Mechanisms of Chemical Degradation of
Cement-based Systems

Mechanisms of Chemical Degradation of Cement-based Systems

Proceedings of the Materials Research Society's
Symposium on Mechanisms of Chemical
Degradation of Cement-based Systems

Boston, USA
27-30 November 1995

EDITED BY

K.L. Scrivener

Lafarge, Laboratoire Central de Recherche,
St Quentin Fallavier, France

AND

J.F. Young

Center for Cement Composite Materials,
University of Illinois, USA

CRC Press
Taylor & Francis Group
Boca Raton London New York

CRC Press is an imprint of the
Taylor & Francis Group, an **informa** business
A TAYLOR & FRANCIS BOOK

CRC Press
Taylor & Francis Group
6000 Broken Sound Parkway NW, Suite 300
Boca Raton, FL 33487-2742

First issued in paperback 2019

© 1997 by Taylor & Francis Group, LLC
CRC Press is an imprint of Taylor & Francis Group, an Informa business

No claim to original U.S. Government works

ISBN-13: 978-0-419-21570-7 (hbk)
ISBN-13: 978-0-367-86471-2 (pbk)

Chapters 9, 22, 37 and 51 cover work done for the US Government and are in the public domain
and are not subject to copyright

Chapters 23 and 31 are © British Crown Copyright 1997. Published by kind permission of
the Controller of HMSO

Publisher's Note
This book has been produced from camera ready copy provided by the individual
contributors in order to make the book available for the conference.

Visit the Taylor & Francis Web site at
http://www.taylorandfrancis.com

and the CRC Press Web site at
http://www.crcpress.com

Contents

Part Two   Corrosion of Reinforcing Steel

Part Three   Microstructure and Mechanisms of Chemical Degradation II

Part Four   **Implications of Curing Temperatures for Durability**

Part Five   **Durability of Non-Portland Cements**

Part Six   Cementitious Waste Forms and Performance of Concrete Barriers for Nuclear Waste Management

Part Seven   Diffusion and Modelling

**Part Eight   Cementitious Waste Forms for Non-nuclear
Applications**

Foreword

Chemical degradation of concrete has been a major concern for engineers throughout the twentieth century. Despite the relative chemical stability of concrete compared to steel - the other major construction material - there are still many diverse forms of chemical attack that must be dealt with such as sulfate attack, corrosion of reinforcing steel, alkali-silica reactions and carbonation. Early solutions to these problems have been successful in controlling the most vigorous forms of attack. Today, however, as structures become older and concrete formulations change, we find that our knowledge about chemical attack is no longer sufficient. New forms of chemical attack have recently been identified, such as internal sulfate attack (delayed ettringite formation), and additional rock types have been recognised as showing alkali-silica reactions. Moreover, we are not able to predict the long-term resistance of concrete to chemical attack.

Now research in materials science and engineering is needed to provide a fresh look at these perennial problems. It is necessary to identify and quantify the links of chemistry and concrete microstructure with chemical attack in order to identify the mechanisms of degradation. The papers presented here are the Proceedings of a symposium held as part of the Materials Research Society's Fall Meeting held in Boston in November 1995, and provide accounts of active research projects in progress around the world. The symposium was financially supported by the US National Science Foundation, the Portland Cement Association, Lafarge and Fondu International.

J.F. Young
November 1996

PART ONE
MICROSTRUCTURE AND MECHANISMS OF CHEMICAL DEGRADATION I

1 THE DIAGNOSIS OF CHEMICAL DETERIORATION OF CONCRETE BY OPTICAL MICROSCOPY

N.THAULOW and U.H. JAKOBSEN
G.M. Idorn Consult A/S, Bredevej, Denmark

Abstract
This paper reviews the diagnosis of chemical deterioration of concrete using optical microscopy. Based on case studies the following mechanisms are emphasised:
- Alkali Silica Reaction (ASR)
- Sulphate Attack
- Delayed Ettringite Formation (DEF)
- Acid Attack

The various mechanisms are mainly diagnosed on the basis of crack pattern, paste texture and secondary mineral phases. It is often found that more than one deterioration mechanism is active, which may accelerate the breakdown of concrete and challenge the skills of the petrographer.

1 Introduction

Diagnosing the deterioration mechanism from the visual appearance of the surface of concrete is almost never possible. Many people do, however, use the surface crack pattern to make their diagnosis which unfortunately often ends up being incorrect. The surface symptoms caused by different breakdown mechanisms may look alike on the surface of the concrete (e.g. map cracking see Figure 1) but when studied in the microscope the microstructure reveals different crack patterns indicative of various deterioration mechanisms.

When diagnosing deterioration mechanisms it is therefore important to conduct a field survey to examine cores taken from relevant areas (including both sound and damaged areas) and to make use of the information which can be obtained by optical microscopic analysis. It is, however, necessary to examine a reasonable number of thin sections as it is important to realise that a single occurrence of alkali silica gel in one

Mechanisms of Chemical Degradation of Cement-based Systems. Edited by K.L. Scrivener and J.F. Young.
Published in 1997 by E & FN Spon, 2–6 Boundary Row, London SE1 8HN. ISBN: 0419215700.

Fig. 1.   Railroad tie showing map cracks on the surface

thin section is a weak foundation for stating that the concrete is damaged by ASR.Unfortunately, for economic reasons the number of thin sections is often quite low.

Another important fact to remember when looking at deteriorated concrete is that the surface crack pattern does not reveal anything about the number of breakdown mechanisms which may be present at the same time. Furthermore, the surface crack pattern provides no information about the sequence of occurrences of the different mechanisms. It is well known [1] that when concrete is first cracked by one mechanism other breakdown mechanisms may follow.

The order in which the deterioration mechanisms occur can under normal circumstances be determined by the use of optical microscopic analysis. The experience of the petrographer is, however, by far the most important factor in assuring a useful outcome of a microscopical investigation. Especially if more than one mechanism is active in the deterioration of concrete (which is often the case!) the petrographer must evaluate the relative importance of the different mechanisms.

Sometimes it is possible to date two generations of cracks that intercept each other: if an empty crack crosses a crack filled with material like alkali silica gel or ettringite the empty crack is the younger of the two.

The most important deterioration mechanism is decided on the basis of the relative amount of cracking caused by the different mechanisms.

ASR and DEF are often found together in heat cured concrete because high temperatures trigger DEF and accelerate ASR. It should be emphasised, however, that ettringite in cracks and voids is commonly found in ASR affected concrete [2], where DEF is not present.

This paper describes four types of deterioration mechanisms, their occurrences and their microscopic appearances and differences: alkali silica reaction (ASR), sulphate attack, delayed ettringite formation (DEF), and acid attack.

2 Microscopic Analysis

Optical microscopy is performed on fluorescent impregnated thin sections using transmitted polarised light, crossed polarised light and fluorescent light (combination of a blue filter and a yellow blocking filter). A 100 watt halogen light source is used as illumination.

The thin sections are made by vacuum impregnating pieces (35 x 40 mm) of concrete with an epoxy resin containing a fluorescent dye. The impregnated pieces are mounted on slabs of glass and ground and polished to a thickness of 0.020 mm (20 μm). The vacuum impregnation of the samples with epoxy causes all voids and cavities in the samples to be filled with fluorescent epoxy. By transmitting blue light through the thin section in the microscope, the fluorescent epoxy in the various porosities will emit yellow light that makes voids, cavities and cracks easy to identify. Aggregates and other less porous phases appear in different shades of green. The fluorescent epoxy also impregnates the capillary pores of the hardened cement paste causing a dense cement paste with low water to cement ratio to appear darker green while a more porous cement paste with a high water to cement ratio appears lighter green. Comparing these green colours to known standards the water to cement ratio (w/c) of the concrete can be estimated with an accuracy of ± 0.02.

Combining information from all 3 types of microscopic illumination usually gives answers to questions such as concrete composition, presence and cause(s) of physical defects and/or chemical deterioration.

3 Alkali Silica Reaction

3.1 Mechanism
Alkali silica reaction (ASR) is a heterogeneous chemical reaction which takes place in aggregate particles between the alkaline pore solution of the cement paste and silica in the aggregate particles. Hydroxyl ions penetrate the surface regions of the aggregate and break the silicon-oxygen bonds. Positive sodium, potassium and calcium ions in the pore liquid follow the hydroxyl ions so that electroneutrality is maintained. Water is imbibed into the reaction sites and eventually alkali-calcium silica gel is formed.

The reaction products occupy more space than the original silica so the surface reaction sites are put under pressure. The surface pressure is balanced by tensile stresses in the centre of the aggregate particle and in the ambient cement paste.

At a certain point in time the tensile stresses may exceed the tensile strength and brittle cracks propagate [3][4]. The cracks radiate from the interior of the aggregate out into the surrounding paste.

The cracks are empty (not gel-filled) when formed. Small or large amounts of gel may subsequently exude into the cracks. Small particles may undergo complete reaction without cracking. Formation of the alkali silica gel does not cause expansion of the aggregate. Observation of gel in concrete is therefore no indication for that the aggregate or concrete will crack.

3.2   Microscopic appearance

Alkali silica reaction is diagnosed primarily by four main features [5]:

- Presence of alkali silica reactive aggregates
- Crack pattern
- Presence of alkali silica gel in cracks and/or voids
- $Ca(OH)_2$ depleted paste

Alkali silica reaction is primarily diagnosed by the presence of cracks in *reactive aggregates* radiating out into the cement paste. Reactive aggregates contain amorphous or microcrystalline silica, SiO_2. Frequently, alkali silica gel is also observed, however, the amount of gel is not a measure of the extensiveness of the reaction but dependent on the type of reactive aggregate. Aggregates containing reactive silica, such as porous flint (Figure 2a), will react rapidly to form alkali silica gel both at the surface and inside the aggregate. More dense polymineralic particles, such as mylonitizised granites (Figure 2b) containing microcrystalline quartz, will react more slowly, the reaction initiated primarily at pre-existing cracks, inhomogeneities, or grain boundaries in the particles that act as pathways for the alkaline pore solution. In such cases only small amounts of gel formation are needed to separate the grain boundaries, leading to expansion and cracking of the aggregate.

The *crack pattern* observed depends on the type of reactive aggregate (Figure 2). Aggregates without preferred orientation of the mineral grains such as porous flint crack creating a triple junction with angles of ideally $120°$. In aggregates with oriented minerals such as mylonites cracks are often parallel to the mineral orientation.

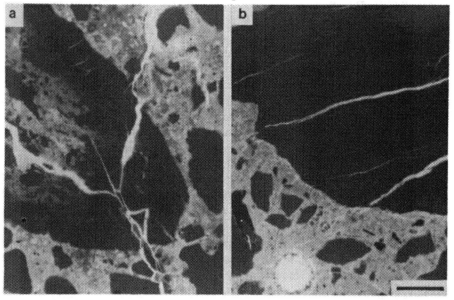

Fig. 2.   Alkali silica reactive aggregates in concrete. (a) cracked porous flint particle, gel appears in the cracks and outer part of the aggregate, (b) cracked mylonite particle, gel appears in cracks. The images are taken in fluorescent light. Scale bar measures 0.5 mm.

In aggregates such as sandstones the cracks may run along grain boundaries.

Typically for all types of aggregates the cracks are wide in the centre and narrow toward the rim of the aggregates. Locally the cracks tend to be perpendicular to the surface of the reactive particle (radial cracks).

If a certain number of reactive particles are present in the concrete a continuous crack pattern is present in the paste. Usually, the cracks follow the line of least resistance perpendicular to the tensile stresses of the concrete. At the surface of the concrete, cracks perpendicular to the surface are commonly seen. Beneath the surface, a large number of cracks tend to be parallel to the surface. These cracks usually run through the cement paste and reactive aggregates, and in rare cases even sound aggregates may be cracked.

Another diagnostic feature of alkali silica reaction is the presence of *alkali silica gel*. Alkali silica gel is a clear, colourless isotropic material with low refractive index (1.46 - 1.53) and a typical shrinkage crack pattern (Figure 3a) [6]. Alkali silica gel may, however, be partly crystallised, showing an orange interference colour in crossed polarised light. However, this latter feature is generally only observed in gel that is situated in cracks inside aggregates or in cracks of rather old concrete [7]. Gel may be observed replacing the outer part of e.g. a porous flint particle (Figure 2a), be situated in cracks and/or in air voids. Ettringite in pores and cracks is commonly found in ASR-affected concrete. The trained petrographer can, however, distinguish alkali silica gel from massive ettringite formation (Figure 3b) by the slight birefringence of ettringite. Furthermore, ettringite exhibits a different crack pattern consisting of almost parallel microcracks. The use of a yellow fluorescent epoxy for preparing the thin sections facilitates the distinction between alkali silica gel and ettringite.

In cases of intensive ASR the $Ca(OH)_2$ of the paste can be dissolved leaving a *black and opaline shining paste* when observed in crossed polarised light. Generally, the

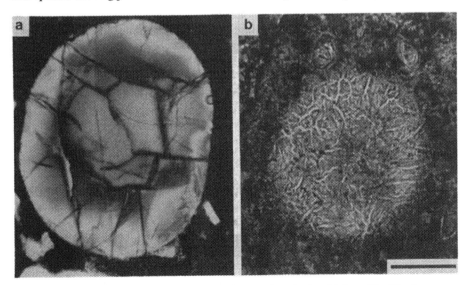

Fig. 3.   Difference between alkali silica gel (a) and ettringite (b) in voids. The images are taken in ordinary polarising light. Scale bar measures 0.1 mm.

dark paste areas are found in a narrow zone around reactive aggregates but are also observed along cracks containing gel.

4 Sulphate Attack

4.1 Mechanism

Sulphate ions may attack components of cement paste. Such attacks can occur when concrete is in contact with sulphate containing water e.g. sea water, swamp water, ground water or sewage water. The often massive formation gypsum ($CaSO_4 . 2H_2O$) and/or ettringite ($3CaO . Al_2O_3 . 3CaSO_4 . 32H_2O$) formed during the sulphate attack may cause concrete to crack.

The formation of gypsum requires a high concentration of sulphate in the ambient water in contact with the concrete. The formation of ettringite occurs by a transformation of the calcium and aluminium containing components in the cement paste. Formation of sulphate phases takes place through dissolution of the cement paste as the sulphate ions consume calcium ions from calcium hydroxide.

4.2 Microscopic appearance

Sulphate attack is diagnosed primarily by four main features:

- Cracks parallel to the surface of the concrete
- Presence of gypsum and/or excessive amounts of ettringite in voids, cracks and paste
- Dissolution of cement paste
- External sulphate source

Sulphate attack is diagnosed when the concrete contains *parallel cracks* (Figure 4a) filled or partly filled with gypsum. The cracks occur parallel to and near the surface of the concrete. The orientation depends on the possible expansion direction of the concrete. The cracks traverse the cement paste and follow aggregate surfaces.

Gypsum is diagnosed by its fibrous texture when observed in crossed polarised light. The interference colour is white to grey. Gypsum is typically observed in parallel cracks and in voids near the surface (Figure 4b). Gypsum is not always recognisable in the optical microscope. If gypsum precipitates in the cement paste scanning electron microscopy (SEM) equipped with an energy dispersive spectrometer (EDS) is useful for positive identification.

Ettringite is identified as needle shaped crystals with low birefringence. It has to be emphasised that the occurrence of ettringite in voids and cracks is common in every mature water-exposed concrete; ettringite by itself is <u>not</u> a diagnostic feature of sulphate attack (or delayed ettringite formation) [2].

To diagnose sulphate attack, near-surface paste expansion forming surface parallel cracks in the cement paste must be present. Chemical analyses of the sulphate content in the surface and the interior is helpful. Furthermore, in order to distinguish sulphate attack from delayed ettringite formation an *outside sulphate source* must be identified.

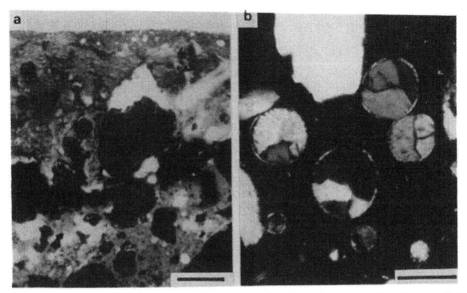

Fig. 4.   Concrete subjected to external sulphate attack (a) gypsum appears in surface parallel cracks and in the cement paste, (b) gypsum filled voids in the upper part of the concrete. The images are taken in fluorescent and crossed polarised light. Scale bars measure 0.5 mm and 0.25 mm, respectively.

5 Delayed Ettringite Formation (DEF)

5.1 Mechanism
Generally DEF is seen as a form of internal sulphate attack. A number of factors such as concrete composition, curing conditions and exposure conditions [8] influence the potential for DEF. Although the fundamental reaction mechanism is still debated among researchers [9].

DEF is believed to be a result of improper heat curing of the concrete where the normal ettringite formation is suppressed. The sulphate concentration in the pore liquid is high for an unusually long period of time in the hardened concrete. Eventually, the sulphate reacts with calcium and aluminium containing phases of the cement paste and the cement paste expands. Due to this expansion empty cracks (gaps) are formed around aggregates (Figure 5). The cracks may remain empty or later be partly or totally filled with ettringite [10].

5.2 Microscopic appearance
DEF is diagnosed primarily by four main features:

- Presence of gaps completely surrounding aggregates
- Wider gaps around large aggregates than around small aggregates
- Absence of external sulphate source
- High temperature heat curing history

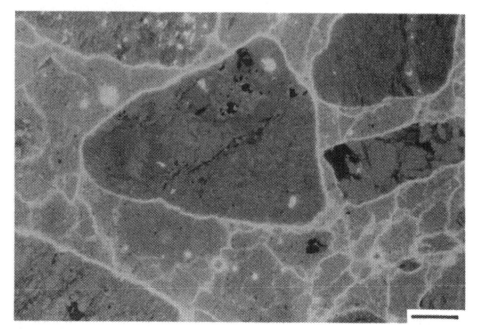

Fig. 5.   Gaps around aggregate particles in an improperly heat cured railroad tie. The image is taken in fluorescent light. Scale bar measures 0.5 mm.

In diagnosing DEF, cracks or *gaps* have to be observed around the aggregate particles (Figure 5). The gaps may be empty, partly filled by needle shaped ettringite or filled with a more massive type of ettringite. In contrast to other researchers [11][12], we do not believe that the crystal growth of ettringite is the cause of the gaps.

The *width of the cracks* are directly proportional to the diameter of the aggregates, resulting in wide cracks around coarse aggregates and narrow cracks around fine aggregates [10]. Under ideal circumstances gaps are only seen around the aggregates. However, in field concrete narrow cracks are also observed perpendicular to the aggregate surfaces (Figure 5). Usually aggregate particles are not cracked by DEF.

DEF is distinguished from external sulphate attack because gypsum is normally not formed. Furthermore, the paste expansion in DEF is not limited to the surface regions of the concrete. It is useful to know the curing conditions of the concrete, as high temperature heat curing is a prerequisite for delayed ettringite formation.

6 Acid Attack

6.1   Mechanism
Concrete is susceptible to acid attack because of its alkaline nature. The components of the cement paste break down during contact with acids. Most pronounced is the dissolution of calcium hydroxide which occurs according to the following reaction:

$$2\,HX + Ca(OH)_2 \rightarrow CaX_2 + 2\,H_2O \qquad\qquad \text{(X is the negative ion of the acid)}$$

The decomposition of the concrete depends on the porosity of the cement paste, on the concentration of the acid, the solubility of the acid calcium salts (CaX_2) and on the fluid transport through the concrete. Insoluble calcium salts may precipitate in the voids and can slow down the attack. Acids such as nitric acid, hydrochloric acid and acetic acid are very aggressive as their calcium salts are readily soluble and removed from the attack front. Other acids such as phosphoric acid and humic acid are less harmful as their calcium salt, due to their low solubility, inhibit the attack by blocking the pathways within the concrete such as interconnected cracks, voids and porosity. Sulphuric acid is very damaging to concrete as it combines an acid attack and a sulphate attack.

6.2 Microscopic appearance

An acid attack is diagnosed primarily by two main features:

- Absence of calcium hydroxide in the cement paste
- Surface dissolution of cement paste exposing aggregates

Acid attack is usually the diagnosis when *dissolution of calcium hydroxide* is observed. Dissolution of $Ca(OH)_2$ makes the cement paste totally black and opaline shiny when observed in crossed polarised light (Figure 6).

Dissolution of the calcium hydroxide is observed in the surface of the concrete and around cracks in contact with the surface. The depth of the dissolution depends on the porosity of the concrete (water/cement ratio) and the type of acid.

Generally, cracks are not produced by the acid attack itself, but instead *exposed aggregates* are observed on the surface, due to the disintegration of the cement paste.

Chemical analyses may be helpful in identifying which acid is present.

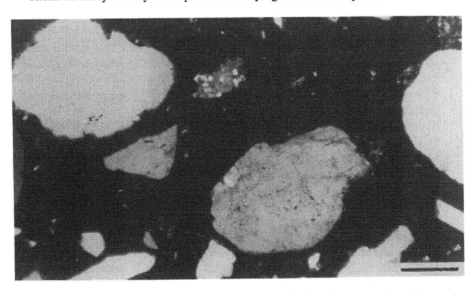

Fig. 6.   Acid attacked concrete. The paste appears black with an opaline shine when illuminated in crossed polarised light. Scale bar measures 0.25 mm.

7 Conclusions

- With an experienced petrographer at the optical microscope thin section microscopy is the principal tool for diagnosing the different chemical deterioration mechanisms in concrete.
- With optical microscopy it is possible to uniquely discriminate between chemical deteriorating mechanisms, such as ASR, DEF, sulphate attack and acid attack.
- Each of the chemical deterioration mechanisms mentioned in this paper can usually be diagnosed by crack pattern, presence and position of various precipitates, types of aggregate and condition of cement paste.
- Optical microscopy can usually detect if several chemical deterioration mechanisms are present at the same time. Furthermore, it is possible to evaluate which was first and what is the main cause of the deterioration problem.
- Optical microscopy should always be supplemented by a field inspection and (if possible) examination of production data. Sometimes it is useful to supplement the optical thin section microscopy with SEM-EDS or chemical analyses. However, optical microscopy should always be performed as the first step before other types of microanalyses because optical microscopy provides both insight and overview.

8 References

1.   Idorn, G.M. (1967) Durability of Concrete Structures in Denmark. *D.Sc. thesis, Technical University of Denmark*, Copenhagen.
2.   Christensen, P., Chatterji, S., Thaulow, N., Jensen, A.D. and Gudmundsson, H. (1981) Filling up pores and fractures in weathered concrete structures. *Proceeding Third International Conference on Cement Microscopy*, Houston, Texas, pp. 298-306.
3.   Svenson, E. (1991) Eigenstresses generated by diffusion in a spacial particle embedded in an elastic medium, *International Journal Mechanism Science*, Vol. 33, No. 3. pp. 211-223.
4.   Thaulow, N. and Andersen, K.T. (1988) New knowledge on alkali-silica reactions (in Danish), *Dansk Beton*, No. 1. pp. 14-19.
5   Andersen, K.T. and Thaulow, N. (1990) "The Study of Alkali-Silica Reaction in Concrete by the use of Fluorescent Thin Sections ASTM STP 1061, (eds. Erlin and Stark) *Petrography Applied to Concrete and Concrete Aggregates*, Philadelphia.
6.   ASTM C-856 (1993) *Standard Practice for Petrographic Examination of Hardened Concrete*, vol. 04.02, pp. 422-434
7.   Thaulow, N., Holm, J. and Andersen, K.T. (1989) Petrographic Examination and chemical analyses of the Lucinda Jetty Prestressed Concrete Roadway, *Proceedings 8th International Conference on Alkali Aggregate Reaction*, Kyoto.
8.   Thaulow, N., Johansen, V. and Jakobsen, U.H. (1996) What Causes Delayed Ettringite Formation ? *MRS, proceedings* (in press)

9. Lawrence, C.D. (1995) Delayed Ettringite Formation: An Issue ? in *Material Science of Concrete IV*, (eds. J. Skalny & S. Mindess), American Ceramic Society, pp. 113-154.
10. Johansen, V., Thaulow, N., Jakobsen, U.H. and Palbøl, L. (1993) Heat Cured Induced Expansion, presented at the *3rd Beijing International Symposium on Cement & Concrete*
11. Heinz, D. (1986) Schädigende Bildung ettringitähnlicher Phasen in wärmebehandelten Mörteln und Betonen, *Thesis*, RWTH Aachen.
12. Heinz, D. and Ludwig, U. (1987) Mechanism of Secondary Ettringite Formation in Mortars and Concretes after Heat Treatment, *Massivbau Baustofftechnologie*, University of Karlsruhe, Vol. 6. pp. 83-92.

2 CONCRETE CHEMICAL DEGRADATION: ANCIENT ANALOGUES AND MODERN EVALUATION

D.M. ROY and W. JIANG
Materials Research Laboratory, The Pennsylvania State University, PA, USA

Abstract
Nobody concerned with concretes in this century can afford not to care about their durability. The international "concrete crisis" of the last decade seems to bring about intensive research efforts. Throughout the history of cementitious materials, excellent resistance to attack by the environment always plays an important role. Ancient analogues provide examples of intelligent use of the available knowledge of technology to properly specify and produce a cement matrix that is durable, which also provides a potential source of information useful for predicting long-term durability of modern cements. The status of the Pyramids in Egypt, the Acropolis in Athens, and the Taj Mahal in India show the effects of environmental impact and pollution. The decay of concrete is closely related to the geologic process of natural rock weathering.
Key words: Ancient analogues, chemical weathering, Roman Concrete, Taj Mahal, Acropolis, Pyramids, pozzolana.

1 INTRODUCTION

Concrete degradation processes can be thought of in two categories: physical action which breaks rocks down into smaller fragments, and chemical action which alters the components of them into different species. However, nature joins these two forces to attack our concretes, many of which owe their very survival to favorable weather conditions. Among mechanism of cement paste degradation due to chemical and physical factors, are: alkali-silica reaction, sulfate attack, chloride reaction, carbonation, leaching, and freezing and thawing[1]. Critics never can be silenced about the deteriorated or deteriorating roads and infrastructure which seem to create a "concrete durability crisis" of the last decade. The question frequently arises, where are those natural, Herculean forces of nature that deteriorate our concretes? Recently the impact of environmental impact on concrete durability has been emphasized[2]. All concrete ages due to the action of the environment. The urban atmosphere of the 20th century is creating special environmental problems to exposed concrete surfaces.

Mechanisms of Chemical Degradation of Cement-based Systems. Edited by K.L. Scrivener and J.F. Young.
Published in 1997 by E & FN Spon, 2–6 Boundary Row, London SE1 8HN. ISBN: 0419215700.

2 ANCIENT ANALOGUES

2.1   Classic examples

Throughout the history of mankind, concern for longevity of buildings and structures is evident, which is implied in the biblical instructions given to Moses on Mount Sinai for construction of the Tabernacle in Jerusalem[3]. Nevertheless, the thousands of surviving ancient buildings and monuments told a lot about their extraordinary stability and durability, which are beyond the legend, when so much dramatically damaged infrastructure built by using modern portland cement now requires rebuilding. The stability and durability of cementitious materials has attracted intensive research interest and contractors' concerns. The analogy of ancient cementitious materials to modern Portland cement could give us some clues to study their stability and durability. A simple classification just divided such materials into two categories: gypsum, and lime and pozzolana, which both could be used to produce durable monuments.

Gypsum was the first binder to be used: The ancient Egyptians used gypsum as jointing mortar to connect the stones of Cheops' pyramid. At the beginning, lime was used in the form of an air-hardening mortar. Later, the Romans discovered that by partially substituting ordinary sand of volcanic origin (the type existing near Pozzuoli), the mortar could be made hydraulic, which is defined as pozzolanic. This discovery made a major contribution to the durability of concrete which has been testified to by very convincing evidence. Until now, much research still focuses on this feature. The pozzolanic effect is principally due to the presence in pozzolana of silica (SiO_2) and alumina (Al_2O_3) that react with lime because of their amorphous, vitreous state, in addition to their high specific surface area. Although the discovery of hydraulic mortars is commonly attributed to the Romans, it is believed that even the Phoenicians and the Israelites in the 10th century BC knew this technique. Prior to the Romans, the Greeks already employed pozzolana. Mallinson studied samples collected from ancient Greece, the city of Camros on Rhodes which was part of a concrete lined, underground, water tank thought to have been concreted around 500 BC[4]. Lamprecht uses the term "Opus caementitium" which indicates Roman concrete technique, his paper illustrated by a number of photos which included the famous Pantheon and Coliseum at Rome, The Pont du Gard, a bridge and aqueduct near the city of Nimes in southern France built in 1st century A.D[5]. Jiang and Roy reviewed some interpretation of stability and durability of ancient cementitious materials, and gave the following clues[6]: 1. Durable lime mortars for surface protection (plasters) may be formed using polishing techniques; 2. Lightweight and air-entrained ancient concrete; and 3. Pozzolana concrete. Chandra and Ohama reviewed ancient polymers utilization in concrete which have proven durability[7].

2.2   Weathering Constructs Finishes

Mostafavi and Mohsen from an architect's point of view in their new book "On Weathering: The Life of Building in Time"[8] used a humorous expression "Finishing ends construction, weathering constructs finishes" which did tell a simple fact. Over time the natural environment acts upon the outer surface of a building in such a way that its underlying materials are broken down, as a result of some chemical reactions. A classic example of how climate affects the rate of weathering was provided when Cleopatra's Needle, a granite obelisk, was moved from Egypt to New York City. After withstanding approximately 3500 years of exposure in the dry climate of Egypt, the hieroglyphics were almost completely removed from the windward side in less than 75 years in the wet, chemical-laden air of New York City. Analysis of the damage to the inscription challenged Winkler to link the damage to its history[9]. He stated that hydration of the entrapped salts caused all the damage, since the action of crystallization would have attacked the stone long ago.

3 MODERN EVALUATION

3.1 Air Pollution

Recently, worldwide general concern was expressed about the state of our natural environment. The alarm was sounded in 1970 by west European countries where "acid rain" destroyed all life in many of their lakes, and attacked buildings and historic monuments in many areas. Winkler described a case of accelerated rate of decay a sandstone sculpture exposed to the industrialized Rhein-Ruhr area of northern Germany since 1702. Two photographs (in 1908 and again in 1969) present the progress of decay[9]. The weathering damage in the first 200 years was relatively mild compared with that suffered from the beginning of this century. The decay of concretes in the engineering structures and monuments is closely related to the geologic process of rock weathering. Most of the decay progresses near or at the ground surface, influenced by the following weathering agents: atmosphere, rainwater, rising ground moisture, stream water, lake water and seawater.

The effects of pollution generated by modern civilization are staggering. Increasing amounts of contaminants entering the air from man-made sources have been found in recent years to be doing far more than occasionally fouling the air and choking the residents of one locality or another: They cause ancient statuary on the Acropolis to crumble, Egypt's pyramids to dissolve[10], and yellowing of the milky white marble of the Taj Mahal due to dust pollution from an oil refinery. Also it is reported that air pollution is causing damage to other historic monuments that were still intact at the beginning of the century, such as the Parthenon in Greece, the Coliseum and Venetian palaces in Italy, and the Cathedral of Strasbourg in France[11]. Those ancient monuments are the historical witness for assessing the potential impacts of global climate change and environmental contaminants.

3.2 Taj Mahal

The modern threats on the Taj Mahal are from environmental pollution, floods, soil erosion, silting and other natural hazards. Since commissioning of an oil refinery at Mathura, 40 km NW of Agra, apprehensions have been growing about the ruinous effects of pollutants on the milky white surface of Taj Mahal. The Archaeological Survey of India, the government agency that repairs and maintains the Taj, has taken adequate measures to protect and preserve the monument. Sharma and Gupta reported the average monthly SO_2 level at the Taj Mahal from 1988 to 1991. In the first half of 1991, they speculated that the unusually high level seemed to be due to the impact of burning of oil wells in the Gulf region during the Gulf war[12].

3.3 Pyramid

The Great Pyramid has given plentiful imagination to the writer, poet, artist, and architect. It also gives the materials scientist a lot to think about. According to Ollier's estimate, the Great Pyramid produced 50,000 cubic metres of weathered debris in 1000 years, indicating an average rate of lowering of 0.2 mm per year over the whole surface of the pyramid; at this rate the pyramids will take ten million years to weather away[13]. The deterioration of pyramids has been accelerated during this century[14]. From a chemical view point the rising water table which is full of salt contacts the limestone monuments, causing a salt crystallization reaction which reduces it to powder. The action of man with the resulting pollution and tourism also plays a very important role. In "The Great Pyramid Debate" (Concrete International, Aug. 1991), the cast-in-place theory of Pyramid construction again raised a point of view; the quality of ancient cementitious materials has led to speculation about secret chemical processes—an ancient mystery. But this fascinating theory is rejected by the opposition who support the "engineering" aspect against the "chemistry" theory, which was recently reviewed by Jiang and Roy[6].

3.4   Acropolis

The Periclean Acropolis is one of the most elaborate examples and perhaps the purest expression of classical form ever created. Rhodes in his interpretive essay[15], when describing Acropolis's ruins, stated that broken columns on an isolated mountaintop, the battered and crumbling wall of a Civil War fort are fascinating because in them every layperson recognizes a direct link with the distant past. In the presence of ruins everyone becomes an active participant in the reconstruction of history: We reconstruct the lines of the building, the lives of its inhabitants, the history of its decay. However from a material scientist point of view, its decay also give us a lot of information about physical, chemical and biological changes.

"The past 40 years have done more corrosive damage to the art treasures of the Acropolis than all of the 2,400 years before." One observer's gloomy assessment of the situation in Athens applies wherever ancient sculpture and architecture have to coexist with industry and automobile traffic[10]. Chemical attack on statuary was studied by a team of chemists from the National Technical University of Athens. Photographs taken in 1955 and 1965 showed the amount of deterioration in a decade. Also photographs taken in 1976 and 1983 showed the amount of deterioration[16].

3.5   Chemical Weathering on Concrete

Following Reich's definition chemical weathering on rock which was rephrased by Loughnan[17], to apply to concrete is proposed to read— Chemical weathering is a process by which atmospheric, hydrospheric, and biologic agencies act upon and react with the constituents of concretes within the zone of influence of the atmosphere, producing relatively more stable, new phases. The chemical changes taking place are essentially (a) the removal of the more soluble components of the concrete constituent and (b) the simultaneous addition of hydroxyl groups and possibly oxygen and carbon dioxide from the atmosphere.

The geochemical carbon cycle[17] governs the transfer of carbon among the land, oceans and atmosphere. When this concept is related to concrete weathering, in which that CO_2 uptake is by concrete weathering, it may be described as follows:

a) Carbonate weathering
$$CaCO_3 + H_2O + CO_2 \longrightarrow Ca^{++} + 2HCO_3^-$$
b) Silicate weathering
$$CaSiO_3 + H_2O + 2CO_2 \longrightarrow Ca^{++} + 2HCO_3^- + SiO_2$$
c) Silicate weathering plus carbonate formation
$$CaSiO_3 + CO_2 \longrightarrow CaCO_3 + SiO_2$$

In the course of chemical weathering, carbonate and silicate minerals in concrete (here represented by $CaCO_3$ and $CaSiO_3$) to produce bicarbonate ions (HCO_3^-), calcium ions (Ca^{++}) and dissolved silica (SiO_2), which were transported and liberated carbon dioxide (CO_2), which eventually escapes to the atmosphere. This is the natural, slow analogue to the burning of fossil fuels.

From the view point of global cycle of silica[18] if sulfate exists, which could be driven from the weathering of $CaSO_4$ minerals (gypsum and anhydrite) in concrete or outside (e.g. acid rain), the process is supplemented by the following:

d) $$2CaCO_3 + H_2SO_4 = 2Ca^{++} + 2HCO_3^- + SO_4^{2-}$$

Atmospheric carbon dioxide diffuses into concrete through empty pores in the binder matrix and reacts with the hydrated cements. The reaction lowers the alkalinity of pore fluid: this can destroy the passive oxide layer on any adjacent steel reinforcement and leaves the steel susceptible to corrosion[19]. The top three on the list of fate of pollutants in the air, both natural and man-made, are sulfate, nitrate and ozone[20]. Sulfate from the combustion of fossil fuel, coal, oil and gas, is the most powerful corrodent of most types of concretes. It usually appears as acid rain or dry fallout. Though also a strong acid, nitric acid is less damaging to carbonate than sulfuric acid and sulfates due the greater reactivity of sulfates with concrete.

4 FIELD OBSERVATIONS AND LAB EXPERIMENT

4.1 A Survey of 227 Pennsylvania Bridges' Chemical Weathering

Pennsylvania, a typical "snow belt" state, maintains 25,000 bridges, the second largest number in the nation. In 1975, Carrier and Cady released a report of 249 four-year old Pennsylvania bridges survey[21]. Twenty years later, in Summer 1994 and 1995, a group of Penn State graduate students involved with the concrete pavements materials made several field trips to revisit 227 of those bridges. All of the observations by the students were made from quantitative measurement rather than qualitative judgments[22]. The information is released for an academic purpose concerning chemical weathering, and more detail will be presented elsewhere. Table 1 gives some chemical weathering deterioration phenomena and related factors.

In 1993/94 winter, a series of fierce storms set a century's record of over 100 inches total snowfall in most Pennsylvania areas. After one season heavy deicing salt use, concrete decks were deeply corroded. From field collected samples, SEM image analyses show the decay process in which the salt apparently acted by crystallization pressure, leading to crumbling and spalling.

The corrosiveness of different atmospheric environments on metal is well known. It appears that metal corrosion approaches the weathering of ferrous-ferric oxides and ferrous-ferric silicate minerals. Larrabee and Mathay reported data giving an indication of the aggressiveness of different atmospheres[23]. An attempt was made by graduate students, who wanted to compare the rate of corrosion of reinforcement and deterioration of concrete of bridges between State College (Penn State) and Pittsburgh. 16 bridges in the Pittsburgh area and 15 in the State College area were chosen. During the summer 1994 they selected objectives and recorded their observations, then in the next summer at same location repeated their observations (table 2 compared with ref.[23]) They show some trend as data reported by Larrabee and Mathay, so It can be concluded that strongly industrial atmosphere will accelerate the corrosion rates of iron.

Table 1 Chemical weathering deterioration—related factors and their source

Factor	Phenomenon	Comments
ASTM weathering indexC-15 and C-62	average annual winter rainfall in inches	bridge location with its climatic condition was key
No. of freeze thaw cycles per years	Damage of south was lighter than north	freeze-thaw environment produce rapid deterioration
Deicer type	$CaCl_2$ more friendly to environment than NaCl	$CaCl_2$ has a great Solubility and low RH
Rate of deicer application	usually 200 to 800 Ib/mile	daily traffic figures affect
No. of deicer application per year	1993/94 winter set a record the most ever	the damage also is the worst ever
Corrosion of reinforcement through crack or pore	by deicers gaining access to steel	a major cause fracture plane and spalls

Table 2 Corrosion rates of iron in different atmospheres

Location	Type of atmosphere	Latitude	Average weight from ref (23)*	loss (g/year) by students**
State College PA	Rural, semihumid	41°	3.75	2.94
Pittsburgh PA	Strongly industrial	41°	9.65	7.18

* iron plate (2x4x1/8 in) ** iron bar (ϕ3/4x8.5 in)

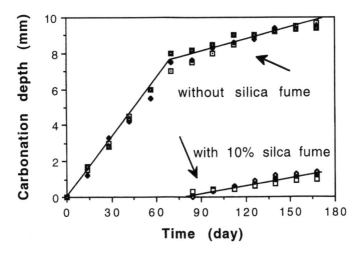

Fig. 1 The carbonation depth

4.2 Pozzolana Put the Durability Back into Concrete

Massazza concluded that pozzolanic cements find their best applications when durability is a priority requirement[24]. In such application, a high resistance to chemical attack, freezing/thawing, and repressing alkali-aggregate-reaction are of concern. Consequently, there is a significant pool of knowledge to draw from in order to optimize the use of pozzolanic cements to produce concretes with properties equal to, or better than those using 100% ordinary portland cement . Roy[25] and others described the effects of silica fume on the microstructural development, and concluded that the ultra-fine particle size of silica fume brings the potential of being much more reactive than other supplementary cementing materials. The microsilica modification of the cement phase can lead to a significant strengthening and resistance to environmental conditions ensuring long-term protection to the reinforcement. The measurements shown in Fig. 1 were assessed by the phenolphthalein neutralization method in a series of concretes at w/c = 0.42, the mix proportion were: cement: ASTM standard sand: aggregate (ASTM C33) :: 1.0:2.75:2.75.

4.3 Artificial Chemical Weathering

The deterioration of a concrete product depends on how and to what extent it interacts with its surroundings. As regards the durability of concretes the weighting which should be given to severe climatic conditions depends on the confidence level required in the performance of them. However, there is as yet no accepted classification of climates for the deterioration of concretes, so more efforts should be made in creating the consciousness of the relationship between concrete durability and the impact of environment. Weatherability is an important characteristic of many commercial new concrete products which are generally put on the market with comprehensive account of their physical and mechanical properties. However there is seldom sufficient information available to assess the length of time the material will fulfill its function satisfactorily in situations on site. Field trials are an important means, but are time-consuming and expensive. Artificial chemical weathering tests are aimed at reducing the 'idle time' by increasing the combined water, temperature, and chemical agents while keeping them properly balanced. At present, a series of systematic artificial chemical weathering tests are being conducted by us, intending to search for reasonably confident answers about reliable prediction of durability.

Fig. 2 Photo of acceleration ageing

4.4 Accelerated Ageing

Compared with trials on site, artificial weathering offers a relatively stable and controllable environment for the systematic study of concrete durability. However, its facility to accelerate weathering processes is its main attraction and some impressive acceleration factors must be identified. A photo of mortar as a function of time under an artificial chemical weathering condition is given in Fig. 2. Mortars were made at w/c=0.35, with ASTM standard sand: cement:slag :: 2.75:0.5:0.5, and control were made by pure cement and sand. They were repeatedly washed by 5% $NaSO_4$ during 6 months (simulating tens of year acid rain attack). The aqueous phase as a transport agent is characterized by the total sulfate concentration in solution, and it is this parameter which controls the chemical state of the system.

5 CONCLUSIONS AND RECOMMENDATIONS

• Ancient analogues provide examples of an intelligent use of the available knowledge of technology to properly specify and produce a cement matrix that is durable. The remarkable longevity of ancient cementitious materials provides a potential source of information useful for predicting long-term stability and durability of concrete.
• Deviation from temperate climatic conditions may be responsible for major differences in the rate of deterioration of a cement matrix in different localities. One of the major concerns of industrialized countries is the pollution of air by emissions resulting from the increased use of hydrocarbon fuels. Development of an in-depth understanding of the impact of the environment on concrete durability is needed, with emphasis on the relevant physics and chemistry. More efforts should made in creating the consciousness of the relationship between concrete durability and the impact of the environment
• More research is need to 'tailor' the use of Pozzolana to ensure durability in concrete. Development is needed of more generally accepted short-term testing methods for assessing the long-term durability of concrete.

ACKNOWLEDGEMENT—This research was partly supported by the National Science Foundation grant MSS-9123239 and EPA grant R 819482.

6 REFERENCES

1. Roy, D.M. (1986) Mechanisms of cement paste degradation due to chemical and physical factors, in Proc. *of 8th Intl. Congress on the Chemistry of Cement.* (Rio de Janeiro, Brasil) Vol. I, pp. 362-380
2. Brantley, L.R. and Brantley, R.T. (1995) *Building Materials Technolgy: Structure Performance and Environmental Impact* (McGraw-Hill, Inc., New York).
3. Maters, L.W. and Brandt, E. (1987) Prediction of service life of building materials and components. *Materials and Structures.* Vol. 20, 55-77.
4. Mallinson, L.G. (1988) Concrete as a barrier for nuclear waste: studies of ancient and old concretes and their implications for long-term durability, in Paper for the First International Conference on Concrete for Hazard Protection, Edinburgh, UK , The Concrete Society, London, UK, pp. 299-318.
5. Lamprecht, H.O. (1986) Opus Caementitium: constructions in Roman concrete. L'Industria Italiana del Cemento, 7-8, pp. 590-605.
6. W. Jiang and D.M. Roy (1994) Ancient analogues concerning stability and durability of cementitious wasteform, in *Scientific Basis for Nuclear Waste Management XVII*, (ed. A. Barkatt, R.A.V. Konynenburg) Mater. Res. Soc. Proc. 333, Pittsburgh, PA, pp. 335-340.
7. Chandra, S. and Ohama, Y. (1994) *Polymers in Concrete*, CRC, Boca Raton,FL.
8. Mostafavi, M. and Leatherbarrow, D. (1993) *On Weathering: The Life of Buildings in Time*, The MIT Press, Cambridge, Massachusetts
9. Winkler, E.M. (1994) *Stone in Architecture: Properties, Durability.* 3rd. Ed. Springer-Verlag, Berlin Heidelberg, Bermany.
10. Allen, O.E. (1983) *Atmosphere*, Time-Life Books Inc., Morristown, NJ.
11. Bequette, F. (1990) Pollution unlimited, in *Environment*, (ed. Allen, J.) The Duskin Publishing Group, Inc., Sluice Dock, Guilford, VA..
12. Sharma, R.K. and Gupta, H.O. (1993) Dust pollution at the Taj Mahal—a case study, in *Conservation of Stone and Other Materials*, (ed. Thiel, M.J.) E & FN Spon, London, Vol. I, pp. 11-18.
13. Ollier, C. (1990) *Weathering and Landforms.* 2nd Ed. MacMillan Ltd., London.
14. Hawass, Z. (1993) The Egyptian monments: problems and solutions. in *Conservation of Stone and Other Materials*, (ed. Thiel, M.J.) E & FN Spon, London, Vol. I, pp. 19-26.
15. Rhodes, F.R. (1995) *Architecture and Meaning on the Athenian Acroplis*, Cambridge University Press, Cambridge.
16. The Committee for the Preservation of the Acropolis Monuments (1985) *The Acropolis at Athens*, Ministry of Culture, Athens.
17. Loughnan, F.C. (1969) *Chemical weathering of the silicate minerals.* (Elsevier Inc., New York).
18. Wollast, R. and Mackenzie, F.T. (1983) The global cycle of silica, in *Silicon Geochemistry and Bioeochemistry* (ed. Aston, S.R.) Academic Press, London.
19 Parrott, L.J. (1991) Assessing carbonation in concrete structures, in *Durability of Building Materials and Components* (ed. Baker, J.M., Nixon, R.J., Majumdar, A.J. and Davies, H.) E.& F.N. Spon, Cambridge, UK, pp. 575-586.
20. Graedel, T.E. and Crutzen P.J. (1989) The changing atomsphere. Sci. Am. 265/9: 58-68.
21. Carrier, R.E. and Cady, P.D. (1975) Factors affecting the durability of concrete bridge decks, in *Durability of Concrete*, SP-47, ACI, Detroit, pp. 121-168.
22. Jiang, W., et al. (1994/95) *Field Trip Report* The Pennsylvania State University.
23. Larrabee, L.R. and Mathay, W.L. (1963) Iron and steel, in Corrosion Resistance of Metal and Alloy (eds. La Quee, FL, Copson, HR) Reinhold, Washington DC.
24. Massazza, F. (1993) Pozzolanic cement. *Cement & Concrete Composites*, Vol. 15, pp. 185-214.
25. Roy, D.M. (1989) Fly ash and silica fume chemistry and hydration, in SP-114 (ed. Malhotra, V.M.) ACI, Detroit., pp. 117-138.

3 PORE SOLUTION ANALYSIS: ARE THERE PRESSURE EFFECTS?

D. CONSTANTINER and S. DIAMOND
Purdue University, West Lafayette, Indiana, USA

Abstract

Pore solution studies have played a prominent role in investigations relating to general cement chemistry and concrete durability problems. Several authors have raised the question of whether the pore solution concentrations obtained in the usual manner (under high triaxial stresses) truly represent the bulk composition of the pore fluid. This report presents parallel determinations of pore solution of mature hydrated cement pastes determined in the usual manner, and separately, in a procedure in which pore solution were recovered at incrementally increasing pressures. Pore solution of mature hcps are predominantly alkali hydroxide, with some SO_4^{2-} ions and very small concentrations of Ca^{2+} ions. There was only a small effect of pressure on the concentrations of the major components, the increase between the ion concentrations at the lowest and the highest pressures used varied from about 7% to 16%. On the other hand, SO_4^{2-} ion concentrations increased significantly with increasing pressure, here the SO_4^{2-} ions increases were on the order of 0.01 to 0.02 N over the full pressure range, insignificant in terms of the total ionic concentrations (0.5 to 0.8N) but representing 65% to 75% of the concentrations obtained at the lowest pressure. It was concluded that pore solutions expressed from hcps over the range of pressures usually employed provide an appropriate measure for the actual concentration of the major ions present in the pore fluid.
Keywords: Pore solution analysis, pressure effects.

---

* Work carried out while the first author was at Purdue University.

Mechanisms of Chemical Degradation of Cement-based Systems. Edited by K.L. Scrivener and J.F. Young.
Published in 1997 by E & FN Spon, 2–6 Boundary Row, London SE1 8HN. ISBN: 0419215700.

1. Introduction

The composition of the liquid phase within the pores of cement paste was the subject of speculation for many years. A method developed about 20 years ago by Longuet, Burglen, and Zelwer [1] and later refined by Barneyback and Diamond [2] has been applied successfully to the expression of the liquid phase and has ended much of this speculation. The liquid phase, commonly referred to as the pore solution, is expressed by means of a high pressure apparatus and then analyzed by conventional analytical techniques. Pore solution analysis of mature portland cement pastes and of mortars has shown, that the solution is mainly composed of alkali hydroxide. Depending on the alkali content of the cement, the concentration of OH⁻ ions can vary from about 0.3 N to as much as 1.0 N.

Concerns have been advanced to question the validity of the pore solution technique [3]. One of the major concerns is that the expressed pore solution might not be representative of the whole, suggesting that this solution primarily comes from the larger pores [4]. In addition, another concern that has been noted is the possibility of altering the ion concentrations due to high pressures necessarily used by this technique.

Lashchencko and Loganina [5], conducted work on portland cement suspensions with w/c of 1. In their experiments, pore solution was expressed as a function of expression pressure, different pore solution removal techniques being used to cover the wide pressure range. Their results indicated that the concentration of alkali ions found at the low and the high pressures was not significantly different. On the other hand, the Ca^{2+} ion concentrations obtained at low pressures were significantly higher (50%) than those obtained at higher pressures.

Silsbee, Malek, and Roy [6] investigated further the effects of expression pressure on the composition of the pore solution. In their experiments, pore solution was collected at individual pressure intervals (0 to 20 ksi, 20 to 30 ksi, 30 to 40 ksi, and 40 to 50 ksi.) from the same specimen, and then analyzed separately. The results obtained were somewhat different then those of Lashchencko et al. and indicated that: a) the concentration of Na^+ increased slightly with increasing pressure, b) the concentration of K^+ first increased slightly with pressure but then dropped substantially at the highest pressure, c) the Ca^{2+} concentrations decreased as the pressure increased -- especially at the highest pressure, and d) the sulfate concentrations increased slightly at low pressures, followed by a more substantial increase at higher pressures.

Recently Duchesne and Bérubé [7] reported on the effects of pressure in the pore solution alkali ion concentrations from paste specimens. In their experiments two portions of solution were obtained and analyzed separately (form 0 to 30 ksi and from 30 to 81 ksi). The data indicated that the alkali concentration in the pore solution obtained at the lower pressure was only slightly different than at the higher pressure (30 to 81 ksi). The difference was usually less than 5%. The concentrations in these two portions where not substantially different than the results from the specimens were the solution was combined.

The work presented here provides additional data to support the thesis that for the major ions present in the pore solution, the expression pressure has only a small impact on the results obtained.

2. Experimental procedure and materials

In this work two ASTM Type I portland cements were used, one of low alkali content (Cement A), the other of moderately high alkali content (Cement B). Chemical compositions are provided in Table I. Cement pastes were prepared at w/c 0.5 and were hydrate in sealed containers for two years.

Table I. Chemical Characteristics of Cements Used in this Study.

Compound	Cement A	Cement B
SiO_2	20.82	20.6
CaO	64.87	61.0
Al_2O_3	5.29	4.0
Fe_2O_3	2.12	3.1
MgO	1.35	4.9
SO_3	3.04	2.8
K_2O	0.58	0.91
Na_2O	0.12	0.25
Na_2O eqv.	0.50	0.85
LOI	1.74	1.90
Insoluble	0.33	0.28

Pore solution were expressed from these mature paste specimens using the apparatus and procedure described by Barneyback and Diamond[2]. The cylindrical specimens are placed in a specially designed high pressure steel die and subjected to triaxial stresses induced by normal stresses applied to the top and bottom flat surfaces. The resulting triaxial stresses remold the specimens and liberate small volumes of pore solution, which are retrieved into a syringe using a vacuum-assisted drain system. Once the solution is obtained, chemical analysis is carried out by conventional means.

In the usual procedure the applied stress is gradually raised to the maximum level, approximately 75 ksi. It is then reduced to ca. 30 ksi and cycled between this reduced level and 75 ksi several times to increase the yield of solution. Such usual determinations were carried out with the present pastes, to serve as controls for the investigation of the pressure effects.

To investigate the effects of pressure, replicate specimens were subjected to a modified procedure, during which the load application was halted and in some cases cycled over specific intermediate applied stress levels. The solution was retrieved and analyzed separately at each increment. The stress level increments, listed in Tables II and III, were slightly different for the two cements. The first increment was from zero to 30 ksi; applied stresses of this magnitude were needed to collect enough pore solution to analyze. The second increment was up to the order of 50 ksi, the third to 60 ksi, and the fourth, to 75 ksi, the maximum design level of the die. Unfortunately, analytical results for the fourth pressure level were not obtained for the Cement A paste, because of the very small yield (0.1 ml). However for the Cement A paste a small amount of solution was collected from the die surface a the conclusion of the experiment and this was analyzed.

The volumes and weights of solution recovered are given in Tables II and III. In these mature cement pastes cycling was needed to recover sufficient solution for analysis at the higher stress increments; an indication that the "easily recoverable" solution had been previously removed.

3. Results and Discussion

The results of the various analyses carried out on the expressed pore solutions are provided in Tables IV and V.

Individual analyses for Na^+, K^+, SO_4^{2-}, OH^- and Ca^{2+} are provided, as are the sums of the respective concentrations for anions and cations (+) and (-), and the net excess charge. The latter are close enough to zero to indicate that the results are mutually compatible.

Table II. Pressure Range, Volume, and Weight of Pore Solution Obtained at Different Pressure Levels from the Two Year Old Cement A Specimen, w/c 0.50.

Sample collection level	Pressure limits used to obtain pore solution ksi	Amount of pore solution collected ml	g
1	0 to 36, monotonically increased	approx. 1	1.0916
2	36 to 52, monotonically increased	approx. 1	1.2187
3	52 to 60, then reduced to 52 and increased to 60	less than 1	0.8621
4	60 to 75, then reduced to 61 and increased to 71	approx. 0.1	0.1170
5	collected from die surface	less than 0.4	0.3721
Total:		approx. 3.3	3.6615

Table III.   Pressure Range, Volume, and Weight of Pore Solution Obtained at Different Pressure Levels from the Two Year Old Cement B Specimen, w/c 0.50

Sample collection level	Pressure limits used to obtain pore solution ksi	Amount of pore solution collected	
		ml	g
1	0 to 32, monotonically increased	less than 1	0.7774
2	32 to 45, monotonically increased	approx. 1.4	1.4430
3	45 to 60, then reduced to 53 and increased to 60	approx. 1	0.9818
4	60 to 74, then reduced to 63 and increased to 69	approx. 0.4	0.4989
	Total:	approx. 3.8	3.7011

Tables IV and V also provide a calculated average concentration for each ion over the range of pressure used, and also a "control" analysis results for the replicate specimen whose solution was expressed in the normal continuous manner. The average value cited a weighted average, with each analysis value weighted by its sample volume.

A comparison of the average values with those of the control values indicates that they are essentially identical for Cement A, and almost so for Cement B. In the latter there is a slight discrepancy for K^+ (0.36N for "average", 0.39N for "control"). Thus the solution obtained in the normal manner is equivalent to the combined solution obtained incrementally. Thus there is no indication of any special effects of high pressure in suddenly breaking down the paste structure.

In comparing the individual values for the different pressure increments, we find that there is indeed a slight increase in alkali and OH^- ion concentrations with pressure. The percentage increase in concentration ranged from about 7% to about 16% from the lowest pressure solution to the highest.

There was an increase in sulfate ion concentrations with pressure. The relative increase was significantly higher than for the other ions. Figure 1 provides plots of concentration vs. pressure. For Cement B, the highest pressure solution had a concentration of 0.047N as compared to 0.027N for the lowest. Cement A appears to show the same pattern. However, it should be pointed out that the general levels of sulfate in the pore solutions are quite low, 0.02N or 0.04N, as compared to the total concentration of either anions or cations, 0.5N or 0.8N. Thus the large proportionate increase with pressure for this ion has little effect on the overall solution concentration.

Table IV. Ion Concentrations in Pore Solution Obtained at Different Pressures for Two Year Old Continuously-Sealed Cement A Specimen, w/c 0.50.

Collection level	Pore solution ion concentrations, N							
	Na^+	K^+	Ca^{2+}	SO_4^{2-}	OH^-	$\Sigma+$	$\Sigma-$	$\Sigma+/-$*
1	0.066	0.369	0.0032	0.0125	0.453	0.438	0.466	-0.028
2	0.068	0.376	0.0036	0.0141	0.458	0.448	0.472	-0.024
3	0.076	0.418	0.0036	0.0185	0.490	0.498	0.509	-0.011
4	NA	NA	NA	NA	NA	NA	NA	NA
Collected from die	0.077	0.418	NA	0.0206	0.490	0.496	0.511	-0.015
Average:	**0.070**	**0.389**	**0.0035**	**0.0154**	**0.468**	**0.463**	**0.483**	**-0.020**
Normal pore solution expression method	**0.070**	**0.398**	**NA**	**0.015**	**0.452**	**0.468**	**0.467**	**+0.001**

*$\Sigma+/- = (\Sigma+) - (\Sigma-)$

Table V. Ion Concentrations in Pore Solution Obtained at Different Pressures for Two Year Old Continuously-Sealed Cement B Specimen, w/c 0.50.

Collection level	Pore solution ion concentrations, N							
	Na^+	K^+	Ca^{2+}	SO_4^{2-}	OH^-	$\Sigma+$	$\Sigma-$	$\Sigma+/-$
1	0.178	0.348	0.0050	0.0268	0.507	0.531	0.534	-0.003
2	0.174	0.350	0.0034	0.0291	0.516	0.528	0.545	-0.017
3	0.189	0.365	0.0034	0.0364	0.532	0.558	0.569	-0.011
4	0.191	0.383	NA	0.0468	0.560	0.574	0.607	-0.033
Average:	**0.181**	**0.358**	**0.0038**	**0.0330**	**0.523**	**0.543**	**0.556**	**-0.013**
Normal pore solution expression method	**0.187**	**0.387**	**NA**	**NA**	**0.530**	**0.574**	**0.530**	**+0.044**

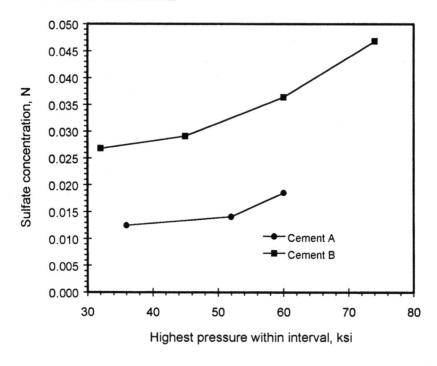

Figure I.   Sulfate Concentrations in Pore Solution as Affected by Expression Pressure.

For Ca^{2+} ions, present at even lower concentration, there seems to be no clear effect of pressure.

In general it appears that despite the modest concentration increases found with higher pressures it is evident that the results obtained by the normal pore solution expression measurement are representative of the pore solution as a whole. It does appear that small amounts of slightly more concentrated solution do exist and can be removed at higher pressures.

4. Conclusions

- For the two year old mature cement pastes examined, with increasing pressure (above the approximately 30-35 ksi needed to express any appreciable amount of pore solution) it was found that the alkali hydroxide concentration increases slightly with pressure. Sulfate ion concentrations showed a proportionally greater increase, but the sulfate levels involved are low. The results for calcium ions were mixed.

• The average concentrations obtained over several pressure increments were essentially identical to the concentration secured by expressing pore solution in the normal manner. Thus no special alteration of ion concentrations due to breakdown of the structure at high pressures was indicated. Nevertheless the results suggest that small amounts of slightly more concentrated solutions exist in the pastes that are removable only at higher pressure increments.

5. Acknowledgments

This work was supported by the N.S.F. Center for Advanced Cement Based Materials as part of its research activities at Purdue University.

6. References

1 Longuet, P., Burglen, L., and Zelwer, A., The Liquid Phase of Hydrated Cement, Rev. des Materiaux de Construction, No.676, pp. 35-41, 1973.
2 Barneyback, R. S., and Diamond, S., Expression and Analysis of Pore Fluid from Hardened Cement Pastes and Mortars, Cement and Concrete Research, Vol. 11, pp. 279-285, 1981.
3 Taylor, H. F. W., Cement Chemistry, Academic Press, 1990, pg. 229.
4 Glasser, F. P., Chemistry of the Alkali-Aggregate Reaction, in The Alkali-Silica Reaction in Concrete, R. N. Swamy Editor, pp. 30-53, London, 1992.
5 Lashchenko, V. A. and Loganina, V. I., Investigation of the Liquid Phase in Hydrated Portland Cement, Journal of Applied Chemistry of the USSR, 47(3), pp. 646-648, 1972 (English translation).
6 Silsbee, M., Malek, R. I. A., and Roy, D. M., Composition of the Pore Fluids Extruded form Slag-Cement Pastes, Eighth International Congress on the Chemistry of Cement, Vol. IV, pp. 263-269, Brazil .
7 Duchesne, J. and Bérubé, M.A., Evaluation of the Validity of the Pore Solution Expression Method from Hardened Cement Paste and Mortar, Cement and Concrete Research, Vol. 24, No. 3, pp. 456-462, 1994.

4 MICROSTRUCTURAL CHANGES AND MECHANICAL EFFECTS DUE TO THE LEACHING OF CALCIUM HYDROXIDE FROM CEMENT PASTE

C. CARDE
Commissariat à l'Energie Atomique, CEN Saclay, France
R. FRANÇOIS AND J-P. OLLIVIER
Laboratoire Matériaux et Durabilite des Constructions, INSA-UPS Génie Civil, Toulouse, France

Abstract

This paper deals with the microstructural changes due to the leaching process of cement based materials.

In order to characterize this effect, we have performed experiments on paste samples made with pure CEM I cement and a mix of CEM I + 30% of silica fume with the same water/binder ratio. Chemical attack and compressive tests are conducted on micro-cylinders samples (10 to 30 mm of diameter). The leaching of calcium is due to the attack of a 50% solution concentrate of ammonium nitrate. The removal of Ca ions is monitored by microprobe analysis which allows to plot the Calcium profile along the samples. The removal of $Ca(OH)_2$ is monitored by the plot of XRD curves at different depths obtained by polish sequences. The results show that the peripheral degraded zone is delimited by the dissolution front of calcium hydroxide and that the leaching process leads to a progressive decalcification of CSH.

The kinetics of leaching is the same for both materials (CEM I paste and paste with silica fume) and is governed by a diffusion mechanism. The same amount of Calcium is removed from both type of materials but in the case of pure CEM I the strength loss is 76% when the degraded thickness reach the radius of the sample whereas the strength loss is only 32% in the case of the paste with silica fume.

The results show that for both materials, the strength loss is a linear function of the degraded area and the residual stress of the degraded zone is the same whatever this thickness and depends on the initial calcium hydroxide content.

Keywords: Leaching, Calcium hydroxide, CSH, silica fume, compressive strength, water porosity, ammonium nitrate.

1 Introduction

The research program in progress tries to characterize the deterioration of the mechanical properties of the concrete surrounding radioactive wastes, due to the water

Mechanisms of Chemical Degradation of Cement-based Systems. Edited by K.L. Scrivener and J.F. Young. Published in 1997 by E & FN Spon, 2–6 Boundary Row, London SE1 8HN. ISBN: 0419215700.

flow during storage. The chemical attack of this small amount of ionized water is essentially a leaching of the calcium hydroxide and a progressive decalcification of the CSH. The slow kinetics of these chemical reactions leads us to increase the aggressively of the environment by using a solution of NH_4NO_3 (the chemical attack due to NH_4NO_3 results also into leaching of the cement paste but the kinetics is quicker than those obtained with water as reactant).

In this paper, the results of an investigation carried out in order to better understand the mechanisms of deterioration of cement pastes in contact with an aggressive solution of NH_4NO_3 are presented. The test program was designed in order to obtain information on both the alteration of mechanical properties and porosity in relation to time and to determine the influence of the dissolution of calcium hydroxide and the decalcification of the CSH.

2 Experimental program

2.1 Materials

Two different types of pastes were investigated. The first one is a pure paste of Portland cement and the second one is a mixture of Portland cement and silica fume. The cement used to make the paste sample is a CEM I 42.5 whose chemical composition is given in Table 1. The W/C ratio is 0.5 for the pure paste. The silica fume used is 30 % substituted of the weight of cement. The W/(C+SF) ratio is 0.45. The high content of silica fume is used to consume all the calcium hydroxide.

Table 1. Chemical composition of the cement

SiO_2	Al_2O_3	Fe_2O_3	CaO	MgO	SO_3	Na_2O
20.2	4.9	3	63.4	0.67	3.2	0.25

The mixtures were cast into cylindrical moulds (11x22 cm). The demoulding was done after 24 hours, then the pure paste samples were cured for 27 days immersed in lime saturated water at 20°C (±1°C). The paste samples with silica fume were cured 75 days to achieve the complete pouzzolanic reaction (fig. 1).

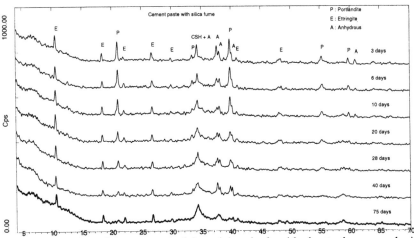

Fig. 1 : Evolution of the removal of the calcium hydroxide due to the pouzzolanic effect in relation to the curing time

2.2 Samples

Because of the slow kinetics of leaching, we have been obliged to work with small sizes samples. The samples used are cylinders whose diameters (φ) are 10, 12, 14, 20 and 30 mm with a h/φ ratio = 2 (h: height of the sample). After curing, the samples are extracted from the test pieces by means of a diamond tipped core lubricated with water. For each sample dimension, two series of samples have been made, the first one which has been immersed in the aggressive solution (treated series), the other one which has been kept in an endogenous environment (control series).

2.3 Leaching process

The aggressive environment used is a 50% concentrate solution of ammonium nitrate NH_4NO_3 (590g/l.). The chemical attack of the cement paste by ammonium nitrate leads to the development of a soluble calcium nitrate, a not very soluble calcium nitro-aluminate of calcium and ammoniac gas NH_3 [1]. This process induces mainly a total leaching of the lime and a progressive decalcification of CSH [2]. The removal of Ca ions is monitored by microprobe analysis which allows to plot the calcium profile along the samples. In the case of the pure cement paste, the removal of $Ca(OH)_2$ is monitored by the plot of XRD curves at different depths obtained by polish sequences (Fig. 2).

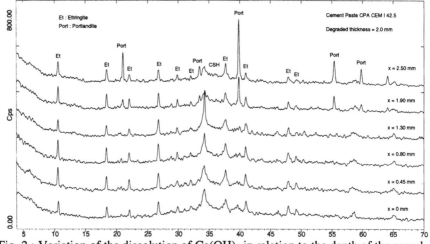

Fig. 2 : Variation of the dissolution of $Ca(OH)_2$ in relation to the depth of the sample

These results show that the peripheral degraded zone is delimited by the dissolution front of calcium hydroxide.

The calcium profile for the pure paste sample is shown on Fig. 3, while in the case of the admixture of silica fume, the profile is plotted on Fig. 4. The results show a difference between the two previous profile: in absence of calcium hydroxide, the variation of Ca is a linear function in relation to the depth of the degraded zone whereas in the presence of portlandite, the profile is linear only on half depth of the degraded zone and almost constant on the other half.

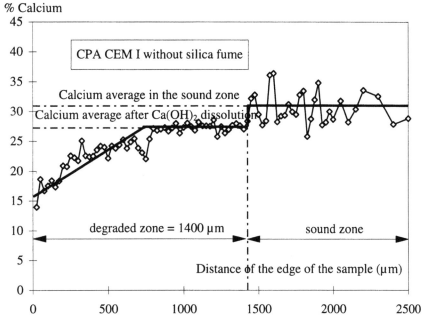

Fig. 3 : Variation of the dissolution of Ca in relation to the depth of the sample for the pure paste sample

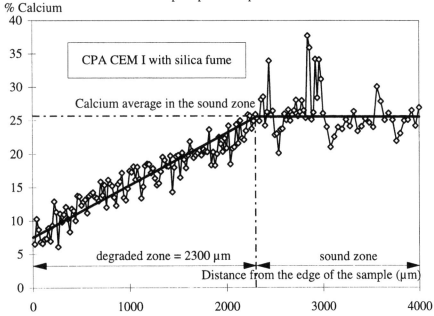

Fig. 4 : Variation of the dissolution of Ca in relation to the depth of the sample for the paste sample with silica fume

2.4 Mechanical tests

Both the treated samples and the control samples have been subjected to a compressive load in order to measure their compressive resistance. The displacement speed is controlled during the load. The force applied on the sample and the longitudinal displacements are measured during the test. The displacement measured is the average of three displacements recorded during the test by means of three transducers fixed on the clamping plate with an angular location of 120°, so the bending effects cannot disturb the measure of the displacement.

The average longitudinal strain εl_{avr} is calculated using the three measured displacements.

$$\varepsilon l_{avr} = \frac{d_1 + d_2 + d_3}{3} \cdot \frac{1}{h} \quad (\mu m/m)$$

where d_1, d_2, d_3 are the displacements ((μm) recorded by means of the transducers 1, 2 and 3, and h is the height of the sample (m).

The compressive strength is evaluated by dividing the maximal load F_{ult} by the cross section of the sample S

The simultaneous recording of the force and the average displacement allow the plotting of the curve $\sigma = f(\varepsilon)$.

$$\sigma = \frac{F_{ult}}{S} \quad (MPa)$$

2.5 Physical tests

Water porosity tests have been performed in order to characterize the microstructure of the degraded zone.

The experimental method used to measure the water porosity consist in drying at 80°C a saturated water sample until it reaches a constant mass. Then, we consider that the loss of mass corresponds to the volume of water contained in the porous system of the material, even if the results obtained by Taylor [3] indicate that some hydrates loss a part of their bound water at this temperature.

3 Experimental results

After treatment, the part of the initial cross section which is degraded and would modify the displacement and the measured irreversibilities is eliminated from the samples.

The values of the average compressive strength measured in the treated group (σ_d) and the control group (σ_T) are plotted on fig. 5. The degradation level is assessed by the ratio of the peripheral degraded area (A_d) over the total area (A_t) of the sample.

To compare the results obtained with the different sizes of samples, the loss of strength is calculated using the following formula where σ_T is the compressive strength of the control sample and σ_d is the compressive strength of the degraded samples.

$$\frac{\Delta\sigma}{\sigma} = \frac{(\sigma_T - \sigma_d)}{\sigma_T}$$

On Fig. 6, results concerning samples which sizes vary from 10 mm to 30 mm are plotted. Because of the only one linear function obtained, these results show that the relative decrease of average stress $\frac{\Delta\sigma}{\sigma}$ is independent of the size of the sample.

When the degradation is completely achieved, there is a residual strength both in the case of pure paste samples which represents about a quarter of the initial strength: $\sigma_d = 0.24\sigma_T$ and in the case of silica fume mixes which represents about 2/3 of the initial strength: $\sigma_d = 0.68\sigma_T$.

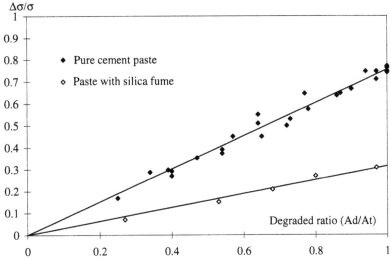

Fig. 5: Variation of the loss of strength in relation to the degraded ratio

The water porosity has been measured on the control group and on degraded samples. The results are presented in fig. 6, where P_t is the total porosity of the sample and P_d is the porosity of degraded samples.

In the case of the pure paste sample, the increase in porosity due to the leaching is a linear function in relation to the degraded ratio and the final increase in porosity due to the total leaching is 19.3%.

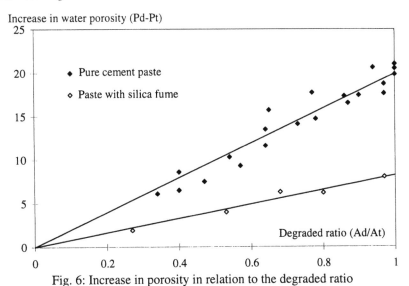

Fig. 6: Increase in porosity in relation to the degraded ratio

With the silica fume containing mixes, the results also reported on fig. 6 show that the increase in water porosity in relation to the degraded ratio is a also linear function. But the slopes are very different, about 2.5 times higher for pure cement paste than the paste containing silica fume. This ratio is almost the same than that found between the residual strength of both materials.

These results confirm that the dissolution of the calcium hydroxide create a macro-porosity which size is about the same than the capillary porosity, whereas the decalcification of the CSH leads to the appearance of a micro-porosity which is the same order than the intrinsic porosity of CSH.

4 Discussion

The size of $Ca(OH)_2$ crystals which is about the same than the capillary pores [4] leads to the formation of a macro-porosity after leaching. This macro-porosity leads to a drop in the strength of the sample and in an increase of the irreversibilities quantified by the plot of the variation of the degree of reversibility [5] (Fig. 7.)

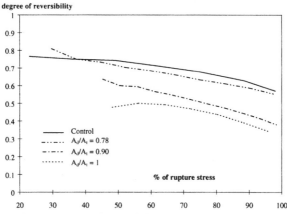

Fig. 7: Degree of reversibility for pure cement paste

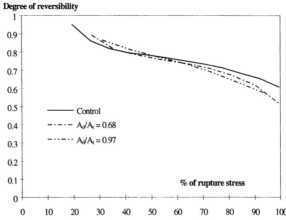

Fig. 8 : Degree of reversibility for paste with silica fume

The removal of Ca in the CSH leads to the appearance of a micro-porosity which is a second order effect compared to the macro-porosity [6] due to the dissolution of portlandite. The fact that the degree of reversibility stays constant in relation to the degraded ratio confirms this result (Fig. 8).

This increase in micro-porosity leads to a decrease of the mechanical resistance but even after total degradation, the material can still resist in both term of mechanical resistance and transfer resistance.

5 Conclusion

The deterioration of the pure cement paste exposed to the action of the ammonium nitrate was manifested by a peripheral zone of less resistance. There is a gradient of C/S ratio in half of the leaching zone but there is a complete dissolution of calcium hydroxide. It seems that, in the case of a paste sample made with CEM I cement which leads to about 20% content of calcium hydroxide, the total dissolution of this calcium hydroxide is the essential parameter governing both decrease in strength and increase in porosity. The loss of strength due to the decalcification of CSH seems weak and could be neglected in relation to the global loss of strength due the dissolution of $Ca(OH)_2$.

In the case of the use of a paste sample with the admixture of silica fume to consume all the calcium hydroxide, the effect of C/S ratio decrease of the CSH reduces the strength of about 32%. Then, the effect of the decalcification of CSH alone is not neglected but the residual strength of the material is important and the material could maintain its mechanical functions.

The initial content of $Ca(OH)_2$ seems to be an important parameter of the ability of cement paste to resist to the leaching process due to ammonium nitrate attack or pure water attack.

6 Acknowledgments

The authors are grateful to the Agence Nationale pour la gestion des Déchets RAdioactifs (ANDRA) for its financial support.

7 References

1 .   F.M. Lea (1965) Magazine of Concrete Research, Vol. 17, n 52, p 115-116.
2 .   C. Carde and R. François - Effect of the leaching of calcium hydroxide from cement paste on mechanical and physical properties. Cement and Concrete Research (to be published).
3 .   H.F.W.Taylor (1990) Cement Chemistry, p. 220, Academic Press, Inc. NewYork.
4 .   G.J. Verbeck and R.H. Helmuth (1968) Structures and physical properties of cement paste. Proc. Of the Vth Int. Symp. On the Chemistry of Cement, Session III-1, Tokyo.
5 .   C. Carde and R. François (1994) Effect of the ITZ on the leaching of calcium hydroxide from mortar, Symposium proceedings vol. 370, MRS, Boston, pp 457-464.
6 .   J-L. Granju and J-C. Maso (1978) Résistance à la compression simple des pâtes pures de ciment durcies, temps de durcissement supérieur à 4 ans. Cement and Concrete research, 8, pp 7-14.

5 MODELLING THE CALCIUM LEACHING MECHANISMS IN HYDRATED CEMENT PASTES

A. DELAGRAVE[1], B. GÉRARD[1-2] and J. MARCHAND[1]

1)Concrete Canada and Centre de Recherche Interuniversitaire sur le Béton, Université Laval, Ste-Foy, Québec, Canada
2)Électricité de France, Moret-sur-Loing, France

Abstract
The leaching of calcium from dissolved portlandite crystals and C-S-H hydrates often contributes to the degradation of concrete structures. Since calcium leaching is a non-linear phenomenon which is influenced by a wide range of parameters, the residual service life of the structures affected by this type of deterioration cannot be predicted satisfactorily on the basis of experimental results alone. In order to provide more fundamental information on this phenomenon, a numerical model was developed. The model considers the thermodynamic equilibrium between the pore solution and the various hydrated phases, and treats the deterioration mechanism from a macroscopic point of view. In order to validate the model, small disks of four different neat cement pastes were soaked for a 6-month period in a pH-controlled solution (pH 7). Test variables included the type of binder and the water/binder ratio. The initial transport properties of the various mixtures were determined by means of a tritiated water diffusion test, and the microstructural alterations of the paste samples were investigated by means of electron microprobe and TGA/DTA analyses. On the basis of the experimental results and the numerical simulations, various aspects of the leaching of calcium are discussed.
Keywords: Calcium leaching, microstructure, durability, numerical modelling, service life prediction.

1 Introduction

The leaching of calcium may be a matter of concern for the durability of concrete [1]. In some cases, it may increase the porosity of the surface layers, and detrimentally affect the resistance of the material to deicer salt scaling and ion penetration [2]. The leaching of calcium may also have a negative influence on the long-term durability of some concrete structures. For instance, it may be a very critical parameter in the durability of nuclear waste barriers for which the required service life ranges from a few hundred years to several thousand years [3, 4].

Over the years, several studies have clearly demonstrated that the leaching of calcium from cement-based materials is essentially a diffusion-controlled phenomenon [4, 5]. Unfortunately, the assessment of the mechanisms of calcium leaching by laboratory

Mechanisms of Chemical Degradation of Cement-based Systems. Edited by K.L. Scrivener and J.F. Young.
Published in 1997 by E & FN Spon, 2–6 Boundary Row, London SE1 8HN. ISBN: 0419215700.

experiments is often difficult and generally time-consuming. Furthermore, since it also affects the pore structure of the material, the overall mechanism quickly becomes non-linear, and a reliable prediction of the evolution of the concrete properties upon leaching cannot be made on the basis of experimental results alone.

Recently, the development of numerical models has provided new tools to investigate the influence of calcium leaching on the evolution of the properties of cement-based materials. BENTZ and GARBOCZI [6], for instance, used a cellular automaton-type digital-image-based model to study the evolution of the hydrated cement paste porosity upon leaching. They showed that calcium hydroxide leaching has a direct influence on the system connectivity and that it can significantly increase the transport properties of hydrated cement pastes.

This paper presents the main features of a new modelling approach that has been specifically developed to study the mechanisms of calcium leaching in cement-based materials. The model considers the thermodynamic equilibrium between the pore solution and the various hydrated phases, and treats the deterioration mechanism from a macroscopic point of view. The degradation characteristics of four different neat cement paste mixtures immersed for 3 months and 6 months in deionized water served as a basis for the validation of the model.

2 Theoretical background

As in any porous solid, the cement paste pore solution is assumed to be in thermodynamic equilibrium with the surrounding hydration products. When a concrete element is immersed in pure water, the chemical potential gradient between the pore solution composition and the external surface causes ions to diffuse in and out of the material. Changes in the chemical composition of the pore solution modify the equilibrium of the system and lead to the dissolution of certain hydrated phases.

The solid/liquid thermodynamic equilibrium is determined by all ions dissolved in the solution [7-9]. However, experimental data show that the equilibrium of the main cement paste hydrated phases can be well predicted on the basis of the calcium concentration of the pore solution [4, 5].

Figure 1 presents the evolution of the CaO/SiO_2 ratio (of the solid phase) as a function of the calcium concentration of the solution. As can be seen, the equilibrium is characterized by a two-step curve, one step being associated with the dissolution of $Ca(OH)_2$, the second with the dissolution of C-S-H. At high concentrations of calcium ions in the pore solution (20 mM/l), the C-S-H and the $Ca(OH)_2$ remain insoluble. As the calcium concentration of the pore solution decreases below a critical level, $Ca(OH)_2$ crystals are first dissolved. When all the portlandite available has been leached, the equilibrium of the system is controlled by the C-S-H which themselves undergo partial decalcification. Beyond a certain critical level (2 mM/l), even the C-S-H becomes totally decalcified, the remaining product being a silica gel without any binding properties [5, 11].

3 Presentation of the model

For the development of our model, we chose to follow a macroscopic approach. This choice was mainly motivated by the fact that the model was developed as part of a larger one aimed at predicting the long-term durability of low-level nuclear waste containers for surface repositories [11, 12]. The main originality of this larger model resides in the fact that its considers the various chemical and mechanical solicitations affecting the concrete structure as well as their eventual couplings. The main features of this larger model are presented in reference 12.

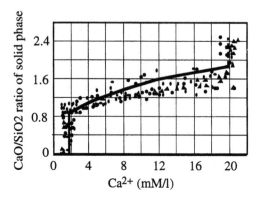

Fig. 1.   CaO/SiO$_2$ ratio of the hydrated cement paste as a function
of calcium concentration in solution [from ref. 10]

We also chose to develop our model of the leaching mechanisms of cement-based materials on the basis of the thermodynamics of irreversible processes. Since according to BERNER [4] and ADENOT [5], the dissolution of the main calcium bearing phases of a hydrated cement paste could well be described by Fig. 1, we chose the calcium concentration of the pore solution as a state variable.

The calcium leaching process was thus modelled by following the modifications of the calcium concentration of the pore solution with time. As previously mentioned, these modifications are mainly induced by the diffusion of calcium ions out of the material (as a result of the chemical potential gradient between the pore solution and the aggressive solution) and by the dissolution of the calcium-bearing hydration product (Ca(OH)$_2$, C-S-H, ettringite,....). The main assumptions of the model include:

- the cement is fully hydrated and all solid phases are in state which can be modelled using the concepts of thermodynamic equilibrium;
- the time required to dissolve a given phase is short compared to transport times;
- when in solution, calcium ions are not reacting to form any new compounds (like CaCO$_3$ for instance);
- the material is and remains saturated over time, and the systems stays in isothermal conditions.

On the basis of these assumptions, the kinetics of the calcium leaching process can be determined by applying the law of mass conservation:

$$\frac{\partial \theta . Ca^{2+}}{\partial t} + \frac{\partial Ca_{solid}}{\partial t} = \text{div}\left[D.\,\text{gräd}(Ca^{2+})\right] \tag{1}$$

where θ is the total porosity and D the diffusion coefficient (m^2/s). The first term of the left-hand side of equation 1 can be neglected considering that the amount of calcium ion initially in solution is very small in comparison with the amount dissolved by the calcium-bearing hydrates during the leaching process. Equation 1 can thus be rewritten:

$$\frac{\partial Ca_{solid}}{\partial Ca^{2+}} . \frac{\partial Ca^{2+}}{\partial t} = \text{div}\left[D(Ca^{2+}).\,\text{gräd}(Ca^{2+})\right] \tag{2}$$

The resolution of equation (2) requires a mathematical relationship for the calcium solid/liquid equilibrium described in Fig. 1. In our model, we chose to approximate the two-step curve of Fig. 1 by the following mathematical equation:

$$Ca_{solid} = a - bx^2 + cx - \left[\frac{e}{1 + \left(\dfrac{Ca^{2+}}{x_2}\right)^N} + \frac{f}{1 + \left(\dfrac{Ca^{2+}}{x_1}\right)^M} \right] \tag{3}$$

where x_1 is the average position of the dissolution front of portlandite (at a calcium concentration of 20 mM/l), x_2 is the average position of the dissolution front of C-S-H at a calcium concentration of 2 mM/l, N and M are constants (N = 70 to 100 and M = 5), and a, b, c, e, and f are determined from the molar fraction of portlandite (S_{por}) and the total calcium content of the hydrated cement paste (S_{tot}):

$$e = S_{por} \qquad f = 0,565(S_{tot} - S_{por}) \qquad b = (S_{tot} - S_{por} - f)/400 \quad (4)$$
$$c = ((S_{tot} - S_{por} - f)/20) + 20b \qquad a = S_{por} + b$$

The molar fraction of portlandite is calculated according to the method proposed by ADENOT [5], and the total calcium content is determined considering the original CaO content of the cement and the water/binder ratio of the paste sample (assuming a full hydration of the cement).

According to the numerical simulations of BENTZ and GARBOCZI [6], the increase in porosity caused by the dissolution of solid phases should significantly modify the transport properties of the paste sample. In our model, this evolution of the diffusion coefficient is linked to the volume fraction of the different solid phases according to the following equation:

$$D = D_0 \left(\frac{D_s}{D_0}\right)^{\frac{\beta V^d_{por} + \alpha V^d}{V^d_{por} + V^i}} \tag{5}$$

where D is the diffusion coefficient of the solid at a given time, D_0 is its initial diffusion coefficient and D_s is the diffusion coefficient of the totally degraded solid (i.e. at the end of the calcium leaching process). V^d_{por} represents the volume fraction of the portlandite, V_d, the volume fraction of the hydrates other than portlandite. V^i represents the volume fraction of the hydrates calculated by subtracting the volume occupied by the portlandite and SiO_2 fractions to the total volume of hydrates in a fully hydrated system.

The coefficient α is determined according to the following equation :

$$\alpha = 1 + (1 - \beta)V^i_{por} / V^i \tag{6}$$

This equation is based on experimental data showing that the silicon content of a hydrated cement paste remains fairly constant throughout the degradation process [13]. β is an empirical coefficient which mainly accounts for the effect of the dissolution of calcium hydrates on the diffusion properties of the material. Figure 2 illustrates the influence of β on the effective diffusion coefficient of calcium.

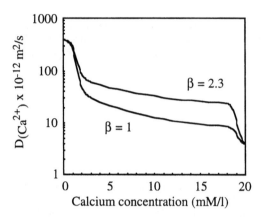

Fig. 2.   Influence of β on the calcium diffusion coefficient

We chose to solve the mathematical equations of the model using the finite element method. The numerical version of the model was thus developed using the 1995 version of CASTEM 2000. A DUPONT 2-type algorithm was used to solve the very non-linear system [14]. To run the model, one has to determine the initial volume and molar fractions of the different solid phases and the initial diffusion coefficient (D_0) of the mixture. One should also estimate the diffusion coefficient of the totally degraded mixture (D_s) and the value of β.

4 Test Program

In order to provide experimental data to validate the numerical model, disks of four different neat cement paste mixtures were soaked in a pH-controlled solution for a maximum period of 6 months. For this research project, only cement pastes were subjected to chemical attack, in order to avoid the influence of the aggregates (particularly that of the interfacial transition zone). The initial transport properties of the neat cement pastes were determined by means of a diffusion test conducted with tritiated water. Test parameters included the water/binder ratio (0.25 and 0.45) and the use of silica fume. Electron microprobe and TGA/DTA analyses were part of the techniques used to study the various samples after removal from the solution. A sufficient number of paste disks were soaked in the aggressive solution to allow measurements to be made after 3 and 6 months of exposure.

5 Materials and Mixtures Characteristics

Four different neat paste mixtures were prepared. The water/binder ratio for the two high performance cement pastes was fixed at 0.25. The cement used was an ASTM type III cement. The C_3S content of this cement was 68.7%, its C_3A content was 7.7% and its Blaine fineness was 5350 cm^2/g. Six percent silica fume was added to one of these two mixtures. The silica fume contains more than 90% SiO_2. A melamine-based superplasticizer was used at a dosage of 2.1% of dry material by mass of cement.

An ASTM type I cement was used in the preparation of the two 0.45 water/binder ratio mixtures. The C_3S content of this cement is 68.7%, the C_3A content 7.4% and the Blaine fineness 4620 cm^2/g. Six percent silica fume was added to one of these two mixtures. The mixture characteristics are presented in Table 1.

Table 1. Mixture characteristics

Mixture	Cement Type	Silica Fume (%)	Water/binder
P25Q	Type III	0	0.25
P25QS	Type III	6	0.25
P45Q	Type I	0	0.45
P45QS	Type I	6	0.45

6 Experimental Procedures

All pastes were cast in plastic moulds (diameter = 7.0 cm, height = 20 cm). The moulds were sealed and rotated for the first 24 hours in order to prevent any segregation of the mixture. At the end of this period, the cylinders were demoulded and immersed in a saturated lime solution for a 2-month period. After curing, the cylinders were cut in 8-mm thick disks. Except for those required for the initial measurements, the disks were soaked in a 50-liter plastic tank filled with deionized water. The solution was maintained at a pH level of 7 throughout the entire test period. The pH level was adjusted every day by adding HNO_3 (0.1N) to the solution. The air above the solution was flushed with nitrogen to prevent any carbonation and the solution in the tank was constantly stirred.

The initial transport properties of the four mixtures were assessed using a simple diffusion cell apparatus similar to the one described by CHATTERJI and KAWAMURA [15]. The test specimens (5-mm thick, 70-mm in diameter) were first vacuum saturated with deionized water and then mounted on the diffusion cells. The upstream compartment of each cell was filled with a lime solution containing 30.8 MBq/l. Tritiated water has been chosen to study the transport properties of the cement paste samples because this tritium has very little interaction with the cement paste hydrates. Two duplicate specimens were tested for each mixture. The experiments lasted approximately 4 months. The effective tritiated water diffusion coefficient was calculated from the steady-state regime according to FICK'S first law:

$$\bar{J} = -D_0 \text{grad} C_{tritium} \tag{7}$$

After 3 and 6 months of exposure, disks of each mixture were removed from the solution. The paste disks were immediately vacuum dried for a minimum period of 10 days to prevent any further chemical reactions. The disks selected for the microprobe analyses were broken into small parts to expose the total internal surface (8-mm thick). They were then impregnated with an epoxy resin, polished, and coated with carbon. Calcium and silicon profiles were performed along lines extending 4000 μm from the external surface in contact with the solution towards the interior of the disks.

The TGA/DTA analyses were performed at the end of the curing period and after 6 months of exposure. Tests were performed on powdered paste samples (passing a 75 μm sieve) representing the full thickness of the disks. In all cases, 40-mg samples were heated, in a nitrogen atmosphere, at a rate of 10° C/min.

7 Experimental Test Results

7.1 Tritiated water diffusion experiments
The test results of the effective diffusion coefficient measurements using tritiated water are summarized in Table 2. As can be seen, the water/binder ratio clearly has a

significant influence on the diffusion property. A reduction of the water/binder ratio led to a significant decrease of the diffusion coefficient of tritiated water by more than an order of magnitude. The use of silica fume has also contributed to reduce the diffusion coefficient. The effective diffusion coefficient of tritiated water is at least 3 times lower for the silica fume mixtures.

Table 2. Diffusion of tritiated water

Mixture	Effective Diffusion Coefficient $(10^{-12} \text{ m}^2/\text{s})$
P25Q	0.63
P25QS	0.16
P45Q	9.83
P45QS	3.79

7.2 Microprobe and TGA/DTA analyses

Figure 3 presents the microprobe analyses of the two OPC pastes after a 3-month immersion period in deionized water. The straight line on each diagram approximates the minimum concentration of calcium in the control specimens. The concentrations above this line correspond to the calcium held in portlandite crystals or in anhydrous cement grains and the concentrations below, to the calcium held in deteriorated cement paste including the C-S-H phases.

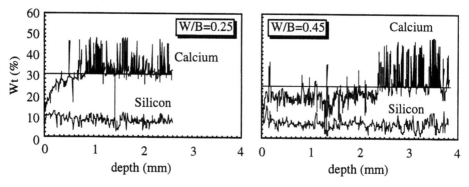

Fig. 3. Microprobe analyses of the two OPC pastes (after 3 months of immersion in deionized water)

In all control samples (i.e. not immersed in the aggressive solution), the concentration of calcium and silicium were found to be constant over the full thickness of the disk. However, after each period of exposure to each aggressive solution, the concentration of calcium near the external part of all disks (in contact with the solution) was lower than in the middle part of the disk. This leaching phenomenon was observed for each of the four mixtures and was found to increase with the time of exposure. The silicium concentration was found to be fairly constant throughout the test period for each mixture and each aggressive solution.

As can be seen in Figure 3, the microprobe profiles for the 0.25 and 0.45 water/binder ratio mixtures after 3 months of immersion are quite different. The calcium content for the 0.25 water/binder ratio mixture is lower than in the control specimen in a zone covering approximately the first 750 µm. In this zone, the C-S-H appear to be partly decalcified. Behind this zone, the calcium profile is similar to that of the control specimen and the C-S-H do not seem to be decalcified (no holes in the calcium profile). The calcium content for the 0.45 water/binder ratio mixture is lower than in the control specimen over almost the full thickness of the disk. There appears to be no more portlandite left in the sample. This result is confirmed by the TGA/DTA analyses. There are signs of C-S-H decalcification at least over the first 2500 µm.

Microprobe analyses also indicate that the addition of silica fume contributes to reduce the calcium leaching process. There is almost no portlandite left in the 0.45 water/binder ratio mixture without silica fume and C-S-H appear to be decalcified practically over the full thickness of the disk. For the 0.45 mixture with silica fume, the decalcified zone is limited to the first 2500 µm and C-S-H are partly decalcified only in the first 1000 µm. The concentration of calcium in the mixture without silica fume is lower than in the control specimen over the full thickness of the disk, while for the silica fume mixture, the calcium content corresponds to the minimum concentration (base line), except for the first 1000 µm. The influence of silica fume on the portlandite content has been confirmed by the TGA/DTA analyses. The positive influence of silica fume is also noticed for the 0.25 mixtures.

8 Numerical Simulations

Table 3 presents the molar and volume fractions of the calcium and portlandite for the four mixtures calculated according to the method proposed by ADENOT [5].

Table 3. Molar and volumic fractions of calcium and portlandite

Mixture	S_{tot} (mM/l)	S_{por} (mM/l)	V_{si}**	V^i	V^i_{por}
P25Q	20158	7553	0.16	0.58	0.25
P25QS	18421	4105	0.20	0.63	0.14
P45Q	14730	5550	0.12	0.43	0.18
P45QS	13558	3020	0.14	0.48	0.10

** Volume fraction of SiO_2

The simulations were run using the effective tritiated water diffusion coefficient as D_0. The value of D_s was first assumed to be equal to the effective diffusion coefficient of calcium in free water (0.8×10^{-9} m^2/s). In a first step, the contributions of the dissolution of both portlandite crystals and C-S-H phases to the diffusion properties are assumed to be equal and the value of β was therefore fixed at 1.

The result of this first numerical simulation for the 0.45 mixture without silica fume is given in Table 4. As can be seen, the depth of calcium leaching is underestimated by such initial parameters. A good prediction of the depth of calcium leaching would be obtained by increasing the values of D_s and β. However, the value of D_s can not be increased because it represents the diffusion of calcium in free water, which is the highest value that parameter can take. β is thus the only variable that can be adjusted to fit the experimental test results.

Table 4.   Numerical simulations for the 0.45 mixture without silica fume
after 3 months of immersion in deionized water

Ca^{2+} at surface (mM/l)	D_s $(10^{-9}\ m^2/s)$	β	Simulated depth of calcium leaching (mm)	Experimental depth of calcium leaching (mm)
0	0.8	1	2.1	2.5-2.7

In fact, the value of D_s is probably lower than that in free water. The residual skeleton of a completely decalcified cement paste still constitutes a very complicated porous network [11]. The high tortuosity of this network certainly modifies the value of the effective diffusion coefficient of calcium. The value of D_s was thus estimated to be equal to $0.4 \times 10^{-9}\ m^2/s$ which represents a value half that of free water.

As can be seen in the microprobe profiles, the amount of calcium at the surface of samples in contact with pure water was never found to be zero. Although the solution in the tank was constantly agitated and regularly renewed, it is probable that a transient zone containing a certain amount of calcium ions was formed at the surface of the sample. The presence of this transient zone surely affected the diffusion of calcium ions out of the samples. In order to account for the effect of this transient zone, the calcium concentration at the surface of the samples was not set equal to 0 mM/l but was rather approximate at 1.5 mM/l.

Table 5 presents the results of the numerical simulations for the 4 mixtures after a 3-month immersion period. As can be seen, the experimental depth of calcium leaching can be well approached with $\beta=2.3$ for the 0.45 mixtures with or without silica fume and with $\beta=1.3$ for the 0.25 mixtures (with or without silica fume).

Table 5.   Numerical simulations after 3 months of immersion in deionized water

Mixture	Ca^{2+} at surface (mM/l)	D_s $(10^{-9}\ m^2/s)$	β	Simulated depth of calcium leaching (mm)	Experimental depth of calcium leaching (mm)
P25Q	1.5	0.4	1.3	0.85	0.8-1.0
P25QS	1.5	0.4	1.3	0.5	0.5
P45Q	1.5	0.4	2.3	2.7	2.5-2.7
P45QS	1.5	0.4	2.3	1.9	1.8-2.2

The values of D_s and β appearing in Table 5 were used to simulate the calcium profiles of the various mixtures after 6 months of immersion. The simulated calcium profiles appear in Figure 4 as solid lines. As can be seen in Figure 4, the simulated profiles fit quite well the calcium distribution in the paste samples obtained with the microprobe measurements.

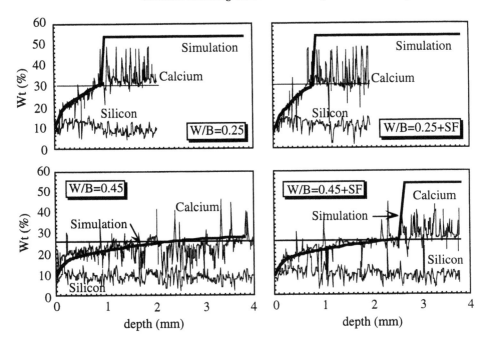

Fig. 4. Microprobe analyses of the four mixtures (after 6 months of immersion in deionized water)

9 Discussion

The experimental test results clearly underline the influence of the microstructure on the chemical degradation. These results are in good agreement with those reported by ADENOT and BUIL [16] and with the numerical simulations of BENTZ and GARBOCZI [6]. The rate at which the paste disks are decalcified depends mainly on the diffusivity of calcium that has to be leached. A reduction of the water/binder ratio and the use of silica fume contribute to the formation of a denser microstructure which clearly modifies the transport properties as shown by the tritiated water diffusion coefficient measurements. A reduction of the effective diffusion coefficient obviously slows down the diffusion of calcium. The calcium concentration in the pore solution remains higher which preserve the chemical equilibrium between the hydrates and the pore solution.

The fact that the numerical simulations could reliably reproduce the chemical degradation of the various samples after an immersion period of 6 months tends to indicate that the hypotheses at the basis of the model are valid. The estimated values of β used for the simulations can also bring forward additional information on the mechanisms that control the leaching process. As previously mentioned, β more or less represents the influence of the dissolution of calcium-bearing hydrates on the diffusion process. An increase in the value of β implies that the dissolution of calcium has a strong effect on the diffusion properties of the material by creating a more open pore network which facilitate the movement of calcium ions.

As can be seen in Figure 2, the increase in the diffusion coefficient (as β increases) is mainly due to the dissolution of portlandite crystals which are in equilibrium in a solution having a calcium concentration of 20 mM/l (see Figure 1). This confirms the numerical simulations of BENTZ and GARBOCZI [6] where the opening of the pore network caused by the dissolution of portlandite crystals was found to strongly influence the transport properties.

It is also interesting to note that the value of β is solely affected by the water/binder ratio, and that the addition of silica fume has no (or a very negligible) influence on this coefficient. The size of portlandite crystals has been found to be affected by a number of parameters [17]. However, as emphasized by SCRIVENER [18], there exists very little quantitative information in the literature on the parameters that influence the distribution of portlandite in the hydrated cement paste matrix. The variation of the value of β tend to indicate that, if silica fume contributes to a reduction of the total amount of portlandite in the matrix, the dissolution of the remaining crystals still has a critical influence on the evolution of the diffusion properties upon leaching.

10 Conclusion

The experimental test results clearly show that the kinetics of degradation is very sensitive to the material microstructure. A reduction of the water/binder ratio and the use of silica fume modify the microstructure of a cement paste, reduce its diffusivity, and significantly improve the resistance of concrete to calcium leaching.

The numerical simulations show that the use of a macroscopic model can be used with a fairly good accuracy to predict the degradation mechanisms of cement pastes by calcium leaching.

The simulations also confirm that the portlandite content of a cementitious system appears to be the key parameter controlling the diffusion of calcium.

11 References

[1]   VERNET, C. (1992), *Stabilité chimique des hydrates. Mécanismes de défense du béton face aux agressions chimiques,* in La durabilité des bétons, Edited by J. Baron and J.P. Ollivier, Presses de l'École Nationale des Ponts et Chaussées, Paris, pp. 129-169.

[2]   MARCHAND, J. (1993), *Contribution à l'étude des mécanismes de détérioration par écaillage des bétons en présence de sels fondants,* Ph.D. Thesis, École Nationale des Ponts et Chaussées, Paris, France, 326 p., (in French).

[3]   ATKINSON, A., EVERITT, N.M., GUPPY, R. (1987), *Evolution of pH in a radwaste repository — Experimental simulation of cement leaching,* Report AERE-R 12594, Materials Development Division, Harwell Laboratory, Oxfordshire, England, 25 p.

[4]   BERNER, U.R. (1992), *Evolution of pore water chemistry during degradation of cement in a radioactive waste repository environment,* Waste Management, Vol. 12, pp. 201-219.

[5]   ADENOT, F. (1992), *Durabilité du béton: Caractérisation et modélisation des processus physiques et chimiques de dégradation du ciment,* Ph.D. Thesis, Université d'Orléans, France, 238 p., (in French).

[6]   BENTZ, D.P., GARBOCZI, E.J. (1992), *Modelling the leaching of calcium hydroxide from cement paste: Effects on pore space percolation and diffusivity,* Materials and Structures/Matériaux et Constructions, Vol. 25, pp. 523-533.

[7]   REARDON, E.J. (1990), *An ion interaction model for determining ion equilibria in cement/water systems*, Cement and Concrete Research, Vol. 20, N° 1, pp. 175-192.

[8]   MA, W., BROWN, P.W., SHI, D. (1992), *Solubility of Ca(OH)₂ and CaSO₄.2H₂O in the liquid phase from hardened cement paste*, Cement and Concrete Research, Vol. 22, N° 4, pp. 531-540.

[9]   ATKINS, M., BENNETT, D.G., DAWES, A.C., GLASSER, F.P., KINDNESS, A., READ, D. (1992), *A thermodynamic model for blended cements*, Cement and Concrete Research, Vol. 22, N° 2/3, pp. 497-502.

[10]   BERNER, U.R. (1988), *Modelling the incongruent dissolution of hydrated cement minerals*, Radiochinica Acta, Vol. 44/45, pp. 387-393.

[11]   GÉRARD, B. (1995), *Vieillissement des structures de confinement en béton pour l'entreposage des déchets radioactifs*, Ph.D. Thesis, Département de Génie Civil, Université Laval/École Normale Supérieure de Cachan, Québec, Canada, (in preparation, in French).

[12]   GÉRARD, B., DIDRY, O., MARCHAND, J., BREYSSE, D., HORNAIN, H. (1995), *Modelling the long-term durability of concrete for radioactive waste disposals*, submitted for publication in Mechanisms of Chemical Degradation of Cement-Based Systems, Edited by K.L. Scrivener and J.F. Young, E & FN SPON, London, England, 14 p.

[13]   DELAGRAVE, A. (1995), *Étude des mécanismes de pénétration des ions chlore dans les bétons conventionnels et à haute performance*, Ph.D. Thesis, Université Laval, Québec, Canada, (in preparation-in French).

[14]   JEANVOINE, E., DE GAYFFIER, A. (1995), *La thermique transitoire et le couplage faible thermique-mécanique dans CASTEM 2000*, C.E.A. Report, DMT/95-134, France, (in French).

[15]   CHATTERJI, S., KAWAMURA, M. (1992), *A critical reappraisal of ion diffusion through cement based materials — Part 1: Sample preparation, measurement technique and interpretation of results*, Cement and Concrete Research, Vol. 22, N° 3, pp. 525-530.

[16]   ADENOT, F., BUIL, M. (1992), *Modelling the corrosion of the cement paste by deionized water*, Cement and Concrete Research, Vol. 22, N° 2/3, pp. 489-496.

[17]   BERGER, R.L., MCGREGOR, J.D. (1973), *Effect of temperature and water-solid ratio on growth of Ca(OH)₂ crystals formed during hydration of Ca₃SiO₅*, Journal of The American Ceramic Society, Vol. 56, N° 2, pp. 73-79.

[18]   SCRIVENER, K. L. (1989), *The microstructure of concrete*, in Materials Science of Concrete, Vol. 1, Edited by J.P. Skalny, American Ceramic Society, pp. 127-161.

6 DIFFUSIBILITY CHANGES DUE TO THE PRESENCE OF CHLORIDE IONS

P.E. STREICHER and M.G. ALEXANDER
Department of Civil Engineering, University of Cape Town, South Africa

Abstract
Various types of concretes, saturated with a high concentration NaCl solution, were tested and then retested after two weeks for their conductivity values. The results showed that the presence of chloride ions at high concentrations substantially reduced the diffusibility of concrete in a short period of time. The magnitude of this effect differed significantly between cement types which, for this work, included ordinary portland cement (OPC) and OPC blends with fly ash and ground granulated blastfurnace slag (GGBS). The results indicate that advantage can be taken of different cement-based systems to control the rate at which chloride ions enter concrete.
Keywords: Chlorides, concrete, effect on diffusion properties, microstructural changes.

1 Introduction

Chloride-induced corrosion of reinforcing steel is a very common cause of deterioration of reinforced concrete structures. Chlorides enter concrete mainly via diffusion through the concrete pore solution. It is well known that cement-based systems can bind chloride ions and thereby reduce the rate of chloride ingress into the material (apparent diffusivity). However, not much is known on how the presence of chloride ions affects the physical resistance of concrete to chloride transport (diffusibility).

Mangat and Molloy [1] have shown that the apparent diffusivity of concrete exposed to a chloride environment reduces substantially over time. This reduction does not necessarily indicate a change in the diffusibility of the material due to the presence of chloride ions, as there are a number of factors that could cause a

Mechanisms of Chemical Degradation of Cement-based Systems. Edited by K.L. Scrivener and J.F. Young. Published in 1997 by E & FN Spon, 2-6 Boundary Row, London SE1 8HN. ISBN: 0419215700.

reduction in the apparent diffusivity of concrete with time:

1. An improvement in concrete quality with depth, ie. lower diffusibility or higher binding capacity at depth would reduce the apparent diffusivity of the material. This is often the case when concrete is poorly cured initially.
2. Additional hydration. Some concrete mixes, particularly those with blended cements, can continue to hydrate for long periods, reducing the diffusibility and hence the apparent diffusivity of the material with time.

A number of researchers (see literature review) indicate that the presence of chloride ions affects the pore structure of cement-based systems. This work was done to investigate to what degree the presence of chloride ions influences the diffusibility of concrete.

2 Review

Midgely and Illston [2] showed with mercury intrusion porosimetry (MIP) tests that the presence of chloride ions (externally derived) reduced the size of the pores of the highest frequency of occurrence. No significant change in the quantity of hydrates could be measured (ie. the presence of chlorides did not accelerate hydration), so it was suggested that the reduction in pore size was due to the formation of calcium chloride on the surface of the CSH. Kayyali and Haque [3] reported that difficulty was experienced with the process of porewater expression for specimens which were immersed in a sodium chloride solution for 90 days compared to similar specimens cured in a fog room, indicating a decrease in water permeability when chloride ions are present.

Suryavanshi et al [4] showed with MIP tests that the addition of chlorides to OPC and SRPC mortars during mixing reduced the total accessible pore volume compared to corresponding chloride-free mortars. They stated that the quantity of chloride ions bound by CSH gel would not alter the hydrate structure sufficiently to cause the observed effects, based on work by Lambert et al [5]. They ascribed the reduction in accessible pore volume to a change in the morphology of the CSH gel. SEM micrographs showed the needle-like structure of the CSH gel for the chloride free mortars, while the mortars in the presence of chloride ions had a much denser structure. It was not clear from the literature whether the ingress of externally derived chloride ions into mature concretes would have the same effect.

Thus in summary, a number of researchers have identified a change in concrete microstructure and pore system due to the presence of chloride ions, but they disagree on the exact mechanisms involved in this change.

2.1 Chloride conduction test
The chloride conduction test developed at the University of Cape Town [6] has been used extensively in our laboratory to characterise the physical resistance of a range of concrete mixes to chloride transport. The test involves vacuum-saturating oven-dried (7 days at 50 °C) concrete samples, 68 mm diameter x 25 mm thick, with a 5 M NaCl solution, before measuring the steady state conductivity of the sample. The

conduction cell was designed to measure conductivity independent of the cell solution concentration.

Atkinson and Nickerson [7] showed that there is a material parameter Q (diffusibility) which can be obtained either by diffusion measurements, or by conductivity measurements (see equation 1). This material parameter is influenced by the tortuosity and the constrictivity of the concrete pore structure.

$$Q = \frac{D}{D_0} = \frac{\sigma}{\sigma_0} \qquad (1)$$

where Q= diffusibility of porous material, D = diffusivity of ion through porous material, D_0 = diffusivity of ion through pore solution, σ = conductivity of porous material, and σ_0 = conductivity of pore solution.

Saturating samples with a highly conductive solution ensures that they have virtually the same pore solution conductivity, unlike normal resistivity tests which are substantially influenced by the pore solution conductivity of the concrete. Different concrete samples therefore yield different conductivities primarily because of differences in their diffusibilities (physical resistance of pore structure to ionic transport). The test is sensitive to important material and constructional factors such as w/c ratio, cement type, and curing (see later). The test procedure allows for concrete samples to be left immersed in the chloride solution, and retested at intervals to determine changes in diffusibility. Factors such as changes in the concrete/mortar pore structure in the presence of chloride ions and additional hydration would reduce the diffusibility of the material.

3 Experimental procedure

A series of Ordinary Portland Cement (OPC) mixes, OPC and fly ash (70/30) blends and OPC and GGBS (50/50) blends with target strengths of 20, 40 and 60 MPa were prepared and tested for chloride conductivity. The cement compositions are shown in table 1, and mix details in table 2.

Aggregates were a fine dune sand (F.M. = 2.0) and a 19 mm greywacke stone. Concrete cubes (100 mm) were cast in steel cube moulds and demoulded after 24 hours.

Table 1. Chemical analyses of cements

%	SiO_2	Al_2O_3	Fe_2O_3	CaO	SO_3	K_2O	Na_2O
OPC	20.9	3.8	3.3	64.6	1.8	0.5	0.2
OPC/FA blend 70/30	30.6	13.2	3.4	46.2	1.4	0.5	0.3
OPC/GGBS blend 50/50	27.7	9.7	2.0	49.7	2.4	0.7	0.3

Table 2. Mix details

Target Strength (MPa)	OPC mixes		Fly ash mixes		GGBS mixes	
	W/C	C/m³	W / (C+FA)	(C+FA) / m³	W / (C+GGBS)	(C+GGBS)/ m³
20	0.83	240	0.76	245	0.83	222
40	0.55	360	0.48	380	0.55	334
60	0.34	465	0.30	495	0.38	398

The mixes were wet cured at 23 °C for either 1, 7 or 28 days (including the first 24 hours in the moulds). The 1 and 7 day wet cured samples were further cured at 60% R.H. and 25 °C until 28 days when samples were removed by coring and slicing, and tested using the chloride conductivity test described above.

Three specimens were tested for each mix and curing combination. The mean of the three test determinations was used to obtain one test result. The coefficient of variation of a single test result (single operator precision) was estimated to be 6%, and the results ranged from 0.1 mS/cm to 5 mS/cm. Clear and consistent trends could therefore be established.

The specimens were kept immersed in the 5 M NaCl solution after the 28 d conductivity test, and retested after two weeks.

4 Discussion of results

Results of the initial (28 day) chloride conductivity tests are displayed in figure 1. Chloride "isoconductivity" lines were interpolated between the measured values. The following trends can be observed.

1. The OPC mixes were not as sensitive to curing as the fly ash and GGBS mixes.
2. All mixes were less sensitive to curing at higher strengths.
3. Substantially lower conductivities were generally obtained using blended mixes.

The percentage reductions in conductivity between the initial test and the subsequent 2 week chloride conductivity test are shown in table 3. Reductions in chloride conductivity over the two weeks were least for the OPC mixes (44% mean), somewhat more for the fly ash mixes (54% mean) and greatest for the GGBS mixes (70% mean).

Trends of chloride conductivity with strength depended on cement type. Generally, greater reductions occurred for the higher strength mixes. There was also a general trend of greater reductions in conductivity for mixes cured for shorter periods.

The reduction in conductivity was mainly ascribed to a change in the physical pore structure (increase in the tortuosity and constrictivity of the pores) due to the presence of chloride ions, and to a lesser degree to:
1. Reduction in pore solution conductivity. Previous work has shown that a 35%

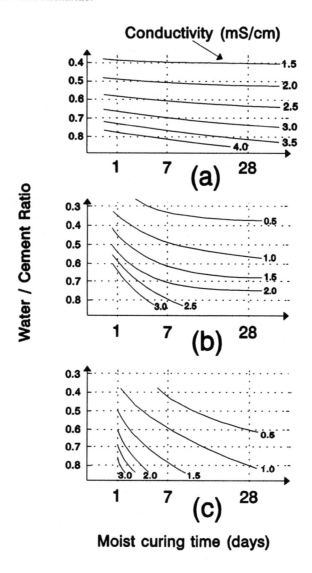

Fig. 1   28 d chloride-isoconductivity graphs for (a) OPC, (b) fly ash and (c) GGBS mixes

reduction in chloride concentration (from 5 M NaCl) results in only about a 14% reduction in conductivity [6].

2. Additional hydration. The differences between samples cured for shorter periods before re-saturation for the conductivity test, and those cured for a full 28 days, are relatively small, except possibly for the GGBS mixes. In contrast, the OPC mixes which were insensitive to initial curing (figure 1) showed a substantial reduction in chloride conductivity. This would seem to indicate that additional hydration from delayed curing has a minor influence.

Table 3. Reductions in conductivity after 2 weeks immersion in chloride solution

Cement type	Wet curing period (days)	Percentage reduction in chloride conductivity (%) Target strength (MPa)			
		20	40	60	Mean
OPC	1	36	45	58	46
	7	28	46	54	44
	28	36	40	54	43
	Mean	33	44	55	44
OPC/fly ash 70/30	1	58	48	64	54
	7	54	47	59	52
	28	57	57	52	56
	Mean	56	51	58	54
OPC/GGBS 50/50	1	53	84	86	77
	7	52	78	70	70
	28	55	66	66	63
	Mean	53	76	74	70

Verification for the above hypothesis also comes from results of another series of tests using a modified version of the Dundee rapid chloride test [8], where the electric potential was maintained for a period of 14 d [9]. A continuous flux of chloride ions flowed through the specimen preventing the depletion of chloride ions due to binding. A reduction in chloride flux occurred with time, as measured by electric current - see figure 2.

Mackechnie [10] performed chloride conductivity tests in our laboratory on a range of specimens immersed in 5 M NaCl solutions (after 28 days moist curing) for various periods up to 98 days - see figure 3. A rapid reduction in conductivity occurs in the first two weeks (28 - 42 days) followed by a more gradual reduction up to 98 days. Mackechnie stated that correlations between chloride conductivity values and two-year apparent diffusion coefficients of marine exposed concretes were improved substantially when 98 day chloride conductivity values were used instead of 28 day values. The improved correlations were probably due to both the effect of the presence of chlorides on the diffusibility of the material, and additional curing (since the on-site exposed concretes also underwent additional curing).

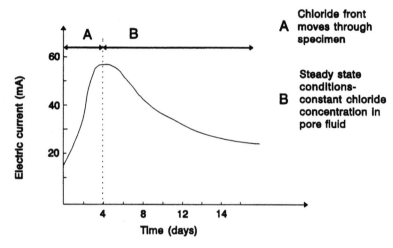

Fig. 2. Reduction in chloride conductivity with constant electric potential applied (10 V).

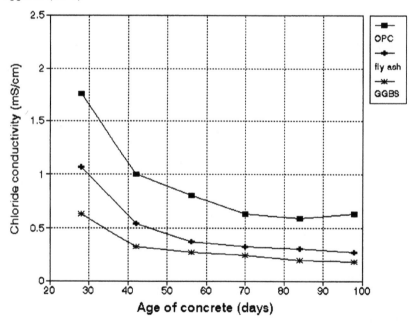

Fig. 3. Reduction in chloride conductivity after immersion in chloride solution at 28d, 40 MPa concretes [10].

5 Conclusions

1. The chloride conductivity of a range of concrete mixes was found to reduce significantly with time (over a two week period) when immersed in a 5 M NaCl

solution. It is suggested that this reduction is mainly due to a change in the physical pore structure affecting the diffusibility of the material. The exact mechanism causing this change is not known, although researchers have suggested that the formation of calcium chloride on the surface of the CSH, or a change in the morphology of the CSH gel occurs in the presence of chloride ions.

2. The reductions in chloride conductivity varied for different cement types. The results indicate that advantage can be taken of different cement-based systems to control the rate at which chloride ions enter concrete.
3. The reductions were only slightly higher for mixes with shorter initial wet curing periods indicating that delayed curing did not have the same benefit as initial curing.
4. Further work is contemplated to determine whether the reduction in conductivity occurs with mature concretes. Pore expression work will also be conducted to measure the change in conductivity in the pore solution as a result of chloride binding. Different concentration chloride solutions will be used to determine to what degree the chloride concentration influences the measured reduction in conductivity.

6 References

1. Mangat, P.S. and Molloy, B.T. (1994) Prediction of long term chloride concentration in concrete, *Materials and Structures*, Vol. 27, 1994, pp. 338-346.
2. Midgely, H.G. and Illston, J.M. (1984) The penetration of chloride ions into hardened cement pastes, *Cement and Concrete Research*, Vol. 14, 1984, pp. 546-553.
3. Kayyali, O.A. and Haque, M.N. (1988) Chloride penetration and the ratio of Cl^-/OH^- in the pores of cement paste, *Cement and Concrete Research*, Vol. 18, pp. 895-900.
4. Suryavanshi, A.K., Scantlebury, J.D. and Lyon, S.B. (1995) Pore size distribution of OPC & SRPC mortars in presence of chlorides, *Cement and Concrete Research*, Vol. 25, No. 5, pp. 980-988.
5. Lambert, P., Page, C.L. and Short, N.R. (1985) *Cement and Concrete Research*, Vol. 15, pp. 675.
6. Streicher, P.E. and Alexander, M.G. (1995) A chloride conduction test for concrete, *Cement and Concrete Research*, Vol. 25, No. 6, pp. 1284-1294.
7. Atkinson, A. and Nickerson, A.K. (1984) The diffusion of ions through water-saturated cement, *Journal of Material Science*, Vol. 19, pp. 3068-3078.
8. Dhir, R.K., Jones, M.R., Ahmed, H.E.H. and Seneviratne, A.M.G. (1990) *Magazine of Concrete Research*, Vol. 42, No. 152, pp. 177-185.
9. Streicher, P.E., Peters, A.G.E. and Alexander, M.G. (1994) Chloride diffusion through fly ash concrete, *2nd International Symposium: Ash - A Valuable Resource*, South African Coal Ash Association, Vol. 2, pp. 623-631.
10. Mackechnie, J.R. (1995) PhD thesis in progress, University of Cape Town.

7 THE DISPERTION OF SILICA FUME

D.A. ST JOHN
Industrial Research Limited, Lower Hutt, New Zealand

Abstract
Agglomeration of silica fume causes the effective mean particle size of silica fume to vary between 1 and 50 µm, much larger than the 0.1 - 0.2 µm extensively quoted in the literature. The distribution and degree of agglomeration is dependent on source, bulk density and age. Ultrasonic dispersion can be used to indicate the dispersivity of silica fume. In practice undispersed agglomerates will always be present in mortars and concrete and under appropriate conditions may lead to alkali-aggregate reaction. It is hypothesised that the fusion of particles into chains is largely responsible for agglomerates that cannot be dispersed.
Keywords: Concrete, dispersion, DSP, silica fume.

1 Introduction

St John et al. [1] reported on the properties of ultra-high strength mortars (DSP) made from silica fume, New Zealand cement and aggregate. Over a period of years it was found that alkali-aggregate reaction developed in test samples exposed to wetting and drying. Petrographic examination identified agglomerates of silica fume as the cause of the alkali-aggregate reaction. He found that some 25% of the silica fume had not been dispersed and was still present as agglomerates exceeding 10 µm in size [2].

1.1 Literature
Bonen and Diamond [3] observed and analysed silica fume agglomerates in laboratory samples of hardened cement paste and found they had only partially reacted with calcium hydroxide. Similarly Pettersson [4] reported that fragments of silica fume granules were often detectable by thin section analysis of mortars. She also reported that granulated, that is fully densified silica fume, caused cracking in mortars tested in 1M sodium chloride

Mechanisms of Chemical Degradation of Cement-based Systems. Edited by K.L. Scrivener and J.F. Young.
Published in 1997 by E & FN Spon, 2–6 Boundary Row, London SE1 8HN. ISBN: 0419215700.

solution which was absent when silica fume slurry was used. Agglomerates in test specimens have also been reported by Shayan *et al.* [5] who found they acted like reactive aggregates at high alkali levels. Sveinsdóttir and Guðmundsson [6] found agglomerates present in a range of Icelandic concretes containing up to 7.5% of silica fume.

There is some confusion over the effective particle size distribution of silica fume. ACI Committee 226 [7] gave a particle size distribution curve quoted as being measured by Fiskaa *et al.* [8] which shows that typically 20% of fume passes 0.05 μm and 100% passes 0.5 μm with a mean size of about 0.1 μm. Examination of the report by Fiskaa *et al.* [8] failed to find the data quoted by ACI Committee 226 and the derivation of that particle size distribution curve is not known. Fiskaa *et al.* present a partial particle size distribution curve of silica fume determined by the Andreason pipette method which is similar to other distribution curves given in this report.

Kolderup [9] investigated silica fume in a dust removal system and found that condensed silica fumes are not really composed of individual microspheres ranging from 0.02-2.0 μm in size, but of agglomerations of these microspheres. Aitcin [10] and De Larrard [11] gave data to show that because of agglomeration silica fume effectively has a mean particle size at least one order of magnitude larger than the 0.1 to 0.2 μm that is widely quoted in the literature. Yonezawa *et al.* [12] and Asakura *et al.* [13] investigated a range of silica fumes. They found that the dispersivity of a silica fume could be measured by plotting the amount of material less than 1 μm against time of ultrasonic dispersion in water. They also found a wide variation in the dispersivity of silica fumes with some indication that those with the lower BET surface areas were more dispersible and less likely to lose dispersivity when stored for up to a year.

1.2 Commercial production of silica fume

The production, composition and many other details of silica fume has been described by Aitcin [10]. When the silica fume is initially collected in the storage silo it typically has a loose bulk density of about 125 to 150 Kg/m^3. Silica fume in this form is difficult to handle and transport and is also a potential health hazard. To overcome these problems silica fume is densified, presumably a combination of both closer packing and agglomeration caused by intermixing. This is achieved by tumbling in a mixer or by blowing air through the silo. One day's aeration raises the bulk density to about 300 kg/m^3 and full densification to 600 kg/m^3 or greater may require at least a week's aeration. The current terms used to describe silica fume as undensified, that is 300 to 400 kg/m^3, and densified, greater than 600 kg/m^{3,} are confusing and would be better described as partially densified and fully densified fumes as used in this report.

2.1 Materials

The following silica fumes were used.

A. Raw fume (128 kg/m^3) without densification, partially densified fume (370 kg/m^3) and fully densified fume (690 kg/m^3). The samples were less than four weeks old at the time of testing. A commercial slurry of unknown age made from this silica fume.

B. Partially densified (370 Kg/m^3) and fully densified (600 kg/m^3) silica fume including a commercial slurry of unknown ages. Manufacturing source as for samples A.

C. A fully densified silica fume (>600 kg/m^3) less than four weeks old.

D. A fully densified silica fume (630 kg/m^3) less than four weeks old.

E. Partially densified Tasmanian silica fume (390 kg/m^3), originally about one to two years old when used [1] and now about ten years old.

F. Fully densified Elkem F100T silica fume (710 kg/m^3) about ten years old.

A 70 x 48 mm core drilled through a concrete slab of a large project under construction. The concrete contained 10% silica fume from the same source as sample A. The age and strength of the concrete at time of sampling was 28 days and 60 MPa respectively.

2.2 Experimental
A Phillips EM 400T transmission electron microscope (TEM) was used to examine the silica fumes which were dispersed in water by 2 minutes of ultrasonic treatment before being mounted on a grid for examination. The BET surface areas were measured on a Quantachrome surface area analyser using a helium/nitrogen mixture and single point determination. Chemical analyses of silica fumes were determined by XRF and weight losses by TG770 Stanton thermobalance using 30 ml/min of dry air at a heating rate of 10 °C/minute.

A Shimadzu SALD-2001 particle size analyser was used to determine the dispersivity of the samples. All the measurements reported were carried out in water using high stirring and circulation speeds. An initial measurement was taken without ultrasonic dispersion and then 30 watts of ultrasonic dispersion was applied to the suspension for fifteen minutes recording particle size distributions on the circulating suspension at regular intervals. In addition, some limited particle size measurements were carried out with a range of superplasticiser additions to the water and also in a saturated solution of calcium hydroxide adjusted to a pH 13 by the addition of sodium hydroxide.

Two complete 70 x 40 mm slices, cut along axial planes of the drilled concrete core, were thin sectioned to a thickness of 25 μm and examined using a petrographic microscope to determine the amount of silica fume agglomerates present and other relevant petrographic data.

3 Results

Table 1 gives the surface areas, including the calculated mean particle sizes, and thermal analysis measurements of the weight loss to 450 °C and from 450 - 950 °C to represent approximate losses due to water, volatiles and carbon present in the silica fumes.

All the silica fume samples examined appeared similar when examined by TEM so only typical examples are given in Figs 1a and 1b. The results of the particle size analyses for fifteen minutes ultrasonic dispersion in water are given in Figs 2 and 3, excluding those for sample B which are similar to sample A. Since neither the addition of superplasticiser nor the use of a pH 13 solution significantly affected the particle size distributions these results are not presented. The data for the particle size analyses versus time of ultrasonic dispersion are given in Fig. 4 as percentage of particles less than 1 μm. This is a convenient method of presenting the large mass of data involved and also shows the time required to obtain maximum dispersion below 1 μm.

Petrographic examination of the thin sections of the concrete core identified 2% of the volume of the hardened concrete consisted of silica fume agglomerates. This equates to

15% of the silica fume remaining as agglomerates most of which were 50 μm or greater in size. No agglomerates less than 30 μm in size were detected and the largest agglomerate seen was 250 μm. Agglomerates varied from globular to irregular in shape and in some cases their interiors were coloured light brown. Crystals of calcium hydroxide inside agglomerates were not identified and obvious interaction of the agglomerates with the pore solution was not visible.

	A-128	A-370	A-690	B-370	B-600	C-600	D-630	E-390	F-710
LOI	4.18	5.76	4.56	4.69	4.96	2.10	2.32	2.09	5.65
-450 °C	0.41	0.53	0.36	1.06	1.75	0.96	0.90	1.52	3.44
+450 °C	2.75	4.44	3.61	2.72	3.36	1.12	0.87	0.68	3.18
BET (m²/g)	29.7	31.0	25.0	23.8	20.6	18.8	21.2	17.5	16.9
Mean size (μm)	0.09	0.09	0.11	0.11	0.13	0.14	0.13	0.15	0.16

Table 1: Results of surface area measurements and thermo-analytical losses of silica fumes.

4 Discussion

The results for the particle size distributions, chemical compositions, surface areas and thermal weight losses are similar to those reported by Aitcin [10]. The silica fumes investigated in this study fall within the typical range of commercially available silica fumes derived from the production of silicon or iron-silicon 90% metals. The mean particle sizes calculated from the surface areas measured in this study are similar to those quoted in the literature although sample A appears to have a higher surface area than previously reported.

4.1 Dispersion of silica fume
The results presented in this study, also confirmed by the literature, clearly indicate that the mean diameter of silica fume is not 0.1 to 0.2 μm. With extended ultrasonic dispersion the reduction in agglomeration only reduces the mean particle size to at the best about 1 μm and it ranges up to at least 50 μm in samples that do not disperse well. This indicates that some portion of silica fume is always agglomerated and the question arises as to the nature of the agglomeration. TEM observations show that silica fume consists of spherical particles which range from 20 nm to about 0.5 μm in size. Studies of sieve residues [14], indicate that small amounts of impurities are also present as particles larger than 1 μm. However, the TEM observations also show that most of the spherical particles present are aggregated together as agglomerates and that some particles are fused to a neighbour and in cases even occur as chains as reported by Kolderup [9]. Individual spheres in the agglomerates will be held together by a combination of Van der Waals forces, fusion of spheres, entanglement of fused spheres and chains which also may entrap individual spheres. This scenario is sufficient to explain the presence of the agglomerates without invoking other forms of chemical bonding between the spheres.

There appears to be little information on the extent of fusion occurring between particles in silica fumes with source of manufacture. Results given in this study and the work of Asakura et al [15] show that the dispersivity of silica fume is not necessarily related to the bulk density or time of storage. This suggests that a range of dispersibilities will apply varying from silica fumes that effectively cannot be dispersed to those which will be easily dispersed. It is hypothesised that the extent to which an agglomerate can be broken up will be a function of

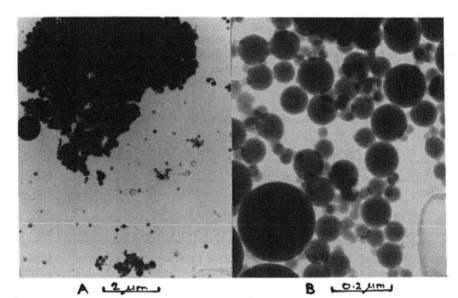

Fig.1.   A. TEM micrograph of fully densified silica fume A showing that very few individual spherical particles are present unattached. The dispersivity of a silica fume will depend on the extent the agglomerates shown can be broken up. B. Detail of densified silica fume A showing that many of the spherical particles are fused together and form chains.

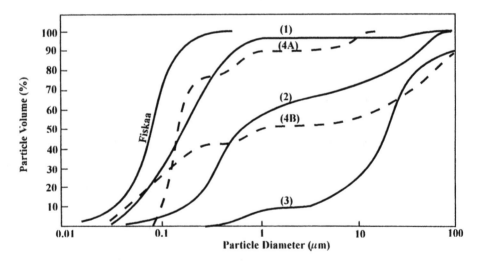

Fig.2.   The particle size distributions of silica fume A after 15 minutes ultrasonic dispersion. Raw fume (128 kg/m³) curve 1, partially densified (350-400 kg/m³) curve 2, fully densified (630 kg/m³) curve 3, commercial slurry curve 4A and commercial slurry without ultrasonic treatment, curve 4B. Fiskaa's curve (ACI Committee 226, 1987) represents a fume where all the spherical particles are unattached and may be considered as the theoretical limit of the possible particle size distribution of a silica fume.

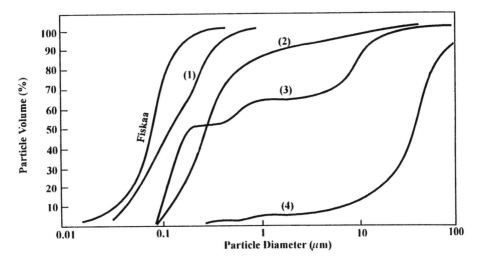

Fig.3. The particle size distributions of silica fumes after 15 minutes ultrasonic dispersion. Densified silica fume C (>600 kg/m³), curve 1, densified silica fume D (600 kg/m³), curve 2, Tasmanian silica fume E (390 kg/m³), curve 3, and Elkem densified silica fume F (710 kg/m³), curve 4.

relative amounts of binding by Van der Waals forces and entanglement. If entanglement of fused spheres and chains predominates it will be difficult break up agglomerates. Formation of smaller agglomerates is more likely than significant release of individual spheres. Dispersion of a silica fume into slurry form may not significantly alter this situation.

The particle size distribution measured by extended ultrasonic dispersion may be a limiting state of dispersion for a silica fume. This assumption has already been made by other workers [12, 13] and it is considered the results obtained in this investigation confirm this assumption. Reference to Fig.4 clearly shows that ease of dispersion of silica fumes varies considerably with some fumes essentially remaining undispersible. The question remains as to the applicability of this assumption to the mixing of silica fume in mortar and concrete.

4.2 Application to concrete

There appears to be a widely held assumption that the agglomerates present in the silica fume are sufficiently dispersed by intermixing in mortars and concrete so as to be able to react as a pozzolan and in the case of DSP to be small enough to pack between the cement grains. There is increasing evidence to show that this assumption is at the best only partially true and in the case of some silica fumes not correct. If a silica fume cannot be de-agglomerated by extended ultrasonic dispersion in water, mixing in a concrete or mortar is unlikely to do better. It is extremely difficult to measure the dispersion of silica fume in plastic cement pastes. The only measurements available of the dispersion of silica fume in freshly mixed mortar suggest that much of the agglomerated silica fume remained undispersed by mechanical mixing [15].

Petrographic examination of hardened mortar and concrete in thin section is another method which is able to indicate the presence of larger agglomerates but is unable to detect agglomerates smaller than 25 μm which either will have been consumed by pozzolanic reaction or remain hidden in the thickness of the thin section. However, this inability of petrography to determine the complete particle distribution of a silica fume in hardened cement paste is not important. It is the detection of agglomerates larger than 25 μm which is critical as these large agglomerates have little pozzolanic reactivity and are potentially reactive with the alkalies in the pore solution.

These points are illustrated by the results obtained from the petrographic examination of the commercial 60 MPa concrete where some 15% of the silica fume remains as agglomerates. The silica fume used will have a particle size distribution similar to that given in curve 2 of Fig.2 which shows that after fifteen minutes ultrasonic dispersion some 15% of agglomerates larger than 50 μm are still present. This indicates the concrete mixing was unable to break up the larger agglomerates.

Similarly petrographic examination of DSP mortars carried out by St John *et al.* [1] indicated that in their DSP mortars some 25% of the Tasmanian silica fume was still present as agglomerates exceeding 10 μm in size. In spite of twenty minutes of the high shear mixing required to make the DSP mortars the agglomerates detected were still clearly visible in thin section even after years of exposure. The particle size distribution of the Tasmanian silica fume (curve 3 in Fig.3) indicates that about 25% of the fume was coarser than 10 μm in size which equates well with the results found from petrographic examination.

The persistence of silica fume agglomerates in concrete has been reported by other workers. Shayan *et al.* [5] also found agglomerates exceeding 30 μm in size and even where the silica fume is interground with the cement agglomerates have still been reported [6]. Pettersson [4] found that silica fume granules were often present in thin section analysis of mortars.

In practice it is possible for a silica fume to be sufficiently dispersed at the lower end of the particle size distribution to provide pozzolanic reactivity and also provide packing between cement grains and for large agglomerates also to be present. This is probably the situation that allowed both the effective fabrication of DSP samples containing the Tasmanian silica fume described by St John *et al.* [1], and also a sufficient number of larger agglomerates to cause AAR. The Tasmanian silica fume used did have a bimodal distribution as shown in Fig.3 which will have led to much more silica fume being required to give workability. This results in an inefficient use of the silica fume and also increases the potential for AAR.

It was initially believed that the extremely low permeability of these ultra-high strength mortars would not allow sufficient water to penetrate to allow AAR to proceed. The expansive cracking of the outdoor exposure samples showed that in spite of the high strength and low permeability the expansive cracking due to AAR was able to proceed and in this respect the DSP behaved like any other concrete.

Where a sufficient number of agglomerates larger than 25-30 μm remain in the hardened cement paste they present a potential for expansive cracking due to AAR provided that sufficient alkali and moisture are present. Both the amount of agglomerates and alkali required to initiate the expansive reaction of silica fume agglomerates remains to be investigated. Some guidance on this topic can be gained from codes of practice. One report [16] recommends particles coarser than 45 μm be limited to 1% and codes of practice for the limitation of AAR generally recommend a maximum alkali limit in the concrete of 2.5-3 Kg/m^3 as Na$_2$O equivalent where reactive aggregates are being used.

The fact that there are few reports of expansive cracking caused by interaction of silica fume with the alkalies in concrete should not be allowed to obscure the potential for AAR to occur with badly dispersed fumes given the appropriate conditions. The history of AAR contains numerous examples where aggregates have initially been considered unreactive but given sufficient time cracking has developed which petrographic investigation has shown to be due to AAR. It is strongly recommended that where long term durability of concrete containing silica fume is required that the ultrasonic dispersivity of the fume be investigated. Where the alkali content of the concrete exceeds 3kg/m^3 the silica fume should not contain more than a few percent of agglomerates larger than 50 µm or alternatively consideration should be given to reducing the alkali content of the concrete to a safer level.

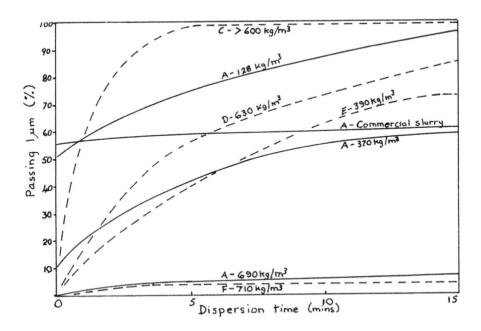

Fig.4   The ultrasonic dispersion of the silica fumes shown as the amount less than 1 µm versus time.

5 Conclusions

1. Because of agglomeration the mean particle size of a silica fume dispersed in water varies between 1 and 50 µm dependent on its source, bulk density and age. The effective mean particle size is not 0.1 to 0.2 µm as extensively quoted in the literature.

2. The ultrasonic dispersion of silica fume in water may represent a limiting case and evidence is presented that indicates that intermixing with concrete or mortar will achieve less dispersion.

3. It is hypothesised that the presence of fused particles and chains are a fundamental property of the silica fume source. Once a silica fume with an unfavourable amount of fusion is agglomerated it will not disperse when mixed in concrete.

4. It is recommended that where a silica fume is used to make DSP, its ultrasonic dispersivity should be measured and if it does not disperse to a mean particle size of 1 µm or less should be rejected as being inefficient.

5. It is recommended that where the alkali content of concrete exceeds 3 kg/m³, the ultrasonic dispersion of the silica fume used should be measured to ensure that only a small percentage of agglomerates greater than 50 µm are present. Alternatively, consideration should be given to reducing the alkali content of te concrete to a safer level if excessive amounts of large agglomerates are detected.

5 References

1. St John, D.A., McLeod, L.C. and Milestone, N.B. (1993) An investigation of the mixing and properties of DSP mortars made from New Zealand cement and aggregates. *Industrial Research Limited Report No.41*
2. St John, D.A., McLeod, L.C. and Milestone, N.B. (1994) *High Performance Concrete Workshop, Bangkok*
3. Bonen, D. and Diamond, S. (1992) Occurrence of large silica fume-derived particles in hydrated cement paste. *Cement and Concrete Research*, 22, No. 6. pp. 1059-1066 (1992)
4. Pettersson, K. (1992) Effects of silica fume on alkali-silica expansion in mortar specimens. *Cement and Concrete Research*, 22, No. 1. pp. 15-22
5. Shayan, A., Quick, G.W. and Lancucki, C.J. (1993) Morphological, mineralogical and chemical features of steam-cured concretes containing densified silica fume at various levels. *Advances in Cement Research*, 5, No. 20. pp. 151-162
6. Sveinsdóttir, E.L. and Guðmundsson, G. (1993) Condition of hardened Icelandic concrete. A microscopic investigation. *Proceedings of the Fourth Euroseminar on Microscopy Applied to Building Materials*. (ed. J.E Lindqvist & B Nitz) Swedish National Testing and Research Institute, Building Technology SP Report 1993:15
7. ACI Committee 226. Silica fume in concrete. *ACI Materials Journal*, 84, No. 2. pp. 158-166
8. Fiskaa, O., Hansen, H. and Moom, J. (1971) Betong i alunskifer. *Publication No. 86, Norwegian Geotechnical Institute*. pp. 1-32
9. Kolderup, H. (1977) Particle size distribution of fumes formed by ferrosilicon production. *Journal of the Air Pollution Control Association*, 27, No. 2. pp. 127-130
10. Aitcin, P. (ed.) (1983) *Silica Fume*. Les Editions de L'universite de Sherbrooke, Canada. pp. 52
11. De Larrard, F. (1992) Ultrafine particles for making very high performance concrete. *High Performance Concrete*, (ed. Y. Malier), E & FN Spon, London. pp. 34-47 (1992)
12. Yonezawa, T., Izumi, I., Okuno, T., Sugimoto, M., Shimono, T. and Asakura, E. (1992) Reducing viscosity of high strength concrete using silica fume. *Fourth CANMET/ACI International Conference on Fly Ash, Silica Fume, Slag and Natural Pozzolans in Concrete*, Istanbul. pp. 765-769
13. Asakura, E., Yoshida, H., Nakato, T. and Nakamura, T. (1993) Effect of characters of silica fume on physical properties of cement paste. *JCA proceedings of Cement and Concrete*, 47. pp. 178-183, (In Japanese, English abstract)
14. Bonen, D. and Diamond, S. (1992) Investigations on the coarse fraction of a commercial silica fume. *Proceedings of the Fourteenth International Conference on Cement Microscopy*. pp. 103-113
15. Nagataki, S., Otsuki, N. and Hisada M. (1994) Effects of physical and chemical treatments of silica fume on the strength and microstructures of mortar. (ed. V.M. Malhotra), *High-Performance Concrete, Proceedings ACI International Conference, Singapore, SP-149*, pp. 21-35
16. Concrete Society. (1993) Microsilica in concrete. *Concrete Society Technical Report No.41*.

8 FIXATION OF ALKALIS IN CEMENT PORE SOLUTIONS

D. CONSTANTINER
Master Builders Inc., Cleveland, OH, USA
S. DIAMOND
School of Civil Engineering, Purdue University, West Lafayette
IN, USA

Abstract

The effects of prolonged air drying followed by vacuum saturation on the ion concentrations found in pore solutions expressed from mature cement pastes was investigated. It was found that air drying followed by vacuum saturation resulted in reduction in both alkali and OH^- ion concentrations to values as low as 40% of those of companion specimens that had never been dried. Attempts to re-equilibrate the pastes by prolonged storage in the saturated condition after vacuum saturation were only partially successful. It appears that air drying 'fixes' a major portion of the dissolved alkali hydroxide in a manner that precludes its easy redissolution upon rewetting. In field concrete, near-surface layers showing this effect after partial drying may undergo much less ASR than the interior concrete.
Keywords: Pore solution, fixation of alkalies, drying, carbonation, alkali silica reaction.

Mechanisms of Chemical Degradation of Cement-based Systems. Edited by K.L. Scrivener and J.F. Young. Published in 1997 by E & FN Spon, 2–6 Boundary Row, London SE1 8HN. ISBN: 0419215700.

1. Introduction

Currently, most pore solution studies of cement, mortar, and concrete systems have been carried out with laboratory specimens which have been kept continuously sealed from the time of casting until the expression of pore solution is carried out. In general the sealed condition can be considered to simulate the environment of portion of field concrete at considerable depth from the surface, and thus isolated from the external environment.

On the other hand, pore solutions can also be obtained from concrete cores obtained from the field. Normally at least a portion of the field concrete contained in such cores has been subjected to atmospheric exposure, wetting and drying, temperature changes and other environmental effects.

Although the laboratory experiments under controlled conditions have helped elucidate various aspects of the chemistry of cement and concrete and of its degradation processes, there are issues related to exposed field concrete that have not been studied. Effects produced as a consequence of field exposure can be particularly important where alkali silica reaction (ASR) concerns may occur.

Laboratory experiments have shown that a significant portion of the alkali hydroxide originally present in the pore solution of sealed specimens can be depleted slowly as a result of ongoing ASR [1]. Thus if low alkali hydroxide concentrations are found in the pore solutions of field concrete produced with normal or high alkali cement, in the absence of conditions indicative of extensive leaching, one might suspect that the low concentrations might have resulted from the effects of ASR.

Such low concentrations, sometimes extremely low concentrations, have been found in pore solutions expressed from field concrete cores by the second-named author.

The work presented here illustrates that the low concentrations sometimes observed are not necessarily the result of ASR. It appears that significant reduction of the alkali hydroxide concentration in concrete pore solutions can also result from combined drying and partial carbonation in exposed field concrete. The process described here is a quasi-permanent fixation of alkali hydroxide that is not recovered by the solution on rewetting. Furthermore, it appears that prior fixation of alkali hydroxide can (and perhaps often does) reduce the subsequent ASR , especially in the outer layer of affected concrete.

2. Experimental Work

The effects of prolonged air drying followed by vacuum saturation on the alkali hydroxide concentrations retained in the pore solutions of paste specimens made with two different cements was investigated in this work.

The chemical compositions of the cements used are given in Table I.

Paste specimens (2 in. in diameter by 2.5 in. in height) were prepared at w/c 0.5 and allowed to hydrate in sealed containers for 1 year; certainly long enough to establish equilibrium pore solution concentration levels. The pore solutions were then expressed and analyzed, following the method described by Barneyback and Diamond [2].

Replicate specimens were also removed from separate containers and allowed to air dry in controlled drying chambers for an additional year, under a gently flowing stream of air conditioned at 45% relative humidity. The CO_2 in the drying chambers was kept at about 0.03%, the normal concentration in the air.

It is normally assumed that the partial drying would reduce the volume of the pore solution and proportionally increase its alkali hydroxide concentration. It is usually further assumed that re-saturating the paste, i.e. providing water to replace that lost in air drying, would return the pore solution to its original concentration levels.

It was not possible to express pore solution from the dried specimens to check the former presumption, but the latter hypothesis could be tested by saturating the paste, allowing a period for equilibration, and then expressing and analyzing the pore solution.

A procedure of vacuum saturation similar to that described in AASHTO T277-83i was used for these purpose. In this procedure the air-dried specimens are placed under vacuum for about 4 hours. Water is then introduced while the vacuum continues to be maintained, until the specimens are completely immersed. The specimens are then are allowed to remain under water at atmospheric pressure for about 16 hours to complete the saturation process.

Table I Chemical Characteristics and Fineness of Cements Used in this Study.

Compound	Cement A	Cement B
SiO_2	20.82	20.6
CaO	64.87	61.0
Al_2O_3	5.29	4.0
Fe_2O_3	2.12	3.1
MgO	1.35	4.9
SO_3	3.04	2.8
K_2O	0.58	0.91
Na_2O	0.12	0.25
Na_2O eqv.	0.50	0.85
LOI	1.74	1.90
Insoluble	0.33	0.28
Compound	Bogue (%)	Bogue (%)
C_3S	58.6	54.1
C_2S	15.6	18.2
C_3A	10.4	5.4
C_4AF	6.4	9.4
$CaSO_4$	5.2	4.7
Fineness Blaine, cm^2/gr	3970	N/A

After the saturation and prior to expressing the pore solution, the specimens were resealed for periods of 2 days, 8 days, 1 month (for cement A only), and 4 months to allow the pore solution to equilibrate; thus any changes occurring in the solutions could be monitored.

3. Results and Discussion

Table II shows the analyses of pore solutions obtained from the 1-year old specimens that had been kept continuously sealed (the reference condition). As expected, the pore solution is mainly composed of alkali hydroxide at concentration levels consistent with the alkali contents of the cements used. The charge balance of the anions and cations determined in the solution is reasonably close to zero.

Table III provides the corresponding pore solution ion concentrations obtained from the companion specimens that were air dried, vacuum saturated, and allowed to equilibrate for 2 days. Comparison of these results with those of Table II indicate that the alkali concentrations found to be present after air drying, saturating, and allowing a 2-day equilibration period are less than half of the expected concentrations if drying had not occurred.

To be complete in the comparison, the alkalis that leached out from the dried specimens during the vacuum saturation should be accounted for and added to the pore solution alkali concentration shown in Table III. The concentrations of Table III were adjusted for this effect, with the results provided in Table IV. The adjusted concentration levels are a little higher than before the correction for leaching, but still very much reduced from those of the reference condition as seen in Table II.

Similar comparisons are shown in Table V for replicate air-dried specimens that were vacuum saturated and resealed -- in the saturated condition -- for increasing periods of time. The results shown in Table V indicate that there is some slight additional recovery of alkali hydroxide concentration with additional equilibration time in the sealed condition. However, the rate of increase is slow, and diminishes with time.

Table II Pore Solution Concentration Results from One Year Continuously-Sealed Paste Specimens Made with Cements A and B, 0.50 w/c.

Specimen made with cement	Pore solution ion concentrations, N					
	Na^+	K^+	OH^-	$\Sigma+$	$\Sigma-$	$\Sigma+/-$
A	0.068	0.393	0.433	0.461	0.433	+0.028
B	0.181	0.358	0.523	0.539	0.523	+0.016

Table III Pore Solution Concentration Results from Specimens Made with Cements A and B, Cured Sealed for 1 Year, Air Dried for 15 Months, Vacuum Saturated, and Resealed for 2 Days.

Specimen made with cement	Pore solution ion concentrations, N					
	Na^+	K^+	OH^-	$\Sigma+$	$\Sigma-$	$\Sigma+/-^*$
A	0.029	0.160	0.174	0.189	0.174	+0.015
B	0.073	0.136	0.180	0.209	0.180	+0.029

$^*\Sigma+/- = (\Sigma+) - (\Sigma-)$

Table IV Alkali Ion Concentration in Pore Solutions from Cement A and B Specimens, Cured Sealed for 1 Year, Air Dried for 15 Months, Vacuum Saturated, and Resealed for 2 Day -- Corrected for Leaching of Alkalis.

Description	Concentration, N				
	Na^+	K^+	$\Sigma+$	OH^-	$\Sigma+/-$
Analysis as determined (specimen made with cement A)	0.029	0.160	0.189	0.174	+0.015
Correction for leaching of alkalis during vacuum saturation	0.004	0.025	0.029		
Net concentration corrected for leaching (specimen made with cement A)	**0.033**	**0.185**	**0.218**		
Analysis as determined (specimen made with cement B)	0.073	0.136	0.209	0.180	+0.029
Correction for leaching of alkalis during vacuum saturation	0.012	0.019	0.031		
Net concentration corrected for leaching (specimen made with cement B)	**0.085**	**0.155**	**0.240**		

Table V   Alkali Ion Concentrations Pore Solutions from Cement A and B Specimens, Cured Sealed for 1 Year, Air Dried for 15 Months, Vacuum Saturated, and Resealed for Various Periods of Time -- Corrected for Leaching of Alkalis.

Specimens made with cement	Resealing time	Concentration, N		
		Na^+	K^+	$\Sigma+$
A	2 days	0.033	0.185	0.218
A	8 days	0.036	0.215	0.251
A	1 month	0.036	0.216	0.252
A	4 months	0.042	0.237	0.280
B	2 days	0.085	0.155	0.240
B	8 days	0.093	0.180	0.273
B	4 months	0.108	0.204	0.312

The alkali concentrations found, as functions of equilibration time, and expressed as a percentage of the alkali concentration of the corresponding reference (undried) specimens, have plotted in Figure 1. It is clear that the 'fixation effect' is not easily reversible.

4. Discussion, Conclusions, and Implications

As shown by the data presented above, a significant and quasi-permanent reduction in pore solution alkali hydroxide concentrations ('fixation') took place on air drying under controlled conditions in these laboratory specimens.

It can be expected that similar fixation will occur as a result of the intermittent air drying that commonly occurs in at least the outer layers in field concrete in many climates.

The effect appears to have the potential to   reduce the alkali concentrations normally expected for concrete produced with high alkali cements to levels associated with low alkali-cement concrete. At these reduced solution alkali levels little if any active ASR is to be expected.

This fixation process appears to be only very slowly reversible after re-wetting. Accordingly, its effects in field concrete can be expected to persist for extended periods of time, even if the concrete is periodically re-wetted.

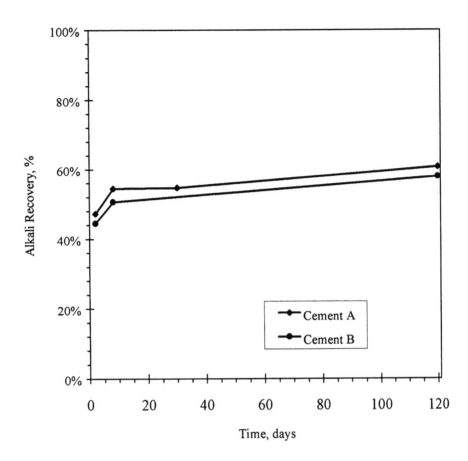

Figure 1 Effects of Resealing Time on the Reincorporation of Fixed Alkalis into the Pore Solution

Additional extensive studies, not reported here, have indicated that some degree of carbonation is fundamental to the fixation of alkalis; air drying in the complete absence of carbon dioxide does not result in significant fixation of alkalis [3].

Exposed field concrete is normally in contact with atmospheric CO_2 and is subjected to drying and wetting cycles. Extensive carbonation (pH lower than 10) can often be detected several millimeters or more into many concretes. The total depth to which the CO_2 - has affected the concrete is usually much greater than the depth indicated by the usual phenolphthalein color change [4].

Given the present findings, it is reasonable to presume that an outer layer of exposed field concrete of significant thickness may often have a significant portion of its original alkali content in the pore solution 'fixed', with the residual alkali hydroxide concentrations much reduced, even in the complete absence of ASR.

The fixation of alkalis conversely may have significance where active ASR is taking place. Susceptible aggregate in the near surface zone of the concrete might be expected to undergo much less actual reaction than corresponding aggregate in the less-affected or unaffected interior. Thus the typical map cracking associated with unrestrained ASR exposures might derive at least in part from this effect, i.e. a differential expansion with the outer shell expanding less than the interior of the concrete. This differential expansion effect has been noted previously [5], but has been explained on the basis of drying shrinkage (modified by intermittent expansion) of the outer layer.

The work shown here suggests that systematic study of the effects of atmospheric exposure on the chemistry of the pore solution in field concrete might profitably be undertaken.

5. Acknowledgments

This work was carried out at Purdue University, and the data are taken from the Ph. D. thesis of the first-named author. The support of the National Science Foundation Center for Advanced Cement Based Materials is gratefully acknowledged.

6. References

1. Barneyback, R. Jr., Alkali-Silica Reaction in Portland Cement Concrete, Ph.D. Thesis, Purdue University, April 1993, 352 pp.

2. Barneyback R. Jr., and Diamond, S., Expression of Pore Fluid form Hardened Cement Pastes and Mortars, Cement and Concrete Research, Vol. 11, No. 2, pp. 279-285, 1981.

3. Constantiner, D., Factors Affecting Concrete Pore Solution: (A) Alkali Fixation Induced by Drying and Carbonation and (B) Alkali Released From Feldspars, Ph.D. Thesis, December 1994, pp. 302-304.

4. Dunster, A. M., A Comparison Between the Carbonation of OPC and PFA Concrete in a Mass Concrete Structure, Advances in Cement Research, Vol. 4, No. 14, pp. 69-74, 1991/92

5. Stark, D., The Moisture Condition of Field Concrete Exhibiting Alkali Silica Reactivity, Durability of Concrete Second Int. Conf., Montreal, 1991, Ed. V.M. Malhotra, ACI SP 126, pp. 973-983.

9   STRESS DUE TO ALKALI-SILICA REACTIONS IN MORTARS

C.F. FERRARIS, J.R. CLIFTON, E.J. GARBOCZI and F.L. DAVIS
National Institute of Standard and Technology, Gaithersburg, MD, USA

Abstract
Alkali-silica reaction (ASR) causing deterioration of mortars and concretes is due to the swelling of gel formed by the reaction of alkali in cement-based materials with reactive silica in aggregates, in the presence of water.   The swelling of the gel generates tensile stresses in the specimen resulting in expansion and cracks.   Most tests designed to detect ASR rely on measurements of the length change.   A new test, designed to measure the stress generated by the swelling of the gel, has a cylindrical mortar specimen placed in a frame under a load cell. The force required to prevent expansion is measured over time while the sample and frame are immersed in a solution of 1 N NaOH at 50 °C.   Along with the design of the apparatus,   some preliminary results are presented. Measurements of stresses showed a strong influence of creep on the mechanical response of the material subjected to ASR. The aggregate influence on the stress and expansion due to the ASR was investigated.
Keywords: Alkali-silica reaction, mortar, stress measurements, Young's modulus, stress relaxation.

1. Introduction

Damage of concrete due to alkali-silica reaction (ASR) is a phenomenon that was first recognized in 1940 by Stanton [1,2] in North America. It has since been observed in many other countries.   Many studies [3,4,5] have been published since Stanton's first paper, but the mechanisms of ASR are not yet clearly understood [6]. Nevertheless, the major factors have been identified. In the presence of water, alkalies in the pore solution react with reactive silica,   found in certain aggregates. Related factors which can play a significant role are environmental relative humidity (RH), porosity of the concrete, and mineral admixtures in the concrete.

Mechanisms of Chemical Degradation of Cement-based Systems. Edited by K.L. Scrivener and J.F. Young. Published in 1997 by E & FN Spon, 2–6 Boundary Row, London SE1 8HN. ISBN: 0419215700.

Most methods in use today to detect ASR in mortars or concretes are based on measurements of the expansion of the sample. In this paper, we will describe a novel technique consisting of the measurement of the expansion stress generated by the reaction. The main advantage in measuring stresses instead of length changes is that the results might be able to be used to design a concrete that resists ASR expansion, i.e., with a tensile strength higher than the stress generated by the reaction.

2. Background

Most researchers agree that the main form of ASR is between certain kinds of silica present in the aggregates and the hydroxide ions (OH^-) in the pore water of a concrete [3,4]. Hydroxide ions from the hydration of portland cement result in a pore solution pH of around 12.5 [7]. The amount of alkalies present in the pore water is related to the amount of soluble alkalies in the cement. The hydroxide ions may attack vulnerable sites exposed in a silica surface. If the silica is well-crystallized the vulnerable sites are few but in the case of poorly-crystallized or amorphous silica, there are many vulnerable sites in the silica structure; in the latter case, alkali attack may lead to complete conversion of the silica to calcium and alkali silicate gel [8,9]. To keep a neutral charge balance, the cations Na^+ and K^+ diffuse toward the hydroxide ions, producing a gel-like material.

The formation of the gel *per se* is not deleterious. The deterioration of the concrete structure is due to the water absorption by the gel and its expansion. If the tensile strength of the system is locally exceeded, cracks will form and propagate in radial fashion around the reaction site. The sites of crack initiation are randomly distributed in the specimen, and there is no preferential direction for cracks to propagate. The crack sites are determined by the location of the reactive silica on the aggregates and the local availability of OH^-.

Most tests available for detecting ASR in concrete are based on measurement of the specimen expansion. The mix design and the condition of testing differs among tests. There are three ASTM tests currently used: ASTM C1260 [10] , ASTM C227 [11], and ASTM C441 [12]. Other tests described in the literature are usually modifications of the above tests. A German test [13, 14] was developed to measure the stress generated from ASR but limited data are available. Recently, Sellier et al. [15] attempted to simulate the stresses and the swelling due to ASR and found good agreement with available measurements. We used a mix design that simulates high performance concrete, with low permeability, low water/cement ratio and high cement content.

3. Experimental Set-up

3.1 Stress measurements
A novel test was designed to measure the stress generated by ASR in mortars. Figure 1 shows a sketch of the apparatus. The specimen, a mortar or concrete cylinder, is placed in the stainless steel frame and connected to a load cell. The load cell, connected to a computer, monitors, at regular intervals, the force generated by the formation of the ASR gel and the specimen expansion. To guarantee that the load cell response (voltage) is uniquely related to a force generated by the specimen, the cell

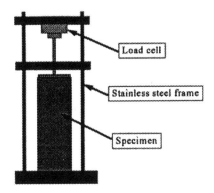

Figure 1: Schematic view of the device used to measure stress due to ASR.

was calibrated, using a dead weight set-up, in the same configuration (frame) as it was used during the experiment.

The frame holding the specimen is immersed in a container with a 1N aqueous solution of NaOH which is the test solution used in ASTM C1260 [10]. The container is then placed in a water bath with a controlled temperature of 50 °C ± 3 °C. The effect of differential thermal expansion between the sample cylinder and the steel legs of the frame was found to be unimportant. Since the experiment was conducted at constant temperature, the initial thermal expansion was easily taken into consideration.

3.2 Specimen preparation

In this initial study, only mortar specimens were tested. Three specimens were prepared for each mix. Table 1 gives the mixture designs used, while Table 2 gives the sand gradation. One sand was selected for its reactivity and one for its lack of reactivity with alkalies.

The specimens were cylinders 38 mm in diameter and 279 mm long (1.5 in. x 11 in.). The cement used had a high alkali content (about 1.2 % Na$_2$O equivalent). As the high alkali content of the cement should result in a high alkali concentration and pH in the pore solution, this cement should promote ASR with reactive siliceous aggregates. Companion cylinders were also placed either in limewater or in 1 N NaOH solution for unrestrained expansion measurements. All the samples were kept at 50 °C ± 3 °C.

Table 1. Mixture design of the mortars

Mixture design	Mix A	Mix B
Water/Cement	0.295 by mass	0.295 by mass
Sand/Cement	1.411 by mass	1.411 by mass
Type of sand	Tecosil[1,3]	U.S. Silica[3]
Sand Gradation	Gradation #1 (Table 2)	Gradation #2 (Table 2)
HWRA[2]	HRWA #1 @ 0.50% by mass of cement	HRWA #2 @ 4% by mass of cement

Table 2. Mortar sand gradation

Gradation simulated	Sand		Mass
	Size range		
	Sieve size ASTM E11	Dimension [μm]	[%]
#1	4-10	4750-2000	15
all sands are from Tecosil[3]	10-20	2000-850	35
	20-50	850-300	25
	50-100	300-150	25
#2	S15	2360-600	35
all sands are from U.S. Silica[3], Ottawa Illinois.	C778 (20-30)	850-600	19
	C778 (Graded sand)	600-300	19
	F95	300-200	25
The sand gradation was selected as a simulation of a smooth size distribution (not gap graded)			

4. Results and discussion

Figure 2 shows the stress-strain plots for the two mixes.   The strain was measured in the free expansion samples, while the stress was measured for the samples confined by the frame. For Mix A, the stress increased rapidly with strain, and then increased much more slowly.   This phenomena is attributed to the occurrence of cracking, which accompanied the large degree of expansion shown by this mix.   The stress for Mix B was always fairly linear with the expansion strain.   If we fit a straight line to the stress-strain graph for Mix B, the "apparent" Young's modulus value obtained is E=3.3 GPa, with an R^2 of 0.86. This value is low by about a factor of 10 compared to concrete or mortar, which usually has a value of E in excess of 30 GPa [16].   For Mix A, since the measured strain in the unrestrained linear expansion specimen consisted mostly of crack opening displacement, the expansion of the cylinder after being removed from the frame was divided into the measured stress to give an estimate of

---

[1] Graded sand provided by C-E Minerals[3], PA USA. The composition is fused silica (amorphous)

[2] by mass of cement; HWRA #1 was supplied by W.R Grace and Co[3]. while HRWA #2 was supplied by Masters Builders[3]

[3] The name of manufacturers are identified in this report to adequately describe the experimental procedure. Such an identification does not imply recommendation or endorsement by the National Institute of Standards and Technology, nor does it imply that the material identified is necessarily the best available for the purpose.

the "apparent" Young's modulus. There was no apparent cracking in this case. The result was nearly the same as that obtained for Mix B. If the strain for Mix B is measured after releasing from the frame an "apparent" Young's modulus of about 18 GPa is obtained. This results seem to confirm the simulation presented by Sellier et al. [15].

Consider now the role of stress relaxation. Suppose the mortars were only linearly elastic. We could then treat the mortar as an elastic composite, which had a small volume fraction of material (ASR gel) that exerted internal stresses by trying to occupy a larger space. The measured stress would then be the effective modulus of the mortar/gel composite multiplied by the measured strain. In this case, for Mix B for example, the measured stress should have been 30 MPa, not 3 MPa, and large scale cracking should have been seen. Similar results should have been obtained for Mix A, when considering the expansion of the cylinder after being released from the frame. This large discrepancy between experimental results and those predicted by linear elasticity implies that stress relaxation played a large role in this experiment in addition to crack opening displacement in Mix A. This finding is in general agreement with the results and conclusion obtained by Stark [17]. Clearly, as the alkali-silica reaction went forward, producing expansive gel, the mortar was able to rearrange itself, thereby relaxing the longitudinal stress to the levels seen. Part of this rearrangement would show itself in lateral movement (bowing and/or increase in diameter), which was indeed observed on the samples. It is known that early-age concrete creeps much more than late age concrete, and since the ASR test started at only 24 hours of hydration, stress relaxation could clearly play a major role. This seems to be the only explanation of the samples' mechanical behavior. Plans have been made to check this explanation by putting an initial compressive stress on a 24 hour old non-reactive aggregate sample, and then measuring the change of stress with time as stress relaxation occurs during the ensuing hydration. However, the problem of stress relaxation induced by internal stresses has not previously been considered theoretically, and is worthy of further investigation.

The two different aggregates had very different effects on the level of the measured expansion and stress. The total free expansion of Mix A was about 10 times greater than that of Mix B (Figure 3). As only the type of aggregates differed, these plots indicate that sand #1 is more reactive than sand #2. However, the stresses

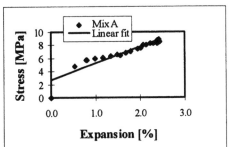 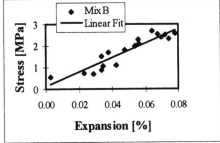

Figure 2   Stress-expansion plots for mortar specimens in NaOH solution at 50 °C.

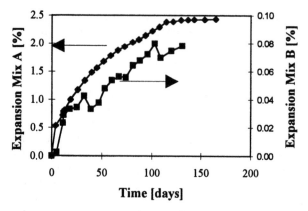

Figure 3   Expansion measurements on the tested mortar.

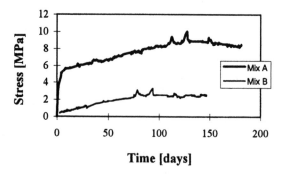

Figure 4   Change of stress with time for the constrained specimens.

measured were not much different (Figure 4).   This result is interpreted in the following way.   The internal stresses in the Mix A sample were able to build up enough so that the tensile strength was exceeded locally, producing the large number of lateral cracks and the large linear expansion seen.   The internal stresses in the Mix B sample never exceeded the local tensile strength, so only a small expansion was seen.   The actual stresses produced could be similar, however, for as long as the stresses in Mix B stayed below the tensile strength, and the stresses in Mix A exceeded the tensile strength, a large difference in the free expansion would be seen. The amount of gel produced in the two samples will be measured in the scanning electron microscope to determine the actual difference in reactivity between the two aggregates.

5. Conclusions

It has been shown, using the new apparatus, that the substantial forces exerted by ASR can be measured, once the role of stress relaxation was made clear.   It now will be possible to study factors affecting the forces for different materials and mixture properties, environmental conditions, and specimen geometries so as to provide data for mixture design and calculation of constrained expansion due to ASR in actual structures.

6. Acknowledgments

This work was sponsored by the "High Performance Construction Materials and Systems" program at NIST. Support for this work was also obtained from the Nuclear Regulatory Commission (NRC). John Winpigler is acknowledged for performing the measurements.

7. References

1.   Stanton, T. E. (1940) Expansion of Concrete Through Reaction Between Cement and Aggregate, *Proc. of the American Soc. of Civil Eng.* Vol. 66, #10, p. 1781-1811.
2.   Frohnsdorff, G., Clifton, J., Brown, P., (1979) *History and Status of Standards Relating to Alkalies in Hydraulic Cements,* Special Technical Publication 663, ASTM, p 16-34.
3.   Hobbs, D.W., (1988) *Alkali-Silica Reaction in Concrete*, Thomas Telford, London.
4.   Diamond, S., Penko, M., (1992) Alkali Silica Reaction Processes: The Conversion of Cement Alkalis to Alkali Hydroxide ,G. M. Idorn Inter. Symposium, *Durability of Concrete* ACI SP-131, American Concrete Institute.
5.   Helmuth R., (1993) *Alkali-Silica Reactivity: An Overview of Research,* SHRP Report C-342. National Research Council, Washington DC.
6.   Capra, B., Bournazel, J.-P., (1995) Perspective nouvelles pour la prise en compte des alcali-reactions dans le calcul des structures, *Materials and Structures*, vol. 28, p. 71-73.
7.   Diamond, S., (1983) Alkali Reactions in Concrete Pore Solutions Effects, Proc. 6th Int. Conf., *Alkalis in Concrete*, Idorn G.M. and Rostam S. eds., p. 155-166. Danish Concrete Association.
8.   Figg, J., (1983) "An Attempt to Provide an Explanation for Engineers of the Expansive Reaction between Alkalis and Siliceous Aggregates in Concrete", 6th Int. Conf. *Alkalis in Concrete*, Copenhagen.
9.   Helmuth, R., Stark, D., (1992) Alkali-Silica Reactivity Mechanisms in *Materials Science of Concrete III*, J. Skalny ed., American Ceramic Society. Westerville OH.
10.   ASTM . (1994) *Standard Test Method for Potential Alkali Reactivity of Aggregates (Mortar-Bar Method)*, ASTM designation C1260-94, Annual Book of ASTM Standards Vol. 04.02.
11.   ASTM. (1994) *Standard Test Method for Potential Alkali Reactivity of Cement-Aggregate Combinations (Mortar-Bar Method)*, ASTM designation C227-90, Annual Book of ASTM Standards Vol. 04.02.

12.   ASTM. (1995) *Standard Test Method for Effectiveness of Mineral Admixtures or Ground Blast-Furnace Slag in Preventing Excessive Expansion of Concrete Due to the Alkali-Silica Reaction,* ASTM designation C441-89, Annual Book of ASTM Standards Vol. 04.02.

13   Kuhlman J., Lenzner D., Ludwig V. (1975) A Simple Method of Measuring Expansion Pressures, Zement-Kalk-Gips 28-12, p. 526-530

14   Lenzner D. (1981) *Untersungen zur Alkali-Zuschlag Reaktion mit Opalsandstein aus Schleswig-Holstein,* Thesis from Rheinisch-West falischen Technischen Hochschule Aachen.

15   Sellier, A., Bournazel, J.P., Mebarki, A., (1995) Une modelisation de la reaction alcalis-granulat integrant une description des phenomenes aleatoires locaux, *Materials and Structures,* vol. 28, pp. 373-383

16   Gutierrez, P.A., Canovas, M. F., (1995) The Modulus of Elasticity of High-Performance Concrete, *Materials and Structures,* Vol. 28, pp. 559-568

17   Stark, D., Morgan, B., Okamoto, P., (1993) *Eliminating or Minimizing Alkali-Silica Reactivity,* SHRP-C343, National Research Council, Washington DC.

10 DURABILITY OF ALKALI ACTIVATED CEMENTITIOUS MATERIALS

R.I.A. MALEK and D.M. ROY
Materials Research Laboratory, Pennsylvania State University, PA, USA

Abstract
In the last two decades, there has been an increasing interest in the use of alkali activated materials. Originally, it was limited to the activation of slag cements using alkali metal hydroxides, but recently the alkali activated cements have extended to include several industrial by-products and wastes. A new class of these materials were developed in this laboratory to reduce CO_2 emission resulting from manufacture of Portland cement. Our recently reported results on some of these materials suggest that zeolite-like materials are being formed. The purpose of this paper is to report on the preliminary results on the nature of these phases and their durability. The effect of these phases on the long-term engineering properties is discussed. It has been demonstrated that the new materials minimize the extent of alkali-aggregate reaction. The mechanism of suppressing AAR is discussed.
Keywords: Alkali-activated cement, alkali-aggregate reaction, aluminosilicate framework, CO_2 emission.

1 Introduction

There is no longer any doubt that we are entering a period in which the CO_2 emission to the environment must be regulated. The cement industry which constitutes the largest volume manufactured materials in

Mechanisms of Chemical Degradation of Cement-based Systems. Edited by K.L. Scrivener and J.F. Young.
Published in 1997 by E & FN Spon, 2–6 Boundary Row, London SE1 8HN. ISBN: 0419215700.

the world is one of the major contributors to carbon dioxide in the atmosphere. This is the result of decomposition of calcite and use of fossil fuel in the kiln during cement manufacture. Developing means to utilize waste materials such as slags and industrial wastes in blended cements for the construction of infrastructure and other building projects should provide significant economic and environmental benefits. Aluminosilicate materials are of known structure and durability. The research work presented here is an attempt to replace Portland cement with synthesized aluminosilicate materials of equivalent or better properties than cement. Because of the difference in charge between Si and Al, a balancing positive charge must be present for each aluminum in the structure. The existence of positively charged ions in their structure offer an advantage to minimize the extent of alkali-silica reaction.

2 Background

The utilization of large amounts of by-products and waste materials plays a very important role in solving ecological problems [1-4]. The use of alkali-activated slags as a binder has significantly increased in recent years. Various national and international standards allow different contents of slag in the blends with cement: up to 65% in Great Britain and USA, 70% (Japan), and 80% (Germany, Russian republics, Czechoslovakia). The chemical activation of slag by different metal salts such as calcium or sodium sulfates or chlorides is sometimes used. Very interesting are the rapid hardening mixes slag + fly ash + gypsum + cement or lime [5], the purpose of which is to economize the building materials costs and the energy for steam curing. Thermal activation of slag without chemical activators is another possibility.

The alumni-silicate binders are formed by mixing some form of alkaline activator with a wide variety of silicate and/or aluminum rich minerals and glasses. Therefore different fine-grained alumni-silicates (clay, ground rock, ash, slag) can be used as raw materials. The possible mechanism involves the chemical reaction of aluminosilicate with alkalis and alkali polysilicate yielding polymeric or crystalline Si--O--Al bonds as shown in Fig. 1.

3 Experimental

A broad range of alkali-activated aluminosilicate frameworks with target composition of variable n and m in the formula:

$$\frac{n}{x} M^{x+} \left[(SiO_2)_m (AlO_2)_n \right] \cdot yH_2O \qquad\qquad m \geqslant n$$

where M is an alkali or alkaline earth metal cation, were prepared. Two systems were studied, namely, K-activated and K, Ca-activated systems.

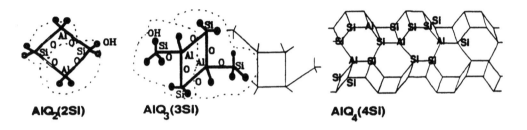

Fig. 1. Typical $AlQ_n(nSi)$ structural units (adopted from [6]).

The K-activated system is composed of metakaolinite, class-F fly ash, potassium silicate and potassium hydroxide. The K, Ca-activated system is composed of metakaolinite, class-C fly ash, slag, potassium silicate and potassium hydroxide. The water to cement ratio was 0.35. Samples were cured at 38° and 90°C. Properties of the samples were monitored, and characterized by various methods. These methods include investigation of mechanical and physical properties. The evolution of the pore system in comparison with plain Portland cement will also be made. Bar expansion tests will be made according to ASTM C227 standard for alkali-aggregate reactivity. Blends with cement were prepared as follows:

   a. Portland Cement 80% + K, Ca-activated cement 20%
   b. Portland Cement 50% + K, Ca-activated cement 50%
   c. Portland Cement 80% + K-activated cement 20%
   d. Portland Cement 50% + K-activated cement 50%

4 Results and discussion

Isothermal calorimetry data of the cement pastes containing 50% of either K-activated or K, Ca-activated blends at 38°C are presented in Figure 2 together with those of neat cement paste for comparison. This figure shows the gradual increase in reactivity as the extent of potassium increases. Compressive strength data at 7 and 14 days for samples cured at 38°C are presented in figures 3 and 4, respectively. In these figures, the

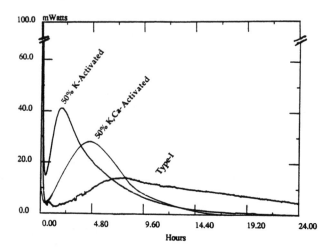

Fig. 2. Isothermal calorimetry data of 50% K-activated and 50% K,Ca-activated materials together with that of neat cement.

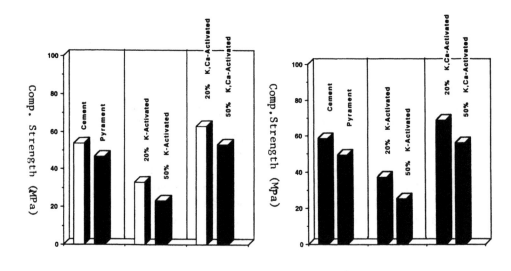

Fig. 3. Comp. strength of K-activated, K,Ca-activated, cement, and pyrament at 7 days, 38°C.

Fig. 4. Comp. strength of K-activated, K,Ca-activated, cement, and pyrament at 14 days, 38°C.

compressive strength data of neat cement pastes and a commercially available aluminosilicate binder (Pyrament) at the same age and temperature are included for comparison. It is evident that the compressive strength values of cement pastes containing 20% and 50% of K, Ca-activated materials are equivalent to those of neat cement pastes and slightly exceed those of the pyrament. The corresponding blends made with K-activated materials are lower in strength but their strength seems increasing with time. The pore structure evolution has been studied by mercury intrusion porosimetry. Results of pore size distributions at 14 days of the blends containing 50% alkali activated materials cured at 38ºC are presented in Figure 5. In the same figure, the pore size distribution of the neat cement paste at the same age and curing temperature is shown. It is evident that the alkali activated systems develop a finer pore structure compared to neat cement pastes. In addition, the blends with K, Ca-activated materials seems developing the finest pore structure. The 28-day mortar bar expansion due to alkali-silica reaction are presented in Figure 6. Results of high alkali cement, cement with potassium hydroxide and pyrament are included for comparison. It is clear that the alkali activated aluminosilicate materials gave the lowest reaction with silica.

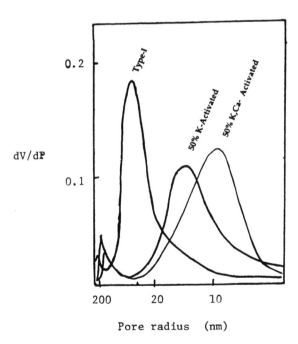

Fig. 5. Pore size distribution of K-activated and K, Ca-activated blends at 38ºC together with those of neat cement paste.

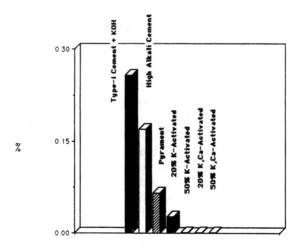

Fig. 6. 28-Days mortar bar expansions due to the alkali-silica reaction.

The durability of aluminosilicate structures arise from the fact that they exist in three dimensional networks that encompass the entire structure through chemical bonding. This structure is more durable than the hydrated cement in which the strength is mostly mechanical through foiled sheets of calcium silicate hydrate. Because of the difference in charge between Si and Al, a balancing positive charge must be present for each aluminum in the structure. The existence of positively charged ions in their structure explains the ability of these materials to minimize the extent of alkali-silica reaction through tie-in of the alkali metal cations [7].

5 Conclusions

New aluminosilicate materials were developed as an attempt to replace Portland cement with building materials of equivalent or better properties. The synthesized aluminosilicate materials provide significant economic and environmental benefits through reducing the CO_2 emission and utilizing industrial wastes for the construction. The aluminosilicate materials are three dimensional durable frameworks formed by chemical bonding. They tie-in alkali metal ions in their structure offering an advantage to minimize the extent of alkali-silica reaction.

6 Acknowledgment

Financial support of the United States Environmental Agency, Grant # R819482 is appreciated.

7 References

1.   Malek, R.I.A. and Roy, D.M. et al. (1986) Slag cement-low level radioactive waste forms at Savannah River plant. *Ceramic Bulletin,* Vol. 65, pp. 1578-1583.
2.   Malek, R.I.A. and Roy, D.M. et al. (1987) Electrochemical stability of embedded steel and toxic elements in fly ash/cement beds, in *Fly Ash and Coal Conversion By-products: Characterization, Utilization, and Disposal III,* (eds. G.J. McCarthy, F.P. Glasser, D.M. Roy and S. Diamond), Materials Research Society, Pittsburgh, PA, Vol. 86, pp. 59-65.
3.   Malek, R.I.A., Roy, D.M. and Licastro, P.H. (1987) Diffusion of chloride ions in fly ash/cement pastes and mortars, in *Microstructural Development During the Hydration of Cement,* (eds. L. Struble and P. Brown), Materials Research Society, Pittsburgh, PA, Vol. 8, pp. 223-231.
4.   Malek, R.I.A. and Roy, D.M. et al. (1995) Structure and properties of alkali activated cementitious materials, in *Novel Cements and Concrete,* 97th. Annual Meeting of the American Ceramic Society, Cincinnati, OH.
5.   Davidovits, J. (1983) Geopolymers II, processing and applications of ultra high temperature, inorganic matrix resin for cast composite structures, molds and tools for RP/C and metal industries. in *PACTEC '83,* Society of Plastic Engineering, Anaheim, CA, pp. 222-230.
6.   Davidovits, J. (1989) *Geopolymers,* University of Tech., Compiegne France, Vol. 2, pp 149-168.
7.   Malek, R.I.A. and Roy, D.M. et al. (1983) Effect of slag cements and aggregate type on alkali-aggregate reaction and its mechanisms, in *Alkalis in Concrete, Research and Practice,* (eds. G.M. Idorn and S. Rostam), Danish Concrete Association, Vester Varimags Gade 31, DK-1606 Kopenhaven V, Denmark, pp 223-230.

11 MICROSTRUCTURAL CHANGES OF HYDRATED CEMENT PASTE DUE TO CARBONISATION

Y.F. HOUST
Federal Institute of Technology, Laboratory for Powder Technology, Lausanne, Switzerland

Abstract
Carbonation of cementitious compounds leads to important microstructural changes. The modification of the microstructure was measured on hydrated cement paste (hcp) by mercury intrusion porosimetry (MIP) and water vapour sorption. This latter technique allowed us to calculated the pore-size distribution of meso- and micro-pores, as well as the BET specific surface area. The total porosity was determined after water absorption under vacuum.

Total porosity is always reduced by carbonation, the water vapour sorption isotherms are always lower than those of the non-carbonated material and the BET specific area is strongly reduced. The effect of carbonation on pore-size distribution depends on the microstructure of non-carbonated material, which is strongly influenced by water/cement (w/c) ratio. Results of MIP show that for well hydrated hcp, the reduction of porosity is higher for low w/c. All pore-sizes are reduced by carbonation, but particularly those of radius ≤ 0.1 μm.

From the durability point of view of cementitious materials alone, carbonation has a beneficial effect: total porosity, through which aggressive gases and solutions penetrate, is reduced; the equilibrium water content, which plays an important role in degradation by freezing-thawing, is also decreased.

Keywords: BET specific surface area, carbonation, hydrated cement paste, mercury porosimetry, microstructure, porosity, water adsorption.

1 Introduction

The carbonation of hcp is a chemical neutralisation process of hydration cement products, i.e. essentially $Ca(OH)_2$ (calcium hydroxide) and C-S-H (calcium silicate hydrate), by carbon dioxide (CO_2) present in the atmosphere at 0.03-0.04% by volume. The anhydrous compounds (calcium sulphates excepted) can also react, but only at high CO_2 concentration or very slowly. The carbonation products of $Ca(OH)_2$ are $CaCO_3$ and H_2O and those of C-S-H are $CaCO_3$, SiO_2 and H_2O. $Ca(OH)_2$ is essentially in

Mechanisms of Chemical Degradation of Cement-based Systems. Edited by K.L. Scrivener and J.F. Young. Published in 1997 by E & FN Spon, 2–6 Boundary Row, London SE1 8HN. ISBN: 0419215700.

crystal form and C-S-H is amorphous. $CaCO_3$ can exist under three crystal forms. The transformation of $Ca(OH)_2$ into $CaCO_3$ causes an increase in volume depending on the crystal form, which is 3% for aragonite, 12% for calcite and 19% for vaterite. Calcite is the stable form under normal temperature and pressure, but the presence of the other two forms in the carbonation products of hcp has been reported. The augmentation of volume decreases the porosity and paradoxically leads to shrinkage.

The most negative effects of carbonation of cement is the decrease of the pH of the pore solution and the loss of protection against corrosion of the steel in reinforced concrete. But, from the macroscopic point of view, carbonation has essentially beneficial effects: strength, elastic modulus, chemical, wear and freezing-thawing resistances are increased. Very few studies on microstructural changes, which could explain the changes of properties, have been reported. A better understanding of the microstructure of carbonated cementitious materials could help to improve these materials.

The present study is a part of a larger project. Papers on carbonation shrinkage kinetics [1], the influence of porosity and water content on the diffusivity of CO_2 and O_2 through hcp [2] and mortars [3], have been already published.

2 Experimental techniques

2.1 Bulk density - Total porosity
The bulk density was calculated by dividing the mass of a dry specimen by its volume, which was determined by the simple method consisting in weighing the specimen in air and in water (buoyancy method). The specimens were dried at 105°C until a constant mass was obtained, weighed in air and in water after saturation under vacuum.

2.2 Mercury intrusion porosimetry
A porosimeter which allows the application of pressures up to 415 MPa was used. This corresponds to a minimum radius of 1.7 nm. After having introduced a number of simplifying assumptions, the radius r of cylindrical pores, which can be penetrated by mercury at pressure P, can be related by the following equation:

$$r = -\frac{2 \sigma \cos \theta}{P} \tag{1}$$

where σ stands for the surface tension of mercury and θ for the contact angle hcp-mercury. A value of 135° for θ was used in this study.

2.3 Water sorption isotherms
The samples crushed to pieces of 3 to 5 mm and dried at 105°C, were first placed in desiccators in which different relative humidities were maintained by saturated salt solutions. After the equilibrium was reached, the water uptake was measured by weighing. The water content as a function of the relative humidity allowed us to construct the adsorption isotherms. The specific surface area was calculated from the well known BET model.

Chemically bound water excepted, water in a cementitious material can be divided into adsorbed, condensed capillary water and, in large capillaries and air pores, free water. All pores can be filled by water when the porous material is immersed into water for a sufficient time. In capillaries, condensation takes place at a lower relative humidity than 100%, depending on the pore radii. This is due to the lower vapour pressure above a concave liquid meniscus than above a plane liquid surface. Kelvin's equation relates the maximum radius r_k of pores filled by water and the relative humidity (p/p0):

$$r_k = -\frac{2\gamma V_w}{RT \ln(p/p_0)} \qquad (2)$$

where γ is the surface tension of water, V_m the molar volume of water, R the gas constant and T the absolute temperature. Kelvin's equation is only valid for capillary condensation, i.e. only for relative humidities where desorption isotherm curve exhibits an hysteresis. In fact, the real radius r_p is higher than the Kelvin's radius r_k. For cylindrical pores in hygro-thermal equilibrium, one has :

$$r_p = r_k + t \qquad (3)$$

where t is the statistical thickness of the adsorbed water layer measured on the plane surface of the same material, but non-porous. Before capillary condensation can take place (at low RH), pore walls are covered by water molecules of thickness t which also reduce the empty porous volume. The t values used in this paper were taken from Badmann et al. [4]. For instance, this thickness is about 0.4 nm at 40% RH and 0.8 nm at 90% RH. Their measurements were carried out on non carbonated hcp. Nevertheless, these data were used since no t value data are available for carbonated hcp.

The method proposed by Barrett, Joyner and Halenda [5], or more precisely the simplest Pierce's method [6] was used for computing the pore volume distribution directly from the adsorption isotherms. This method is based on:

- the Kelvin's equation,
- the thickness t of the adsorbed layer as a function of the relative humidity,
- a model of cylindrical pores.

More precise methods have been reported, but this method is largely sufficient for the comparison of the micro- and mesoporosity of carbonated and non carbonated hcp.

3 Materials

Portland cement paste cylinders (hcp) of 138 mm diameter were prepared. Such large samples were especially prepared for gas diffusion measurements [2]. After six months of curing in lime water, discs of about 3 mm thickness were cut off. In order to avoid sedimentation and bleeding, at high water cement/cement ratio (W/C), it was necessary to assure that cement particles were kept in suspension before setting. This procedure was based on previous work by Sereda and Swenson [7]. Due to the bigger size of our cylinders, it was necessary to introduce some modifications. Cement and water were mixed and placed under vacuum and the cylindrical mould was then placed on a roller device for rotating the mix during setting and hardening, usually for a period of 48 to 72 hours. This method allowed us to prepare hcp with a W/C ratio from 0.3 to 0.8. The cement was a Swiss Portland cement corresponding approximately to ASTM type I. Then, the specimens were artificially carbonated at 76% RH in an atmosphere of 80% to 90% CO_2 until complete carbonation, checked by the phenolphthalein test and X-ray diffraction, was obtained. Pieces of the same discs were used for bulk density-porosity, mercury porosimetry and water sorption measurements.

4 Results and discussion

First of all, it must be emphasized that carbonation of hcp was carried out at a very high CO_2 concentration (80%-90%) and at the most favourable humidity (76% RH) for quick carbonation. These conditions were necessary to carbonate fully our samples in a reasonable time. The rate of carbonation is not proportional to the CO_2 concentration at high concentration, it depends on the porosity of the material. The carbonation reaction produces water which can block the pores and thus limit the diffusion of gaseous CO_2 [8]. It was reported [9] that accelerated carbonation leads to a higher degree of carbonation and it is possible that the changes induced in the microstructure by carbonation could also depend on the CO_2 concentration as long as moisture diffusion out the material is not the limiting factor. But, this affirmation is very difficult to prove, because it is in general not possible to obtain complete carbonation of a dense material at the natural CO_2 concentration level. In any case, one has to keep in mind that the modifications observed could be affected by accelerated carbonation at high CO_2 concentration level.

The influence of carbonation on the microstructure was quantitatively measured by mercury intrusion porosimetry. The results are given in fig. 1 and 2, except for the carbonated hcp specimen of W/C = 0.3, since even after more than five years of exposure in the atmosphere of CO_2, discs are carbonated only to a depth of 0.2 to 0.3 mm. The porosity is significantly reduced (see also table 1). This reduction is greater for low W/C. All the pores of the hcp of W/C = 0.4 are affected by carbonation, but in particular those with radii below 0.1 μm. It is essentially those below 0.1 μm of the hcp of W/C = 0.5 which are reduced. The very porous hcp with W/C = 0.8 reveals a increase of the amount of pores with radii between 2 μm and 0.04 μm and a reduction of the finest pores. The coarsening of the pore structure has been already reported [9] for Portland cement blast furnace slag cement paste with a large amount of silica gel formed by carbonation, but not for Portland cement paste. According to Eitel [10], the carbonation of the C-S-H forms a porous silica-gel with coarse pores with radii of about 0.26-0.4 μm. In our opinion, the formation of large amount of silica-gel with coarse pores cannot explain the present results, because the coarsening of pores is not observed with the other hcp's. This result could be explained by the formation of microcracks due to carbonation shrinkage in the more porous hcp.

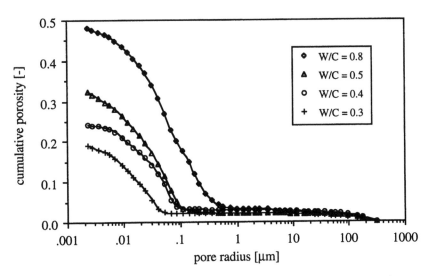

Fig. 1. Porosity of non carbonated hcp measured by mercury intrusion porosimetry.

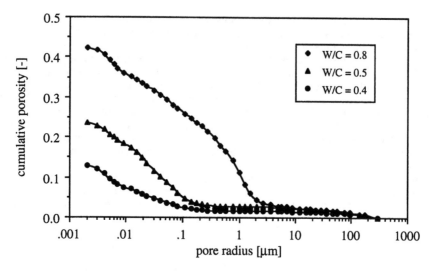

Fig. 2. Porosity of carbonated hcp measured by mercury intrusion porosimetry.

The reported effects of carbonation on the pore size distribution measured by MIP are somewhat inconsistent. It seems established that it leads to a reduction of the total porosity [9,11,12]. Bier [9] concludes that carbonation of hcp made with Portland cement leads to finer capillary pores and a coarsening of the pore structure of blast furnace slag cement (75% slag). However, Philajavaara [11] found that essentially pores between 0.1 µm and 0.02 µm were reduced and Ying-yu and Qui-dong [12] observed on mortars essentially a reduction of the pores < 0.63 µm. The different observed behaviours could come from the age of the specimen at the time of the exposure to CO_2 (only 7 days for Bier). The hcp can continue to hydrate during carbonation; this can influence the microstructure.

The water adsorption isotherms of the different samples of hcp, measured on non-carbonated and carbonated samples, are shown in fig. 3. Carbonation causes a decrease of adsorbed water as it was already observed [11]. The higher the equilibrium moisture content in a non-carbonated hcp, the higher the decrease of moisture content in an identical but carbonated hcp. The adsorption isotherms of carbonated hcp of W/C between 0.4 to 0.8 do not vary to a large extent. For example, the equilibrium moisture content of these hcp at 97% RH runs from about 8.5% to 11%. The BET specific surface area was calculated from the data of the adsorption isotherms between 9% and 44% RH and the results are reported in table 1. The BET specific surface area of non carbonated hcp increases with W/C. It is an indication of an increase of the degree of hydration of the cement. Carbonation strongly reduces the specific surface area to an approximately constant value of 57 to 62 m^2/g for the three higher W/C.

The cumulative volume of pores ≤ 30 nm were computed for the different hcp from equation (3), according to the Pierce method. The results are given in fig. 4 and 5. The effect of the degree of hydration appears in fig. 4: the higher the degree of hydration or W/C, the higher the porosity. The amount of hydration products increase with W/C, but the assembly of the xerogel particles is less compact and this leads to an increase of the porosity. The derivative of the curves of fig. 4 present a maximum at about 2 nm. According to Wittmann and Englert [13], the height of this maximum is a measure of the amount of xerogel. The height of this maximum increases also with W/C in the present study. The effects of carbonation appear in fig. 5. The volume of micropores is strongly reduced and the maximum of the derivative curves at 2 nm is also reduced to an approximately constant value. This shows that the microstructure of the xerogel is significantly modified by carbonation.

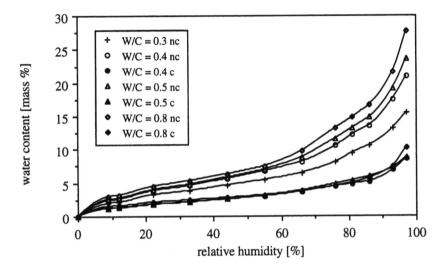

Fig. 3. Water vapour adsorption isotherms (18°C) of non carbonated (nc) and carbonated (c) hcp. The W/C ratio was chosen to be 0.3, 0.4, 0.5, and 0.8.

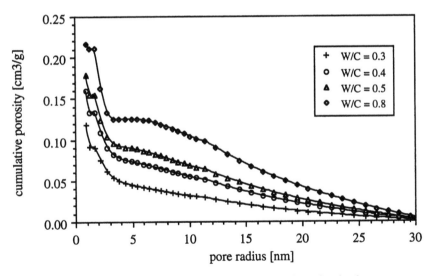

Fig. 4. Porosity of non carbonated hcp computed from adsorption isotherms.

Total porosities and bulk densities measured by MIP and water saturation are given in table 1. The porosity measured by MIP is lower than that calculated after water saturation. This is because MIP does not measure pores with a radius > 300 μm. The smallest pores (gel pores) are only partly measured by this technique. But, the difference between non carbonated and carbonated hcp are coherent for the two techniques. The increase of the bulk density shows the densification due to carbonation.

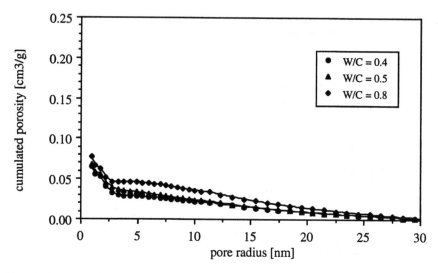

Fig. 5. Porosity of carbonated hcp computed from adsorption isotherms.

The pretreatement of hcp (duration of hydration, type of drying, carbonation, etc.) has a major influence on the microstructure and it is difficult to compare precisely the results from specimens with different histories. Furthermore, these results were obtained on well hydrated hcp (at least 6 month of curing in water) of Portland cement. This can explain certain differences in comparison with other studies.

The analysis of the specimens by Fourier transform infrared spectroscopy (FTIR) always shows the presence of calcite only. So no microstructural difference can be explained by differences in structural form of calcium carbonate.

Table 1. Porosity, bulk density and BET specific surface area.

W/C	porosity [vol. %] (MIP)		porosity [vol. %] (water saturation)		bulk density [kg/m³] (MIP)		BET specific surface area [m²/g]	
	non carb.	carb.	non carb.	carb.	non carb.	carb.	non carb.	carb.
0.3	19.05	-	23.2	-	1828	-	107	(75.7)
0.4	24.05	13.4	31.7	21.0	1595	2026	130	57.4
0.5	32.3	23.0	36.7	27.1	1448	1805	134	60.0
0.8	48.0	42.3	-	46.8	1092	1364	143	62.4

5 Conclusions

The accelerated carbonation of hcp leads to large modifications of the porous system. The lower W/C, the lower the reduction of porosity. All pore sizes are affected by carbonation. In hcp of W/C ≤ 0.5, it is essentially the pores with a radius < 0.1 μm which are diminished. As capillary condensation takes place essentially in these pores, certain capillary effects are reduced.

The adsorption isotherms curves of carbonated hcp are much lower than those of non carbonated hcp and depend little on W/C. BET specific surface area is consequently reduced by carbonation to an approximately constant value of 60 m²/g. The micropore and mesopore size distribution computed from adsorption isotherms is also strongly affected by carbonation. The volume of micropores is reduced and especially that corresponding to the xerogel particles.

Carbonation leads in general to a decrease of the equilibrium water content and of the capillary porosity. Such a material is certainly more resistant to chemical and environmental aggressivity. Its strength and resistance to wear are also improved. Artificial carbonation could be used to improve the performance of non steel reinforced cementitious materials.

6 References

1. Houst, Y.F. (1993) Influence of Moisture on Carbonation Shrinkage Kinetics of Hydrated Cement Paste, in Creep and Shrinkage of Concrete, (Ed. Z.P. Bazant and I. Carol), E & FN Spon, London, pp. 121-6.
2. Houst, Y.F. and Wittmann, F.H. (1994) Influence of Water Content and Porosity on the Diffusivity of O_2 and CO_2 Through Hydrated Cement Paste. Cement and Concrete Research, Vol. 24, No 6, pp. 1165-76.
3. Houst, Y.F., Sadouki, H. and Wittmann, F.H. (1993) Influence of Aggregate Concentration on the Diffusion of CO_2 and O_2, in Interfaces in Cementitious Composites, (Ed. J.C. Maso), E & FN Spon, London, etc., pp. 279-88.
4. Badmann, R., Stockhausen, N. and Setzer, M.J. (1981) The Statistical Thickness and the Chemical Potential of Adsorbed Water Films. Journal of Colloid and Interface Science, Vol. 82, No 2, pp. 534-542.
5. Barrett, E.P., Joyner, L.G. and Halenda, P.P. (1951) The Determination of Pore Volume and Area Distribution in Porous Substances, I. Computation from Nitrogen Isotherms. Journal of the American Chemical Society, Vol. 73, pp. 373-80.
6. Pierce, C. (1953) Computation of Pore Sizes from Physical Adsorption Data. Journal of Physical Chemistry, Vol. 57, pp. 149-52.
7. Sereda, P.J. and Swenson, E.G. (1967) Apparatus for Preparing Portland Cement Paste of High Water Cement Ratio. Materials Research & Standards, Vol. 7, pp. 152-4.
8. Arliguie, G. and Grandet, J. (1991) Représentativité des résultats d'essais accélérés de carbonatation sur éprouvettes de béton, in The Deterioration of Buiding Materials, UIT, La Rochelle, pp. 245-253.
9. Bier, Th. A. (1987) Influence of Type of Cement and Curing on Carbonation Progress and Pore Structure of Hydrated Cement Pastes, in Microstructural Development During Hydration of Cement, (Ed. L.J. Struble and P.W. Brown), Materials Research Society, Pittsburgh, Symposia Proceedings Vol. 85, pp. 123-34.
10. Eitel, W. (1954) The Physical Chemistry of Silicates. The University of Chicago Press, Chicago.
11. Philajavaara, S.E. (1968) Some Results of the Effect of Carbonation on the Porosity and Pore Size Distribution of Cement Paste. Materials and Structures, Vol. 1, 521-26.
12. Ying-yu, L. and Qui-dong, W. (1987) The Mechanism of Carbonation of Mortars and the Dependence of Carbonation on Pore Structure, in Concrete Durability - Katherine and Bryan Mather International Conference, (Ed. J.M. Scanlon), American Concrete Institute, Detroit, Vol. 2, pp. 1915-43.
13. Wittmann, F.H. and Englert, G. (1967) Bestimmung der Mikroporenverteilung in Zementstein. Materials Science and Engineering, Vol. 2, pp. 14-20.

12 CORROSION OF CEMENTITIOUS MATERIALS IN ACID WATERS

G. HEROLD
Institut für Massivbau und Baustofftechnologie, Universität Karlsruhe, Germany

Abstract

Experiments were performed on corrosion mechanisms likely to occur between aqueous acid solutions and solid phases in cementitious systems. The experiments carried out included the influence of preconcentrated attacking solutions on the dissolution behaviour of solids, the influence of exposure time, acidity and temperature of solutions.

Different elements (Ca, Mg, Fe, Al, Si, Na, K) in the solutions were analyzed, mainly by atomic absorption. The dissolved mass of various elements provided quantitative information on the dissolution behaviour of mineral phases. From the results equations are obtained which allow the calculation of corrosion depth from the examined parameters.

Keywords: acidity, corrosion depth, dissolution behaviour, exposure time, paste phases, reaction kinetics, temperature

1. The principle of the reactions

All phases in cementitious systems react with acid solutions by consumption of H^+-ions [1]. In case of pure dissolving attack producing soluble Ca-salts (hydrochloric and nitric acid) dissolution of paste phases occurs either congruently (portlandite) or incongruently (CSH-phase). Incongruent dissolution behaviour of CSH causes residual layers [2] on solid surfaces.

In stationary systems a marked deflection is observed in the course of dissolution curve of heterogeneous reactions. The reasons for deflection of the curve after a very short time are:

Mechanisms of Chemical Degradation of Cement-based Systems. Edited by K.L. Scrivener and J.F. Young. Published in 1997 by E & FN Spon, 2–6 Boundary Row, London SE1 8HN. ISBN: 0419215700.

1. The growth of thickness of the residual protective layer. Therefore, a rapid change from a reaction-controlled to a diffusion-controlled process takes place.
2. The gradual approach of the temporary concentration in solution to final concentration. An increase of the temporary concentration causes a depression of the dissolution behaviour of paste phases. Hence, these experiments were performed at very low temporary concentrations in the attack solutions and without erosive conditions so that protective layers were preserved.

2. Materials and experimental method

The experiments were carried out in 4-litre wide neck reaction vessels (with tempering jacket). A pH control device (set for electrometric titrations) and a thermostat were used to keep the pH-value and temperature constant.

Material: hardened OPC-paste with w/c = 0.5 (OPC: SiO_2 20.65%, CaO 62.08%, MgO 2.43%, Al_2O_3 5.85%, Fe_2O_3 3.44%, K_2O 1.00%, Na_2O 0.13%, SO_3 2.90%).

Performance: frequent exchange of solution (hydrochloric acid) with a high V/A-ratio of 500 ml/cm² (V = volume of solution, A = surface area). Samples of 40 ml from the total volume of solution were taken out at regular intervals and analyzed for some paste-specific elements. Total volume of aggressive solution, volume of sampling and volume of diluted acid supply allowed the calculation of dissolved element mass upon time of exposure. These data are converted to corrosion depths with the aid of element-specific densities (element mass/volume of solid).

3. Results

3.1 "Preconcentration" of the acid solutions with paste-specific elements
In pure acid hydrous systems the pH-value will control the amount dissolved from solution and decomposition rate of paste is very high. But in preconcentrated acid solutions of the same pH-value dissolution of paste phases is depressed. Increasing (pre)concentration (from 1 to 7 in table 1) in attack solutions results in reduced amounts dissolved from the solid. This is true for all measured elements (Fig. 1). In highly preconcentrated solutions (No. 7) only about 20% of the total amount in pure acid systems will be dissolved (compare No. 0 with No. 7). At lower acidities (pH 3 or 5) also much lower preconcentrated solutions depress the "regular" dissolved amount.

Hence also the actual concentration (generated by decomposition of the solid) in the attacking acids will influence the dissolution behaviour of the paste phases. Therefore to study the "real" dissolution kinetics of paste minerals, the actual concentration of components in solution should be low and far out of chemical equilibrium. For this reason in stationary systems without flow of acid water a frequent exchange of solution and/or a high V/A-ratio (A = attack surface; V = volume of solution) is an absolute necessity. Otherwise, the dissolution rate constants of paste phases and therefore also the depth of corrosion depend on experimental conditions (for example V/A-ratio, flow rate of solution).

Table 1. Composition of preconcentrated acid solutions

Element	Acid solution HCL/pHl				[ppm]			
	0	1	2	3	4	5	6	7
calcium	0,00	904,00	1300,00	1944,00	2604,00	3328,00	3950,00	4590,00
silicon	0,00	35,80	37,60	60,20	72,10	91,50	111,50	118,00
aluminium	0,00	31,00	44,20	63,70	86,00	108,50	128,40	142,60
magnesium	0,00	28,50	39,50	58,70	80,50	101,10	120,00	135,80
iron	0,00	41,40	58,50	89,60	117,00	150,00	171,00	202,00

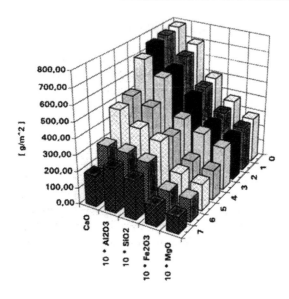

Fig. 1. Dissolved amount at increasing preconcentration of acid solutions (HCl/pH1/1 day)

3.2 Dependence of dissolved components upon time of exposure

When the dissolved components were entered as a function of the time in a log-log-diagram (here for example pH 1 experiments), it was noted that the curves obtained are straight lines with similar slopes (Fig. 2). Only the amount of alkali shows an exception with markedly higher slopes for all examined pH-values. The evaluation of these data allows the calculation of individual corrosion depths, in which the components are theoretically completely extracted (Fig. 3). Alkalis are preferably leached and this results in a broad alkali-free zone far in front of calcium reaction zone. Increasing pH leads to an increase of the distance between the depth of both zones. Ratios of constituents in solution in comparison with those ratios in solids and the calculated corrosion depths of different elements lead to conclusions about the dissolution behaviour of paste phases:

The similar corrosion depth for the alkaline earths is a strong argument that congruent dissolution of one phase (portlandite) occurs. The same slope for the dissolved components calcium, aluminium and iron as well as identical corrosion depths for aluminium and iron suggest that the aluminate phase is leached nearly congruently at pH 1-experiments. Decomposition of AFm/AFt occurs at a small

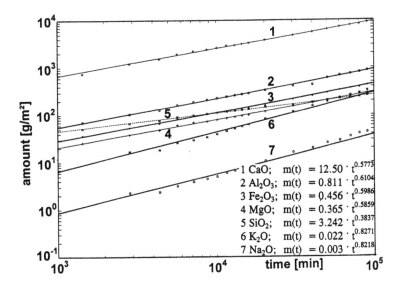

Fig. 2. Dissolved amount of different elements versus time

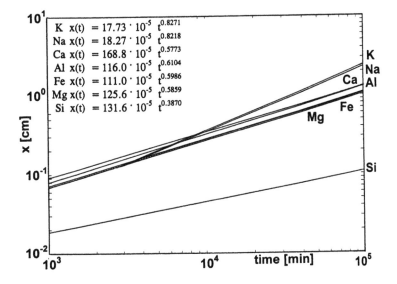

Fig. 3. Corrosion depth of different elements versus time

distance behind the calcium dissolution front. The corrosion depth of calcium exceeds that of all other elements and therefore determines the effective corrosion depth. The small theoretical zone depleted of silicon confirms deposition of this element as a residual layer on the surface of solids formed by incongruent dissolution of CSH.

3.3 Influence of reaction time and acidity

The acidity of aqueous systems, i.e. the availability of H^+-ions, strongly influences their reactivity versus hardened pastes. This is easily understood if the mechanism of reaction is taken into account: increasing acidity favours the formation of dissolved products. It was found that the dependence of dissolved amount m upon the reaction time t of acid solution can be expressed by the formula:

$$m(t) = K * t^b \qquad\qquad K, b = \text{constants} \qquad (1)$$

The individual formulae expressing the influence of time for each concentration (pH-value) of acid solution are presented in Fig. 4. The average value of the exponents b is approximately 0.56.

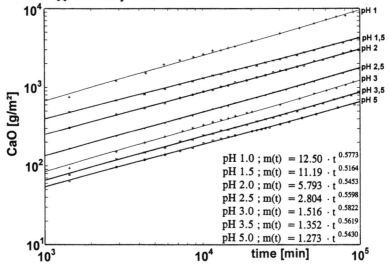

Fig. 4. Amount of CaO dissolved at different pH-values versus time

The graphic presentation of the dissolved calcium mass m versus an increasing concentration c of acid solutions at different exposure times (t = const) is given in Fig. 5. It is obvious that the measured values at pH 5 do not fulfil the curves described by the other pH-values. This confirms that at lower acidities the dissolved amount only slightly depends on pH variations. The course of the individual curves can be expressed by the formula (with exception of pH 5-values)

$$m(t) = K * c^d \qquad\qquad K, d = \text{constants} \qquad (2)$$

The average value of the exponent d is 0.4074.

Therefore, in the pH range from 1 to 3.5 the final comprehensive formula expressing the influence of H^+-concentration c of acid water and the time of its exposure t was found to be

$$m(t,c) = K * t^{0.5572} * c^{0.4074} \qquad (3)$$

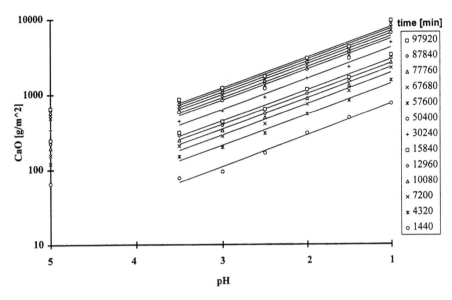

Fig. 5. Amount of Ca at different times versus concentration

The curves calculated according to this relationship and the measured experimental values are presented in Fig. 6. From the formula (3) and the density of calcium ions in the paste the corrosion depth x is calculated.

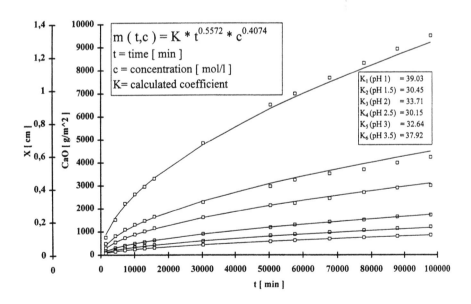

Fig. 6. Dependence of dissolved Ca (corrosion depth) upon concentration and time

3.4 Influence of temperature

The disintegration of hardened paste phases depends not only on solution composition (pH-value or "preconcentration") and reaction time but also on temperature. The plot of measured dissolution rates k versus 1/T leads to a straight line (Fig. 7), hence the equation of Arrhenius is valid:

$$\ln k = \ln k_0 - \frac{E}{RT} \quad or \quad \ln k = \frac{-a}{T} + b \tag{4}$$

E = activation energy, R = gas constant, T = absolute temperature

Generally, the concentration of dissolved species increases with increasing temperature, but the ratio of dissolution rates is independent of temperature and pH. Therefore, it is possible to determine the reaction rate at any temperature, if the dissolution rate at one temperature and the pH is known.

The corrosion depth and also the dissolved amount versus temperature is an only weakly bended e-curve (Fig. 8), which can be considered in a first approximation as a linear plot. However, regression analysis indicated that by the application of an e-function the coefficient of determination is closer to one than by the application of a linear relationship. This confirms the fundamental applicability of the physico-chemical motivated Arrhenius equation for the heterogeneous reactions under consideration. For a quick estimation of the temperature influence the following rule of thumb can be assumed: an increase of temperature of $\Delta T = 40$°C causes a doubling of reaction rate.

The effective activation energy E (table 2) was calculated from the reaction rates:

$$\ln \frac{k_2}{k_1} = \frac{E}{R}\left(\frac{1}{T_1} - \frac{1}{T_2}\right) \tag{5}$$

k_1 (k_2)   reaction rate at temperature T_1 (T_2)

Table 2. Activation energy of dissolution reaction

$[H_3O^+]$-concentration	T-range T1/T2	activation energy
[Mol/1]	[°K]	[kJ/mol]
10^{-5}	275/333	11,8
10^{-3}	275/333	11,6
10^{-1}	275/323	11,0

The following point is noteworthy: The potential energy barrier for dissolution reaction in strong (10^{-1}M) and weak (10^{-5}M) hydrochloric acid is low and almost equal.

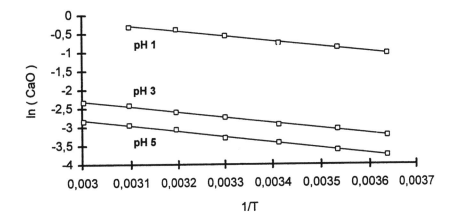

Fig. 7. Dissolution rate k [g/m²·min] of CaO versus 1/T (2 days experiments)

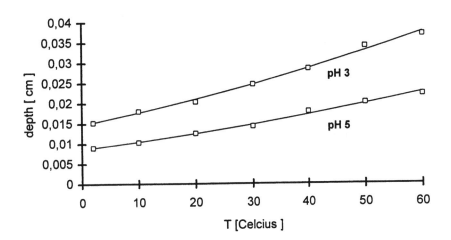

Fig. 8. Corrosion depth versus temperature (after 2 days)

References

1. Mchedlov-Petrosyan, O.P. and Babuschkin, V.I. (1974) *Thermodynamics and thermochemistry of cement*, VIth Int. Congr. Chem. Cement, Moscow.
2. Herold, G. (1994) *Charakerisierung der Korrosionsschichten von Zementstein nach Einwirkung mineralsaurer Lösungen*. European Journal of Mineralogy, Vol. 6, Beiheft No. 1, pp. 104.

13 IMPROVEMENTS IN THE DURABILITY OF CELLULOSE REINFORCED CEMENTITIOUS COMPOSITES

X. LIN, M.R. SILSBEE and D.M. ROY
(Intercollege) Materials Research Laboratory, The Pennsylvannia State University, PA, USA

Abstract
The use of cellulose fibers derived from wood as reinforcement in cement based systems has often been hindered by the poor resistance of the composites to wet dry cycling. Degradation of the fibers in the strongly alkaline environment found in hydrated cement based systems has also been a concern.

An approach has been developed that has resulted in a significant improvement in the resistance of cellulose fiber reinforced composites to wet-dry cycling. The results of monitoring the mechanical properties of the composites for two years indicate that the reinforcement is stable in the alkaline matrix. The approach has been found to be equally effective whether virgin fibers or fibers derived from recycling newspaper or Kraft papers were used as reinforcement.
Keywords: Cellulose fibers, cement, wet-dry cycling, durability

Introduction

Cellulose or wood fibers possess a set of properties that make them unique among the materials that are used as reinforcement in a cementitious system. They are flexible as opposed to rigid glass fibers. Cellulose fibers are inexpensive compared to many other fibers. Recent studies[1-3] have suggested that fibers derived from recycling sources may be effectively used as reinforcement in cement matrices. When dry the cellulose fibers are in the form of flat ribbons (Fig. 1). However, when exposed to moisture, as for instance when placed in a moist cement paste (Fig. 2) they swell and resemble hollow tubes.

Cellulose fibers have been found to be effective in enhancing the flexural strength of cementitious matrices and for improving the toughness and hence the

Mechanisms of Chemical Degradation of Cement-based Systems. Edited by K.L. Scrivener and J.F. Young. Published in 1997 by E & FN Spon, 2–6 Boundary Row, London SE1 8HN. ISBN: 0419215700.

impact resistance.   The possible mechanisms by which the fibers contribute to the mechanical performance of the composite are as follows:

1) Fiber fracture without failing of interfacial bond, which contributes little to the toughness of the composite, but may increase the strength since the tensile strength of the wood fibers is 900 MPa. much higher than the tensile strength of cement at 3.7 MPa. [4].

2) Energy may be consumed as fibers pull out from the matrix contributing to the toughness and the deformation capacity of the composite.

3) Only a portion of the fiber may be parted from the matrix.   Crack paths may be dislocated as a result, halting crack propagation.

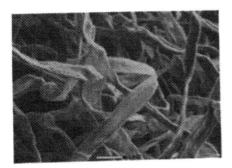

Fig. 1 - Environmental scanning electron micrograph of virgin (first use) softwood fibers.   In a dry state, as shown here, the fibers are in the form of a flat ribbon.

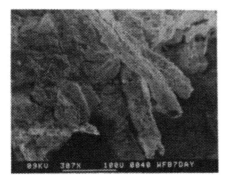

Fig. 2 - Scanning electron microscope of cellulose fibers embedded in a cement matrix.   The fibers as compared to the ribbons shown in Fig. 1, clearly show their hollow shape.

Wood fibers are sensitive to moisture.   Cellulose fibers are bonded together by hydrogen bonds that are water sensitive.   Wet or dry, a wood fiber will have about the same tensile strength but its stiffness is about ten times greater when dry. The Si-OH and Ca-OH groups will, in the high pH environment found in cement pore fluids, react with C-OH groups in the fibers to form hydrogen bonds. As water molecules are inserted the hydrogen bonds can be broken. Coutts[5] reported a higher percentage of fibers failing by pull out in wet composites as opposed to fracturing in dry composites. The physical structure of the fibers can affect the degree of mechanical bonding found in the composites. Microfibrils often less than 0.1 µm in diameter may interlock with the hydrating cement matrix. The interlocking of the fibers and the matrix have been found to improve the mechanical performance in the composites.

Approach

Wood fibers can be treated by water soluble chemicals, polymers and coupling agents. Filling the lumen of the fiber will promote adhesion of the fiber's internal wall.

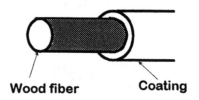

Wood fiber   **Coating**

Fig. 3 - Schematic illustration of a coated wood fiber.

Treatment of the fibers, shown schematically in Fig. 3, can make the fibers stiffer, more durable and more stable. The coatings can have the effect of 1) limiting the movement of water soluble organics (i.e. sugars, lignosulfonates, etc.) from the fiber into the hydrating cement matrix; 2) limiting contact between the fiber and the highly alkaline cement pore fluid; 3) improving the interfacial bond between the cement matrix and 4) improving the dispersity of wood fibers during mixing. The materials that have been found to be useful in the pretreatment of fibers can be broadly divided into three groups, 1) inorganics, such as $CaCl_2$, alkali hydroxides or silicates[6,7]; 2) organic polymers and monomers such as carbowaxes, polyethylene glycol[8,9], vinyl monomers and acrylonitrile[10] and acrylics[1-3].

In this case the fibers were soaked in deionized water at room temperature for more than 1/2 hour. After soaking and while still in the water the fibers were passed through a Valley Beater. After vacuum filtration the flakes of the fibers were removed. After air drying the fibers were removed from the flakes by milling. All fibers were passed through a 2 mm sieve. Treatments were applied to the fibers by soaking the fibers in an aqueous solution containing the desired treatment agent for 1/2 hour; followed by hand pressing to remove the liquid, filtering, air drying and milling as described above. Figs. 4 and 5 show the surface of a wood fiber after treatment with solutions containing 1 and 10 weight percent PVA. The Figs. show that the polymer has to a certain extent coated the surface of the fibers. Small globules of the polymer can be seen adhering to the surface of the fibers.

Sample Preparation

Table 1 shows the formulations used in this study. The dry wood fibers, cement and powdered superplasticizer were pre-mixed. The dry mixture was then placed into water in the bowl of a Hobart mixer (meeting the requirements of ASTM C109). The paste was then mixed at slow speed for 1/2 minute when using treated fibers, 1 minute when using untreated fibers. After scraping down the sides of the bowl and the mixing paddle the paste was mixed at medium speed for another minute. The specimens (either 25.4x25.4x127mm bars or 25.4 mm cubes) were demolded after 24 hours and then cured over water at the designated temperature until testing began. The flexural strengths were determined according to ASTM C1018 procedures.

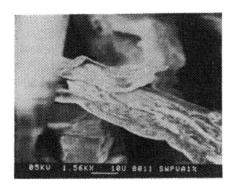

Fig. 4 - Soft wood fiber after treatment in a solution containing 1 wt. % PVA.

Fig. 5 - Surface of a softwood fiber after treatment with a solution containing 10 wt. % PVA.

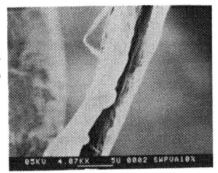

Table 1 - Formulations employed in this study.

Sample	Wt. % of Components			
	Cement	Water	Fibers	Superplasticizer
Control	75.99	22.80	-.-	1.21
Composite	67.38	26.95	4.24	1.43

Results

The results shown in Table 2 are intended to compare the flexural strengths found in wood fiber reinforced composites prepared with different untreated fibers with that found in an unreinforced control paste. The results show that in this case all the composites exhibited flexural strengths higher than those found in the control paste. Fig. 6 tracks the flexural strengths observed in a series of fiber reinforced composites for two years. The results indicate that especially at the later ages that the fiber reinforced composites had significantly higher average flexural strengths as compared to the control. However, while the toughness[1-3] of the composites was improved at early ages, at later times little improvement was observed. For instance, the I_5 toughness (calculated according to ASTM C1018) of the composites prepared with untreated fibers averaged approximately 3.5 at 28 days as compared to near 1 at 90 days. (Note: A toughness of 1 would be found in the unreinforced control paste.)

Table 2 - The Effect of fiber type on flexural strength development in wood fiber reinforced composites. The cement used was an alkali activated blended cement for which the composition has been previously presented[11]. The samples were cured at 38°C in high humidity.

	Flexural Strength (MPa.)
Control	5.1
Virgin Hardwood	12.8
Recycled Kraft	8.8
Recycled Newsprint	9.6

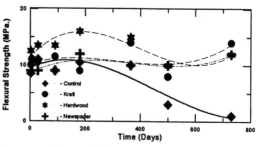

Fig. 6 - Flexural strength development as a function of time in fiber reinforced composites .

This indicates that scenario (1) fiber fracture without failing of interfacial bond, which contributes little to the toughness of the composite, but may increase the strength since the tensile strength of the wood fibers is much higher than the tensile strength of cement, was occurring here.

During this study the effect of pretreating the cellulose fibers with several different polymers was examined. The effects of PEG, PVAc, silane and acrylic polymers were examined. Little or no improvement in the flexural strength or toughness of the composites occurred as the result of pretreating the fibers. However, as shown in Fig. 7, the pretreatments do have the ability to improve the compressive strength of cellulose cement composites. Apparently the pretreatments improved the

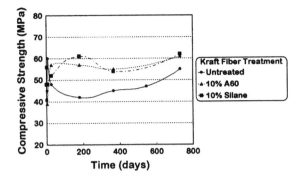

Fig. 7 - The effect of fiber treatment on compressive strength development in fiber reinforced composites. (A60 was an acrylic polymer in an aqueous emulsion from Thoro System Products, Inc., Silane was an alkylalkoxysilane in water from Hydrozo Incorporated, Inc.)

interfacial bonds between the fibers and the cement matrix resulting in improved compressive strengths.

Cycles of wetting and drying can cause considerable drop in the flexural strength and toughness of fiber reinforced composites, in particular when natural fibers are employed. The loss in strength has been attributed to the attack on the lignins in the fiber by the alkaline pore fluids found in cement. Cycles of wetting and drying are thought to accelerate the degradation of the fibers by facilitating the movement of the pore water to the surface of the fiber and of the degradation products away from the fiber[12]. In this case either cellulose fibers obtained by a chemical pulping process or recycled fibers were employed. The lignin content of these fibers can be expected to be reduced as a result of processing. However, other mechanisms can result in the degradation of the fibers. The hollow inner cores of the fibers may gradually fill with hydration products leading to embrittlement of the fiber. In extreme cases the hydration products may swell the fiber causing it to burst and lose integrity. Densification of the cement matrix in the vicinity of the fibers can lead to the embrittlement of the composites.

To simulate the effects of repeated wet-dry cycles on the composites, a wet-dry cycling test was developed. The procedure consisted of soaking the sample bars (25.4x25.4x127mm) in 25°C water for 7 days followed by 7 days at 40°C and 29-34% relative humidity. This results in a typical composite cycling between 30 and 90 percent of maximum absorbed water. The weight and linear dimensions of each specimen were measured after each drying and wetting cycle. Details of the wet -dry cycling experiments are given elsewhere[1]. Fig. 8 compares the performance of an unreinforced control sample, a composite prepared with untreated hardwood fibers and treated hardwood fibers. As can be seen in Fig. 8 both the unreinforced paste and the paste reinforced with the untreated hardwood fibers began to expand soon after the wet dry cycling begin. Cracking of the matrix during the cycling lead to the expansion. However, the composite reinforced with the treated hardwood fibers shrank during the cycling. Some shrinkage during curing is typical of cementitious materials, in this case apparently the treated fibers were able to bind the matrix together tightly enough to prevent expansion due the cracking of the matrix phase. The unreinforced paste failed catastrophically after 5 cycles, cracking into small pieces. The composite prepared using the untreated fibers failed after 23 cycles, normally with a single crack across the narrow dimension of the bar. The composite prepared with the treated fibers failed at 54 cycles, after nearly two years of continuous wet dry cycling. Again, failure in the treated fiber composite normally consisted of a single crack across the narrow dimension of the bar.

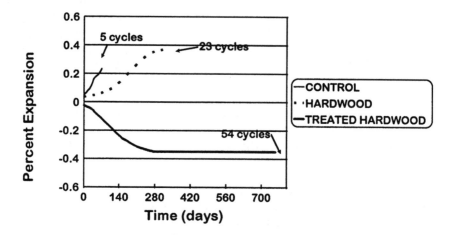

Fig. 8 - Dimensional fluctuations in during wet-dry cycling for an unreinforced cement paste (control), untreated fiber reinforced composite (hardwood) and a composite prepared using hardwood fibers that had been treated with a 10% silane solution. The measurements were taken at the conclusion of the wet cycle before drying.

Summary

Cellulose fibers both virgin (first use) and recycled can be an economical route to producing relatively lightweight cementitious composites. The performance of cellulose fibers in cementitious composites seems to be strongly influenced by the nature of the interfacial bond between the fiber and the cement matrix and the characteristics of the cement matrix in the vicinity of the fiber. The fiber reinforced composites examined during this study exhibited lower densities, modestly improved flexural strengths, lower compressive and significantly improved resistance to wet-dry cycling as compared to the unreinforced control samples. Pretreatment of the cellulose fibers with relatively dilute solutions (1-10 wt. %) of polymers has been found to be an effective route to improving the performance of cellulose fiber reinforced composites. Composites prepared using treated fibers showed improved compressive strengths compared to composites using untreated fibers. The most dramatic effect was in the performance in wet/dry cycling the resistance to failure of the treated fiber composites to wet dry cycling was twice than of composites using untreated fibers and ten times that of an unreinforced cement paste.

Acknowledgments

The authors would like to acknowledge the technical contributions of Darlene Wolfe-Confer and Maria DiCola to this study. The authors also appreciate the cooperation of Dr. Paul Blankenhorn of Penn State's Forest Resource Laboratory. The financial support of the U.S. Department of Agriculture and Environmental Protection Agency is also appreciated.

References

1. Lin, X. (Dec. 1994) High Performance Cementitious Composites Reinforced With Wood Fibers and Polymers M.S. Thesis, The Pennsylvania State University, University Park, PA.
2. Lin, X., Silsbee, M.R., Roy, D.M., Kessler, K., and Blankenhorn, P.R. (1995) By-Product Wood Fiber Reinforced Cementitious Composites and Chemical Pretreatments of Fibers. Second CANMET/ACI International Symposium on Advances in Concrete Technology, pp. 805-824.
3. Lin, X., Silsbee, M.R., Roy, D.M., Kessler, K., and Blankenhorn, P.R. (1994) Approaches to Improve the Properties of Wood Fiber Reinforced Cementitious Composites. Cem. Con. Res. 24 [8] pp. 1558-1566.
4. Bentur, A. and Mindess (1990) S., Fiber Reinforced Cement Composites, New York, McGraw-Hill, Inc., pp. 1-11.
5. Coutts, R.S.P., "Fibre-matrix Interface in Air-Cured Wood-Pulp Fibre-Cement Composites," J. Mat. Sci. Letters, 6 pp. 140-142 (1987).
6. Guthrie, B.M., and Torley, R.B., "Composite Materials Made from Plant Fibers bonded with Portland Cement and method of Producing Sam," U.S. Patent, No. 4,406,703: Sep. 27, (1983).
7. Chen, Y., Park, C.K., Silsbee, M.R., and D.M. Roy (1992) The Use of Natural Fibers in Cementitious Composites. Mat. Res. Soc. Symp. Proc. 245 pp. 229-234.
8. Stamm, A.J. (1956) Dimensional Stabilization of Wood with Carbowaxes. Forest Prod. J. 6(5) pp. 201-204.
9. Stamm, A.J. (1959) Effect of Polyethylene Glycol on The Dimensional Stability of Wood. Forest Prod. J. 9(10) pp. 375-381.
10. Ellwood, E.L., Gilmore, R.C. and Stamm, A.J., "Dimensional Stabilization of Wood with Vinyl Monomers," Wood Sci. 4(3) pp. 137-141 (1972).
11. Roy, D.M. and Silsbee, M.R. (1990) New Rapid Setting Alkali Activated Cement Compositions in Specialty Cements With Advanced Properties Proc. The Materials Research Society, 179, 203-218 Ed., B.E. Scheetz, et. al.
12. Marikunte, S. and Soroushian, P., (Nov.-Dec. 1994) Statistical Evaluation of Long-Term Durability Characteristics of Cellulose Fiber Reinforced Cement Composites ACI Mat. J. 91 [6].

14 CORROSION OF TOBERMORITE IN STRONG CHLORIDE SOLUTIONS

W. KURDOWSKI, S. DUSZAK AND B.TRYBALSKA
University of Mining and Metallurgy, Cracow, Poland

Abstract
The corrosion of tobermorite was followed under SEM and X-Ray. The diffusion of Cl^- and Mg^{2+} ions led to the formation of basic magnesium chloride and basic calcium chloride. As these compounds were expansive, the sample underwent constant expansion. At the same time the amorphisation of tobermorite was found to occur, but the relicts of this phase contained only small amount of magnesium. The samples were destroyed relatively quickly after 390÷420 days.
Keywords: calcium chloride, cement paste, chloride solution, corrosion, C-S-H, magnesium chloride, tobermorite.

1. Introduction

The durability of concrete in strong chloride solutions, especially those of magnesium chloride, is of great practical importance. The waters in the salt mines are generally very strong chloride solutions and the content of magnesium chloride is variable and often very high. For example in one Polish mine the content of $MgCl_2$ lies between 70 g/l and 270 g/l. To ensure good durability of concrete for underground works in this mine, and among other uses in the construction of dams and mine wall stabilization, is a big problem.

Over several years in our laboratory we have been studying the durability of pastes prepared from different cements, namely HAC, slag cements including alkali activated slag cements [1,2]. For better understanding of the mechanism of corrosion processes we also studied monophase samples of C-S-H [3] and tobermorite. This paper describes the results of investigations of the samples rich in tobermorite.

Mechanisms of Chemical Degradation of Cement-based Systems. Edited by K.L. Scrivener and J.F. Young. Published in 1997 by E & FN Spon, 2–6 Boundary Row, London SE1 8HN. ISBN: 0419215700.

2. Preparation of the samples

The C-S-H of molar ratio CaO/SiO_2 = 5/6 was obtained by hydration of $3CaO.SiO_2$ with adequate addition of amorphous silica (Aerosil, Degussa). From this paste small prisms of dimensions $20 \times 20 \times 100$ mm were formed, and cured 14 days in water at $20 \pm 2°C$. After two weeks the samples were autoclaved at $150°C$ in saturated water vapour, for 72 hours. Fig. 1 presents the X-ray diffraction pattern of the product, which was rich in tobermorite, and Fig. 2 is an SEM micrograph of it. Apart from tobermorite the sample contained a small quantity of C-S-H (peak at 0.302 nm) and a small amount of xonotlite (peaks at 0.290; 0.261; 0.253 and 0.207 nm). A field relatively rich in C-S-H is shown in Fig. 3.

After autoclaving, the samples were cured for 48 hours in water at $20 \pm 2°C$ and then immersed in a solution rich in $MgCl_2$. The composition of this solution, which corresponds to an example of salt mine water, was as follows:

$MgCl_2$ - 272 g/l   KCl - 10 g/l
$CaCl_2$ - 164 g/l   KBr - 4 g/l
NaCl - 14 g/l

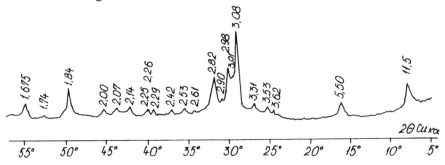

Figure 1. X-ray diffraction pattern of tobermorite sample.

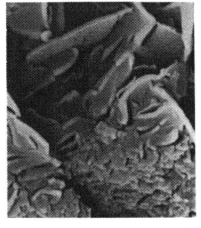

Figure 2. SEM of tobermorite sample

Figure 3. SEM of tobermorite sample, field rich in C-S-H.

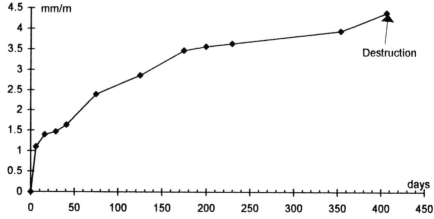

Figure 4. Expansion curve of the prism.

The volume ratio of leachant to cement was approximately 20:1. The solutions were totally renewed at approximately two monthly intervals, as the main goal was to keep conditions close to those in the salt mine.

The porosity of the samples before immersion was relatively high, 0.44 cm^3/g for the pores in the 1.7÷500 nm range, and the average pore diameter was 14 nm.

3. Results of experiments

From the beginning of the immersion the samples revealed an expansion, which was at first very quick, but later became slower (Fig. 4). Shortly before the prisms disintegrated at ages of 390÷420 days an acceleration of the expansion was noted. This acceleration of expansion took place at 40 to 90 days before the disintegration.

The X-ray study of the samples immersed only 230 days in the chloride solution showed a total amorphisation of the tobermorite, which then gave no diffraction peaks. The strong amorphisation of the sample is seen also in the SEM (Fig. 5). On the X-ray pattern (Fig. 6) only three diffuse bands can be seen at 0.444, 0.257 and 0.154 nm, which correspond to the strong peaks of basic magnesium chlorides, namely:

$Mg_2(OH)_3Cl.2H_2O$: 0.4425 - I = 100, 0.1572 - I = 50; [PDF: 7-419]
$Mg_3(OH)_4Cl_2.4H_2O$: 0.2597 - I = 60. [PDF: 12-116]

In a few fields examined by SEM relicts of tobermorite crystals were found (Fig. 7). EDAX analysis revealed that the content of magnesium in these relicts was relatively very low and that of chlorine high (Fig. 8).

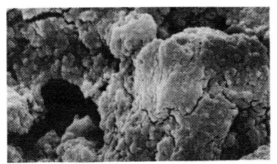

Figure 5. SEM of tobermorite sample after 230 days of immersion in chloride solution.

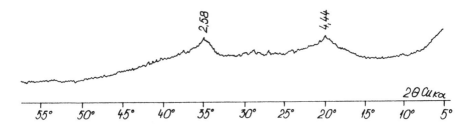

Figure 6. X-ray diffraction pattern of the tobermorite sample after 230 days of immersion in chloride solution.

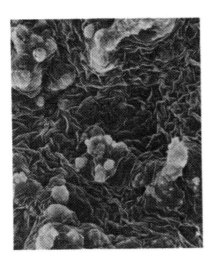

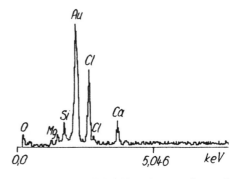

Figure 7. SEM of the sample after 230 days of immersion, showing relicts of tobermorite.

Figure 8. EDAX of a relict of tobermorite.

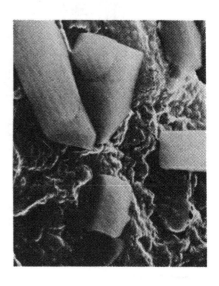

Figure 9. SEM of sample after 420 days of immersion, showing an altered external layer.

Figure 10. SEM, showing crystals of basic magnesium chloride.

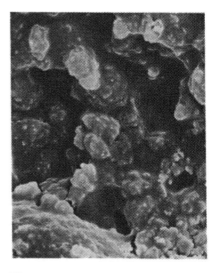

Figure 11. SEM, showing crystals of basic calcium chloride.

Figure 12. SEM of the field of irregular grains in an amorphous matrix.

Figure 13. SEM of porous region of the sample.

Figure 14. SEM of a dense "grain".

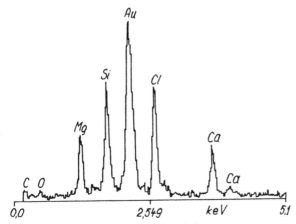

Figure 15. EDAX of a dense "grain".

The external layer of the prism (Fig.9) was much more compact than the core, but its chemical composition was very similar (Table 1). Only the content of chlorine is higher. In this external layer greater crystals of basic magnesium chloride appeared sporadically (Fig. 10) and in other regions very small crystallites of this phase were present. In a few regions crystals of basic calcium chloride were also found (Fig. 11).

Immediately below the external layer a zone of irregular grains embedded in an amorphous matrix was observed (Fig. 12). EDAX of this region showed them to have the same chemical composition as the dense grains.

The third zone was composed of more loosely packed grains and the porosity of this region was higher (Fig 13). In this zone individual dense grains were visible (Fig. 14). They seemed to be of uniform composition and their chemical compositions were very similar to those of the loose grains i.e. they were high in Si and Cl, and lower in Mg and especially in Ca (Fig. 15).

The core of the sample was relatively compact; incipient formation of dense grains was seen.

Table 1. Chemical composition of the prism after 420 days of immersion

Chemical composition (weight %)	Core	External layer
LOI	26.25	26.85
SiO_2	27.17	27.05
CaO	14.83	12.37
MgO	16.03	16.03
Cl	14.35	16.04
Na_2O	0.56	0.60
K_2O	0.36	0.39

4. Discussion of results

The process of corrosion of tobermorite is relatively very quick. It contrasted with that for the pure C-S-H samples [3], which after 28 months of immersion in a solution of similar composition, but of lower $MgCl_2$ content (77 g/l), gave on the X-ray diffraction pattern relatively strong lines typical for this phase. With the tobermorite, the peaks of that phase disappeared after only 230 days. The products included a basic magnesium chloride, which gave diffuse bands in the X-ray diffraction pattern, and crystals of which could be seen in the SEM. Crystals of basic calcium chloride could also be found by SEM, but the content of this compound was below the threshold for detection by X-ray diffraction. All these basic salts are expansive [4] and their formation explains the expansion of the prisms.

The chemical compositions of the several forms of the amorphous products of the decomposition of the tobermorite were very similar. The high contents of chlorine and silica were associated with relatively smaller contents of magnesium and calcium. The chemical compositions of the dense grains were practically the same as those of the areas of higher porosity. The results obtained do not permit an explanation of phase composition of these amorphous grains. It is very probable that they are a mixture on the nanometric scale of several amorphous phases, namely a C-S-H gel, basic magnesium chlorides, brucite and magnesium silicate hydrate.

The observation that the content of magnesium in the tobermorite relicts is very low is important information. It seems that the tobermorite crystals are relatively quickly decomposed with the formation of a C-S-H poorer in calcium together with silica gel. Increased formation of silica gel leads to the crystalization of basic calcium chloride.

It is clear that the process of corrosion is based on the diffusion of chloride and magnesium ions towards the core of the sample and of OH groups and calcium ions in the opposite direction. In this context one can argue that, without any doubt, the relatively great porosity of the samples and the high content of crystalline phases are major reason for the high rate of decomposition. A small content of gel causes absence of the shrinkage that was observed in case of cement pastes containing much gel [1,3], for example C-S-H [2,3] or $Al(OH)_3$ in the case of HAC [1]. The shrinkage causes densification of the structure, which retards diffusion of the ions from the solution into the interior of the immersed samples. It seems that the porosity and the absence of gel are among the main reasons for the quick destruction of tobermorite samples in strong magnesium chloride solution. It is apparent from the chemical compositions of the external layer of the sample and of the core (Table 1) that the diffusion of magnesium ions is quick but that the amount of calcium in the sample was only slightly lower. It seems that most of the calcium remains in the sample but in the form of an amorphous basic calcium chloride. The crystalline form of this phase is very rare.

5. Conclusions

1. Tobermorite is very unstable in a strong solution of magnesium chloride
2. Under the influence of magnesium chloride tobermorite decomposes with the formation of amorphous C-S-H and basic magnesium chloride
3. The amorphous C-S-H undergoes a process of gradual decalcification, so that the end components are: basic magnesium chloride, basic calcium chloride, magnesium silicate hydrate and silicic acid

Acknowledgment

The authors are grateful to the Committee of Scientific Research in Warsaw, Poland, for financial support under Grant no. PBO 613/S2/92/03.

References

1. Kurdowski W., Taczuk L. and Trybalska B. (1989) Long Term Study of Different Cements in Chlorides Solutions; in "Durability of Concrete", Aspects of admixtures and industrial by-products. 2nd International Seminar, June 1989. Swedish Council for Building Research, Stockholm, p. 85.
2. Kurdowski W., Duszak S. and Trybalska B. (1994) Corrosion of Slag Cement Paste in Strong chlorides Solution, "Alkaline Cements and Concretes", Proc. First Int. Conf. Kiev, Ukraine, 11-14 October 1994 Ed. by V. Krivenko, Vol. II, p. 961.
3. Kurdowski W. and Duszak S. (1995) Changes of C-S-H gel in strong chloride solution, Advances in Cement Research, Vol. 7, No. 28. pp. 143÷150.
4. Smolczyk H. G. (1968) Chemical Reactions of Strong Chloride-Solutions with Concrete, 5th Int. Symp. on the Chemistry of Cement, Tokyo, vol. III, p. 274.

PART TWO
CORROSION OF REINFORCING STEEL

15 MONITORING THE CORROSION OF REINFORCING STEEL IN CEMENT-BASED SYSTEMS USING IMPEDANCE SPECTROSCOPY

S.J. FORD AND T.O. MASON
Department of Material Science and Engineering, Northwestern University, Evanston, IL USA

Abstract

The underlying microstructure and chemistry in cementitious systems greatly influence the corrosion of reinforcing steel. Impedance spectroscopy allows the separation of responses from distinct features of the microstructure. Responses from the bulk paste, the paste/steel interfacial zone, and a passive iron oxide/hydroxide film on the steel surface are observed for hydrating cement-based systems.

From the bulk paste and interfacial responses, transport properties, such as permeability and diffusivity, can be obtained. These properties are important in understanding the mechanisms of ingress of corrosive species, and the effect of these species on corrosion of the reinforcement.

The impedance response of the passive film provides a measure of both the oxide thickness and the corrosion rate of the steel. Therefore, the effects on corrosion, caused by changes in the steel and cement chemistry, can be monitored. Various cement-based systems will be discussed with emphasis on changes in the interfacial microstructure, cement chemistry, and composition of the steel.

Keywords: Cement-based materials, corrosion, impedance spectroscopy, in-situ characterization, microstructure/property relationships, reinforcing steel.

1 Introduction

Corrosion of reinforcing steels in cementitious systems has become an important consideration both economically and in terms of public safety, causing billions of dollars in damage each year [1]. Impedance spectroscopy (IS) is well suited for the study of corrosion in reinforced cement-based systems and the characterization of underlying microstructural features. IS reveals the condition of the paste/steel

Mechanisms of Chemical Degradation of Cement-based Systems. Edited by K.L. Scrivener and J.F. Young. Published in 1997 by E & FN Spon, 2–6 Boundary Row, London SE1 8HN. ISBN: 0419215700.

interface, as well as information concerning the microstructure of the bulk cement paste [2]. Because of this ability to separate responses, impedance spectroscopy allows study of both the overlayers and interfaces associated with reinforcing steel.

Typical impedance spectra of cement pastes, over 10 orders of magnitude in frequency from mHz to MHz, exhibit three arcs in the complex plane as shown in Figure 1. The assignments of individual features to bulk, charge transfer/double layer effects, and a passive oxide film are discussed further below.

2 Experimental Procedure

ASTM Type I ordinary Portland cement (OPC) was used for all experiments. A.I.S.I. C-1018 low carbon ground flat stock was used to simulate the reinforcing steel, providing more compositional control than would be allowed by construction grade rebar specimens. The chemical composition of the steel was (weight percent): Carbon 0.15 - 0.20, Manganese 0.60 - 0.90, Phosphorus < 0.040, and Sulfur < 0.050. Synthetic pore solutions were composed of reagent grade potassium hydroxide (KOH) and sodium hydroxide (NaOH) with distilled water.

Cement paste specimens were prepared with Type I OPC and distilled water. Appropriate weights of OPC and distilled water were mixed in a Hobart planetary mixer for fifteen minutes at low speed. The paste was then cast in rectangular Plexiglas molds with dimensions 2.54 x 2.54 x 10 cm^3, and cured at 100% relative humidity in a chamber attached to the front of the impedance analyzer. C-1018 steel electrodes were cut with dimensions 0.25 x 1.90 x 3.85 cm^3. These electrodes were

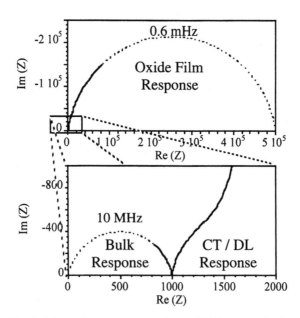

Fig. 1. Typical Impedance spectrum for an OPC paste of w/c = 0.4 at 50hrs with C-1018 steel electrodes.

then polished to 6 μm, rinsed with acetone and distilled water, and immediately cast in the cement pastes perpendicular to the long axis of the bars. An area of 4.85 cm² was immersed in the paste.

Synthetic pore solutions (SPS) were made to simulate the pore solution of an OPC paste of w/c = 0.5 at 100 hours [3]. The main constituents of the pore solution for this study were 100 mM NaOH and 175 mM KOH, corresponding to a pH = 13.5. Minor constituents, such as sulfur, silicon, calcium, and aluminum ions were ignored because of their low concentration (μM) in OPC paste pore solution at these hydration times. Studies of their influence on passivation, especially at early hydration times, are in progress.

Impedance measurements were made with a Schlumberger 1260 Frequency Response Analyzer, and Z60 data collection software [4]. Alligator clips and coaxial cables were used to make electrical connections to the specimens, and the cables were kept as short as experimentally possible to reduce immitance effects from the leads (10 cm). Excitation amplitudes were kept to a minimum (< 25 mV), in an attempt to keep the response in the linear regime, and to reduce inductive contributions from the cables. A nulling procedure was employed to correct for lead and connector immitances, with a more detailed description of this procedure given elsewhere [3]. Also, computer software was used for deconvolution when relaxation times for various processes were not vastly different [5], along with the implementation of equivalent circuits which idealize and model the response of the systems. A more detailed description of the use and interpretation of equivalent circuits is given elsewhere [6].

3   Results

The complete equivalent circuit proposed to model the impedance response of the cement paste/steel system is presented in Figure 2. The bulk response is modeled as a single resistor, although there are currently several more detailed microstructural models in the literature [2,7-11]. Concerning the corrosion processes in these materials, only the bulk resistance, R_{bulk}, is of importance to the present study, which is given by the low frequency intercept of the bulk arc with the real axis.

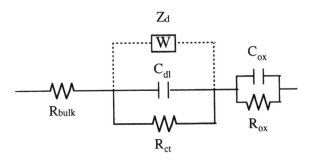

Fig. 2. Equivalent Circuit Model for the cement paste/reinforcing steel system. Circuit elements are defined in the text.

The intermediate frequency response is modeled by a parallel RC circuit comprised of the ionic double layer capacitance, C_{dl}, and a charge transfer resistance or diffusional impedance. Under normal circumstances, the charge transfer resistance, R_{ct}, dominates the intermediate frequency polarization resistance, but upon addition of microstructure altering components such as silica fume, or a decrease in the water to cement ratio (w/c), a diffusional element may replace the double layer capacitance in the model (Z_d). Finally, for passive systems the low frequency arc is modeled by the parallel combination of the oxide resistance and capacitance (R_{ox} and C_{ox}).

The bulk response of cementitious materials is well characterized. From the impedance spectrum, the bulk resistivity can be established [9], and the interconnectivity of the pore structure can be determined [12]. The roles of admixtures on microstructural development can, therefore, be monitored [2,3,13]. Using the Nernst-Einstein and Katz-Thompson equations, ion diffusivity and permeability can be estimated, respectively, each of which has implications for the ingress of corrosive species to the surface of the reinforcing steel. Several sources exist in the literature which propose microstructural models, so that a detailed analysis will not be presented here [2,7,9,11,14]. Regardless of the model, the value of the bulk resistance from which the transport properties of these species can be determined, is reliable.

At frequencies below approximately 100 Hz, two impedance arcs are observed in the complex plane plots of cementitious systems. The values of the capacitance for these responses, being in the mF range, indicate that the processes responsible for the low frequency arcs are characteristic of interfacial reactions at electrolyte/electrode interfaces [15]. Evidence is presented below which supports the theory that the intermediate frequency arc is associated with a charge transfer/ionic double layer process, modified by the effective surface area of capillary pores in contact with the electrode. The low frequency arc is ascribed to the response of an adherent oxide layer on the steel surface, with implications for film thickness and corrosion rate determination.

Several researchers have described the existence of an intermediate frequency arc in corroding systems [16-20]. Previous reports have suggested that this arc is the response of corrosion products formed in the pores of cement systems, or of a calcium hydroxide layer at the interface [19]. Figure 3 shows the impedance response of C-1018 steel in a calcium-free synthetic pore solution simulating the aqueous phase of cement paste. It can be seen that the intermediate frequency arc is evident in the

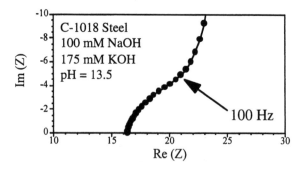

Fig. 3. Impedance spectrum for low carbon steel in synthetic pore solution (SPS).

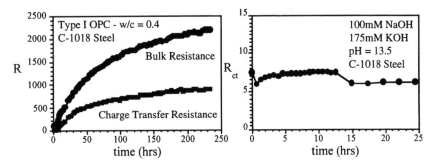

Fig. 4. Charge transfer resistance vs. time for a normal OPC paste and in SPS.

absence of the paste, which suggests that it is associated with the charge transfer/double layer response, since no paste matrix is present, and no corrosion products would be expected in this highly passivating environment. Also, capacitance values are on the order of $20\mu F/cm^2$, consistent with previously reported values for ionic double layers [21,22].

Another feature of the intermediate frequency response is that its resistance parallels the resistance versus time behavior of the bulk. As can be seen in Figure 4, the resistance of the intermediate frequency response follows the same trend with hydration as does the bulk resistance. In contrast, the charge transfer resistance in synthetic pore solutions remains essentially constant with time. The microstructure of the paste in the near interfacial zone has a direct effect on the intermediate frequency arc, which may explain why prior workers have attributed the intermediate frequency arc to corrosion products formed near the interface. Since the impedance response of the system is measured through the conductive pore phase, as the volume fraction of capillary porosity decreases with hydration, the effective area of pores adjacent to the interface also decreases. This reduction in the effective electrode area would explain the increase in the charge transfer resistance with hydration of the paste matrix, giving a qualitative explanation for the observed trends.

Support for this model is also obtained with the addition of silica fume (SF) to the paste. SF is known to reduce the volume fraction of capillary porosity, and therefore the ability of the matrix to transport aqueous species [13,23,24]. A diffusional element, characterized by a 45° line in the complex plane and an n-value of 0.5, becomes apparent in these systems, replacing the semicircular charge transfer/double layer response as seen in Figure 5. We have also altered the intermediate frequency response of steel/OPC samples from arc-like (Fig. 5a) to diffusion-like (Fig. 5b) behavior by dip-coating the steel in SF prior to immersion in fresh paste. With respect to the corrosion of reinforcing steel in these systems, the tighter microstructure indicated by the presence of a diffusional element, should reduce the transport of corrosive species to the interface.

It is well known that in the high pH environment of cement pastes, a passive oxide film forms on the steel surface [25-28]. This adherent film acts as a barrier to corrosion, slowing the process by several orders of magnitude [26,29]. Using the Stern-Geary equation in conjunction with Faraday's law, under the assumption of

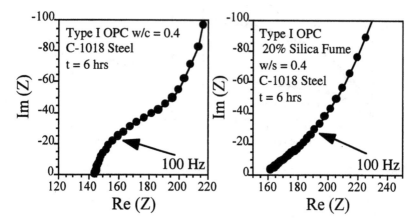

Fig. 5. Examples of complex plane plots for systems with a double layer capacitance response (normal OPC) and a diffusional response (20% silica fume).

uniform corrosion, the corrosion rate of the steel can be estimated from IS measurements by relating the corrosion current, I_{corr}, and the polarization resistance, R_p [30,31]. Therefore, distinguishing active vs. passive corrosion is facilitated. For example, when NaCl is added to the solution in Figure 6a, the oxide arc shrinks dramatically within the first hour until the situation in Figure 6b is obtained (At the same time, the intermediate frequency arc is largely unchanged). This demonstrates how IS can be used to monitor the onset of pitting corrosion.

Figures 7a and 7b show preliminary results for the corrosion of various steels in cement paste with varying cement chemistries. As seen in these figures, the steel chemistry plays a much more dominant role in limiting its corrosion than does the cement chemistry. Changes in steel chemistry, from stainless steel to low carbon

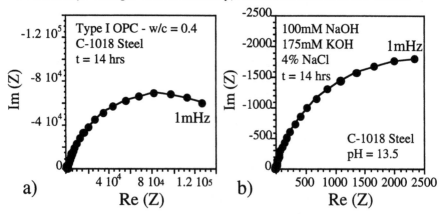

Fig. 6. Passive vs. active corrosion. Note the change in magnitude of the arc size.

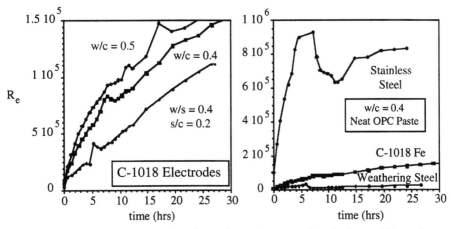

Fig. 7. Oxide resistance vs. time for various a) cement chemistries and b) steel chemistries.

steel, alter the oxide resistance and therefore the corrosion rate by orders of magnitude. In contrast, the variations in cement chemistry result in little change in the corrosion resistance of the low carbon steel.

IS also allows an estimation of the passive layer thickness based upon oxide capacitance, C_{ox}. C_{ox} stabilizes rapidly with time to a value corresponding to a thickness of approximately 15 nm (assuming a relative dielectric constant of 10) consistent with prior reports [27-29].

4 Summary

In conclusion, the use of impedance spectroscopy for the study of corrosion in cement-based materials is an effective way to monitor the process in-situ and non-destructively. The proposed circuit model explains the available experimental data well, and with verification, will be of great use in the study of these systems. Each of the microstructural features described have implications for the corrosion processes in cement-based materials, so that a view of the system as a whole is necessary for complete understanding of these materials.

5 References

1. Borgard, B., Warren, C., Somayaji, S. and Heidersbach, R. (1990) *Mechanisms of Corrosion of Steel in Concrete*. Berke, N.S., Chaker, V. and Whiting, D., ed. Corrosion Rates of Steel in Concrete. American Society for Testing Materials, Philadelphia, ASTM STP 1065, pp. 174-188.
2. Christensen, B.J., Coverdale, R.T., Olson, R.A., Ford, S.J., Garboczi, E.J., Jennings, H.M. and Mason, T.O. (1994) Impedance Spectroscopy of Hydrating

Cement-Based Materials: Measurement, Interpretation, and Application. *Journal of the American Ceramic Society*, Vol. 77, No. 11. pp. 2789-2804.

3. Christensen, B.J. (1993) *Microstructure Studies of Hydrating Portland Cement-Based Materials using Impedance Spectroscopy.* Doctoral Thesis. Northwestern University, Evanston, IL.

4. Scribner Associates (1994) *Z60 / ZVIEW for Windows.* 1.1d ed. Charlottesville, VA 22901.

5. Boukamp, B.A. (1988) *Equivalent Circuit.* 2nd ed. University of Twente, Netherlands.

6. Boukamp, B.A. (1986) A Nonlinear Least Squares Fit Procedure for Analysis of Immitance Data of Electrochemical Systems. *Solid State Ionics*, Vol. 20, pp. 31-44.

7. Coverdale, R.T., Christensen, B.J., Jennings, H.M., Mason, T.O., Bentz, D.P. and Garboczi, E.J. (1995) Interpretation of Impedance Spectroscopy of Cement Paste via Computer Modelling. Part I: Bulk Conductivity and Offset Resistance. *Journal of Materials Science*, Vol. 30, No. 3. pp. 712-719.

8. Gu, P., Xu, Z., Xie, P. and Beaudoin, J.J. (1993) Application of A.C. Impedance Techniques in Studies of Porous Cementitious Materials (I): Influence of Solid Phase and Pore Solution on High Frequency Resistance. *Cement and Concrete Research*, Vol. 23, No. 3. pp. 531-540.

9. McCarter, W.J. and Brousseau, R. (1990) The A.C. Response of Hardened Cement Paste. *Cement and Concrete Research*, Vol. 20, No. 6. pp. 891-900.

10. Xie, P., Gu, P., Fu, Y. and Beaudoin, J.J. (1994) A.C. Impedance Phenomena in Hydrating Cement Systems: Detectability of the High Frequency Arc. *Cement and Concrete Research*, Vol. 24, No. 1. pp. 92-94.

11. Scuderi, C.A., Mason, T.O. and Jennings, H.M. (1991) Impedance Spectra of Hydrating Cement Pastes. *Journal of Materials Science*, Vol. 26, pp. 349-353.

12. Garboczi, E.J. (1990) Permeability, Diffusivity, and Microstructural Parameters: A Critical Review. *Cement and Concrete Research*, Vol. 20, No. 4. pp. 591-601.

13. Christensen, B.J., Mason, T.O. and Jennings, H.M. (1992) Influence of Silica Fume on the Early Hydration of Portland Cements Using Impedance Spectroscopy. *Journal of the American Ceramic Society*, Vol. 75, No. 4. pp. 939-45.

14. Xu, Z., Gu, P., Xie, P. and Beaudoin, J.J. (1993) Application of A.C. Impedance Techniques in Studies of Porous Cementitious Materials (III): ACIS Behavior of Very Low Porosity Cementitious Systems. *Cement and Concrete Research*, Vol. 23, No. 5. pp. 1007-1015.

15. Irvine, J.T.S., Sinclair, D.C. and West, A.R. (1990) Electroceramics: Characterization by Impedance Spectroscopy. *Advanced Materials*, Vol. 2, No. 3. pp. 132-138.

16. Lemoine, L., Wenger, F. and Galland, J. (1990) *Study of the Corrosion of Concrete Reinforcement by Electrochemical Impedance Measurement.* Berke, N.S., Chaker, V. and Whiting, D., ed. Corrosion Rates of Steel in Concrete. American Society for Testing Materials, Philadelphia, ASTM STP 1065, pp. 118-133.

17. John, D.G., Searson, P.C. and Dawson, J.L. (1981) Use of AC Impedance Technique in Studies on Steel in Concrete in Immersed Conditions. *British Corrosion Journal*, Vol. 16, No. 2. pp. 102-106.
18. Reinhard, G., Rammelt, U. and Rammelt, K. (1986) Analysis of Impedance Spectra on Corroding Metals. *Corrosion Science*, Vol. 26, No. 2. pp. 109-120.
19. Hachani, L., Carpio, J., Fiaud, C., Raharinaivo, A. and Triki, E. (1992) Steel Corrosion in Concretes Deteriorated by Chlorides and Sulphates: Electrochemical Study Using Impedance Spectrometry and "Stepping Down the Current" Method. *Cement and Concrete Research*, Vol. 22, No. 1. pp. 56-66.
20. Hachani, L., Fiaud, C., Triki, E. and Raharinaivo, A. (1994) Characterisation of Steel/Concrete Interface by Electrochemical Impedance Spectroscopy. *British Corrosion Journal*, Vol. 29, No. 2. pp. 122-127.
21. Flis, J., Dawson, J.L., Gill, J. and Wood, G.C. (1991) Impedance and Electrochemical Noise Measurements on Iron and Iron-Carbon Alloys in Hot Caustic Soda. *Corrosion Science*, Vol. 32, No. 8. pp. 877-892.
22. Yilmaz, V.T., Sagoe-Crentsil, K.K. and Glasser, F.P. (1991) Properties of Inorganic Corrosion Inhibition Admixtures in Steel-Containing OPC Mortars Part 2: Electrochemical Properties. *Advances in Cement Research*, Vol. 4, No. 15. pp. 97-102.
23. Byfors, K. (1987) Influence of Silica Fume and Flyash on Chloride Diffusion and pH values in Cement Paste. *Cement and Concrete Research*, Vol. 17, No. 1. pp. 115-130.
24. Mindess, S. and Young, J.F. (1981) *Concrete.*Englewood Cliffs, NJ, Prentice-Hall, Inc.,
25. Miley, H.A. and Evans, U.R. (1937) The Passivity of Metals. Part VIII. The Rate of Growth of Oxide Films on Iron. *Journal of the Chemical Society*, Vol. 33, No. 2. pp. 1295-1298.
26. Pourbaix, M.J.N. (1949) *Thermodynamics of Dilute Aqueous Solutions.*Edward Arnold & Co.,
27. Mayne, J.E.O. and Pryor, M.J. (1949) The Mechanism of Inhibition of Corrosion of Iron by Chromic Acid and Potassium Chromate. *Journal of the Chemical Society*, Vol. 45, No. 3. pp. 1831-1835.
28. Kruger, J. (1989) The Nature of the Passive Film on Iron and Ferrous Alloys. *Corrosion Science*, Vol. 29, No. 2. pp. 149-162.
29. Uhlig, H.H. and Revie, W.R. (1985) *Theories of Passivity* Corrosion and Corrosion Control: An Introduction to Corrosion Science and Engineering. 3rd ed. New York: John Wiley & Sons, 69-72.
30. Stern, M. and Geary, A.L. (1957) Electrochemical Polarization I. A Theoretical Analysis of the Shape of Polarization Curves. *Journal of the Electrochemical Society*, Vol. 104, No. 1. pp. 56-63.
31. Mansfeld, F. and Oldham, K.B. (1971) A Modification of the Stern-Geary Linear Polarization Equation. *Corrosion Science*, Vol. 11, pp. 787-796.

16 MICROSTRUCTURAL EXAMINATION OF THE DEVELOPMENT OF CORROSION IN REINFORCED CONCRETE

A.G. CONSTANTINOU and K.L. SCRIVENER
Imperial College of Science, Technology and Medecine, London, U.K.

Abstract
Carbonation of the concrete is slow but once the carbonation front reaches the reinforcement and enough humidity is present, the corrosion of the reinforcement will initiate. This will lead to cracking and spalling of the concrete cover due to the stresses exerted on the concrete cover by the formation of corrosion products which are higher in volume than the original steel.

In this investigation, two batches of samples of w/c = 0.56 and 0.65 respectively, were fully carbonated in 100% CO_2 and 65% RH. After carbonation, the samples were placed in the humidity regime of 90% RH.

Samples were removed from the humidity regimes after six, twelve and eighteen months and prepared for microstructural examination. The samples were examined using the scanning electron microscope (in the backscattered mode). EDS microanalysis was employed in order to detect any changes in the concentration of the corrosion products with respect to time and water / cement ratio.

A montage was created for the interfaces studied and the corroded areas were traced out using tracing paper. The total corroded areas were then calculated using a planometer. The corrosion layer thickness for the samples was also calculated using a ruler.

Keywords: accelerated carbonation, corrosion, EDS microstructure, montage, reinforcement.

1 Introduction

With regard to reinforcement corrosion, carbonation of concrete is not generally considered to be as damaging as chloride contamination. However, although the effects of carbonation may take longer to develop, once the carbonation front has reached the steel reinforcement corrosion will proceed if the humidity is high enough.

Mechanisms of Chemical Degradation of Cement-based Systems. Edited by K.L. Scrivener and J.F. Young. Published in 1997 by E & FN Spon, 2–6 Boundary Row, London SE1 8HN. ISBN: 0419215700.

The corrosion of the steel reinforcement results in the cracking and eventual spalling of the cover due to the internal stresses created as a result of the increase in volume associated with the transformation of steel to rust [1].

In carbonated concrete, the corrosion process is thought to be generalised and homogeneous producing, over the long term, a reduction in the cross-sectional area of the steel and a significant amount of oxides which may crack the cover or diffuse through the pores to the surface of the concrete. However, the relationship between the cracking of the concrete cover, the spatial distribution of corrosion and the deposition of the corrosion products are not well understood.

Good quality concrete will carbonate more slowly and will have a lower permeability when carbonated. However, once corrosion is initiated, the initial quality of the concrete has no effect on the corrosion rate [2]. The major factors affecting the quality of the concrete are the w/c ratio, the degree of curing and the reinforcement cover.

In this investigation, concrete samples were prepared with two different w/c ratios and cured for 1 day under damp sacks. After carbonation they were placed in a high humidity environment (90% RH). The samples were studied microscopically (using a scanning electron microscope) with respect to time in the humidity regime.

2 Experimental

The samples were cast with w/c ratios of 0.56 (cement content 315 kg/m^3) and 0.65 (cement content 275 kg/m^3) in cylindrical moulds 100 mm diameter by 200 mm in length, with steel reinforcement bars (8 mm diameter) at cover depths of 11 mm and 20 mm. The samples were demoulded after one day and then air cured. The samples were then carbonated in 100% CO_2 and 65% RH at 20 °C. The progress of carbonation was checked regularly by spraying phenolphthalein indicator on parallel control samples cast without reinforcement.

After full carbonation, the samples were placed in 90% RH. One sample from each batch was removed from the tank for microscopical examination after six, twelve and eighteen months of being in the 90% humidity regime.

Sections of the reinforced samples were cut using a Buhler Isomet Plus precision saw at a speed of 3800 rpm. and a load of 380 grams with a high concentration diamond wafering blade. The sections were then resin impregnated, lapped with 9 µm alumina powder and polished down, using diamond paste, to ¼ µm grit size and carbon coated. A JEOL-35CF scanning electron microscope (in the back scattered mode) was used to examine the samples. EDS analysis was done on the corrosion products present at the steel / concrete interface as well as further into the paste.

Adjacent images were collected around the interface (between 50 - 60) to make up a montage of the whole interface at a magnification of x200. The interfaces were examined, the amount of corrosion products traced (with tracing paper) and the relative corroding area measured using a planometer. The maximum thickness of the corrosion layer has also been measured for all the interfaces examined.

3 Results and discussion

3.1 Montage results

Corrosion products first started to form on the side of the steel which was closer to the atmosphere, i.e. the side of the concrete cover. With time, corrosion products developed on the other side of the steel but most of the corrosion products continued to form on the concrete cover side. Figure 1 shows a schematic diagram of the montage for the sample with w/c = 0.65, after one year in 90% RH.

A common feature of the corroded samples was the existence of local areas of corrosion which arised probably from defects on the steel which were present on the surface before the steel was cast into the samples. The initial state of the steel is therefore variable and hence the results on the total corroded area of the steel would depend on were in the steel the section was taken from. Corrosion products also formed in areas of high porosity in the paste. Figure 2 shows one such area were corrosion was initiated from defects usually in the form of pits. This is also an area of high porosity in the paste.

The corrosion products formed two distinct layers. The first corrosion layer is dense (higher concentration of iron oxides) and forms in the paste adjacent to the steel surface. The second corrosion layer forms adjacent to the first corrosion layer, appears less dense and consists of iron oxides interspersed with the cement hydrates.

Figures 3, 4 and 5 show the steel / concrete interface in a severely corroded region of the sample after six, twelve and eighteen months exposure to 90% RH respectively. The two layers of corrosion products can be seen in all the micrographs.

Figure 6 shows the steel / concrete interface of a sample with w/c = 0.56 after one year in 90% RH. Again, two corrosion layers can be distinguished, a thick layer directly in contact with the steel reinforcement and a second layer, which appears less dense and is therefore free to diffuse into the paste.

Figure 7 shows a graph of the total corroded area of the sample with w/c = 0.65 as calculated by the planometer method with respect to time in the humidity regime. A general increase is observed with the exception of the sample which was placed in 90% RH for one year which appears to have a higher total corroded area than the equivalent sample after 18 months in the humidity regime. In the case of the sample with w/c = 0.56, the total corroded area was half of the one for the equivalent sample with w/c = 0.65 at 515 cm^2.

The result for the sample with w/c = 0.65 after one year in the humidity regime, leads to the conclusion that the original state of the reinforcing bars is important in the initiation of corrosion. Therefore, in order to obtain accurate measurements, a set of sections should be studied from each of the samples. This would be a very time consuming and cumbersome process.

Figure 8 shows the maximum thickness of the samples with respect to time in the 90% humidity regime. A linear relationship is observed which leads to the conclusion that the samples are corroding at a constant rate. The maximum thickness of the corrosion layer for the sample with w/c = 0.56 after one year in the 90% RH was identical to the equivalent sample with w/c = 0.65 (at 60 mm).

3.2 EDS Results

Figure 9 shows the EDS results for the two corrosion layers of the samples with w/c = 0.56 and w/c = 0.65. The results of the two individual corrosion layers are very similar for the two water / cement ratios.

The first corrosion layer is dense with a high concentration of iron oxides, therefore, the Ca/Fe ratio tends to lower values. The second corrosion layer has a lower concentration of iron oxides and the iron ions in the layer are mobile and free to move into the paste through a concentration gradient in order to form iron oxides and precipitate in the paste.

An interesting feature however, is the fact that the corrosion layers for the sample with w/c = 0.56 are consistently denser than the equivalent ones for w/c = 0.65. This is due to the fact that in the lower w/c, less space is available for the corrosion products to deposit after they form so more products deposit in the same space. The corrosion products would therefore appear to have a higher concentration of iron oxides.

4 Conclusions

1. Two corrosion layers can be distinguished with differences in their iron oxide concentration.
2. A linear relationship exists between the thickness of the corrosion layer with time in the humidity regime, which indicates constant corrosion rate.
3. The original state of the steel reinforcement may be important in the amount of corrosion products formed.
4. EDS results indicate slightly denser corrosion layers for the sample with the lowest water / cement ratio.

5 References

1. Rosenberg, C.M. Hansson and C. Andrade (1989) Mechanisms of corrosion of steel in concrete. *Materials Science of Concrete-Vol.1*, (Ed. J.S. Skalny), The American Ceramic Society, pp. 285-313.
2. Constantinou, A.G. and Scrivener, K.L. (1995) The corrosion of steel reinforcement in carbonated concrete under different humidity regimes *Microstructure of cement based systems / bonding and interfaces in cementitious materials* (Ed. S.Diamond, S. Mindness, F.P. Galsser, L.W. Roberts, J.P. Skalny and W.D. Wakeley) Vol.370, Materials research society, pp.471-478.

6 Acknowledgements

The authors would like to thank Dr. N.N. Dioh for help with the planometer measurements. AGC would like to thank the Building Research Establishment for financing the project.

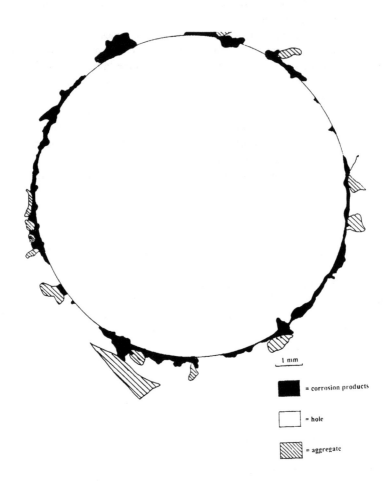

Figure 1: A schematic diagram of the entire steel / concrete interface for the sample with w/c = 0.65, after 1 year in 90% RH.

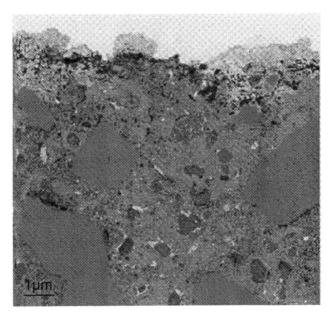

Figure 2: The formation of corrosion products in areas of prorous concrete with defects on the steel surface

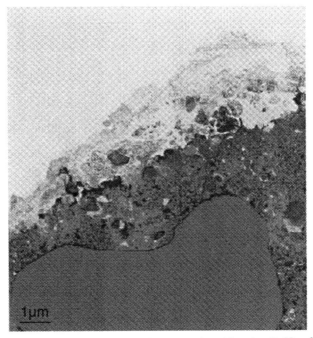

Figure 3: The steel/concrete interface of the sample with w/c=0.65, after 6 months in 90% RH

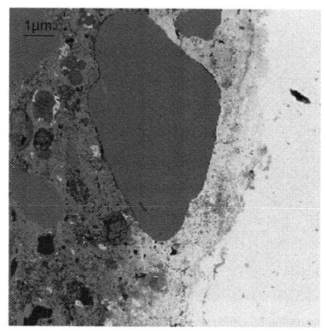

Figure 4: The steel/concrete interface of the sample with w/c=0.65, after 12 months in 90% RH.

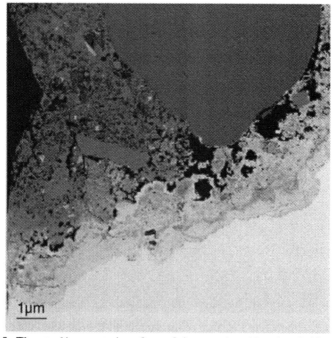

Figure 5: The steel/concrete interface of the sample with w/c=0.65, after 18 months in 90% RH.

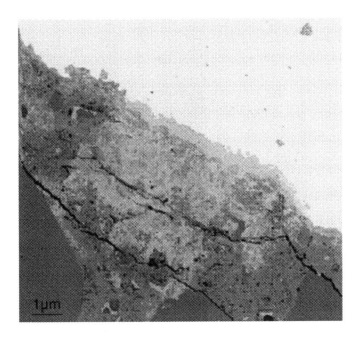

Figure 6: The steel / concrete interface of the sample with w/c = 0.56, after 12 months in 90% RH.

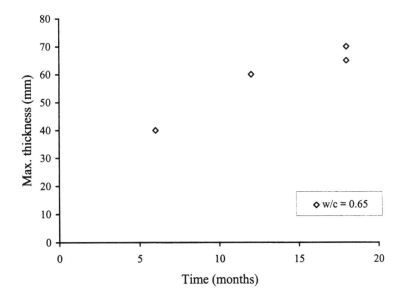

Figure 7: The maximum thickness of the corrosion layer for a sample with w/c = 0.65 with respect to time in 90% RH.

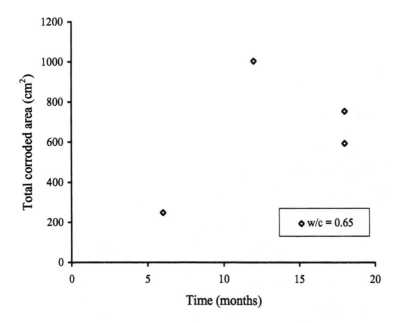

Figure 8: The total corroded areas with respect to time in 90% RH for the sample with w/c = 0.65.

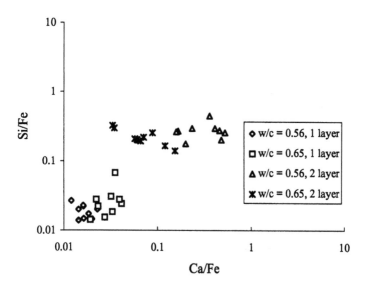

Figure 9: The composition of the two corrosion layers in the samples with w/c = 0.56 and 0.65 after 1 year in 90% RH.

17 BEHAVIOUR OF STEEL IN SIMULATED CONCRETE SOLUTIONS

H.G. WHEAT, J. KASTHURIRANGAN
Center for Materials Science and Engineering, University of Texas, Austin, USA
C.J. KITOWSKI
Ford Motor Company, Dearborn, Michigan, USA

Abstract
The behavior of steel in chloride-free and chloride contaminated simulated concrete solutions was investigated in order to study the corrosion protection provided by such solutions. The chloride concentration was increased by means of incremental additions of NaCl as well as by diffusion of NaCl. The corrosion behavior of the steel was studied electrochemically and changes in the steel surfaces were examined visually as well as by scanning electron microscopy and Auger electron spectroscopy. A comparison of the results of this investigation with the behavior of steel in salt-contaminated concrete is presented.
Keywords: concrete, corrosion, simulated pore solution, steel.

1 Introduction

Corrosion of steel in concrete is a well known problem with severe economic consequences. Steel in concrete is expected to be protected from corrosion due to the barrier that concrete provides and/or the passive and protective film (s) formed on the steel due to the alkaline environment associated with the concrete [1]. This protection can be jeopardized when critical chloride concentrations are achieved [2], however, the mechanism by which this takes place is not clearly understood.

During cement hydration, the aqueous phase rapidly acquires a high pH value [3]. The liquid phase initially contains hydroxides and sulphates of calcium, sodium and potassium, but the sulphate ions are rapidly precipitated. The remaining liquid becomes increasingly concentrated with respect to sodium and potassium hydroxides, and as the pH rises, the concentration of calcium ions decreases in accordance with the solubility product of calcium hydroxide [4]. This is reflected in the fact that, after a few weeks of hydration, pH values well in excess of 13 are frequently recorded for Portland cement pastes of water/cement ratios that are commonly used in construction practice [3].

Some researchers [5] have attributed this protection to the formation of a thin oxide film that protects the steel. It is also feasible that the passive film consists of several phases, in response to changing oxygen gradients across the interface [6]. Still others have attributed this protection to the lime-rich interfacial layer associated with the concrete[7].

Mechanisms of Chemical Degradation of Cement-based Systems. Edited by K.L. Scrivener and J.F. Young.
Published in 1997 by E & FN Spon, 2–6 Boundary Row, London SE1 8HN. ISBN: 0419215700.

Studies by Leek and Poole [8] on the mechanism of depassivation show that the chloride ions attack the film/substrate bond, but do not chemically dissolve the film. Other researchers [9] have indicated that breakdown occurs by the chemical dissolution of the passive film. Foley [10] has also addressed the processes by which chloride ions may influence the steel-concrete interface. The theories proposed are the adsorption theory and the oxide film theory. The adsorption theory proposes that chloride ions can get preferentially adsorbed onto the metal surface in competition with dissolved oxygen and hydroxyl ions. It also considers the possibility of chloride ions displacing the passivating species eventually leading to pitting. The oxide film theory suggests that chloride ions possibly penetrate the existing oxide film more easily than other anions. Penetration occurs through pores and other defects in the film, which again leads to pitting. Both mechanisms suggest the formation of localized high concentrations of chloride ions leading to local acidification and pitting [11].

It is frequently desirable to predict the corrosion behavior of steel in concretes made from different mixes. In order to do this, reinforced concrete specimens are subjected to exposure to sodium chloride solutions and electrochemical information, such as Ecorr as well as polarization resistance (Rp) values are monitored as a function of time. It has been observed that corrosion initiation is associated with a sharp decrease in Ecorr accompanied by an increase in 1/Rp (which is related to the corrosion rate of the steel). An example of such a plot is shown in Figure 1 [12]. Values of Ecorr more positive than -0.220 V vs SCE have been associated with passive behavior while values more negative than -0.270 V vs SCE have been associated with active behavior [13]. Moreover, values of 1/Rp greater than 25 $\mu S/cm^2$ have been associated with visible corrosion activity [14].

Some researchers [15-19] have attempted to minimize the time required for testing reinforcing steel in concrete by simulating the liquid phase in concrete. For example, Hausmann [15] tested steel in saturated calcium hydroxide solution and determined that 700 ppm of chloride was required to destroy the initial passivity of the steel. It is commonly accepted that corrosion initiation in actual concrete structures occurs at chloride levels in excess of about 0.9 kg/m^3 of concrete [2].

The aim of this project was to study the electrochemical behavior of steel rebar in simulated concrete solutions in order to gain information about the evolution and mechanism of the corrosion process of steel in concrete.

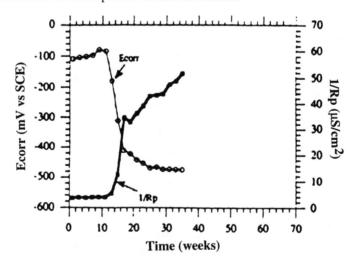

Figure 1. Variation of Ecorr and 1/Rp vs time for a concrete having w/c = 0.60.

2 Experimental Procedure

All electrochemical tests (Ecorr vs time, polarization resistance, and in some cases anodic polarization) were performed on #5 grade 60 reinforcing steel meeting the specifications outlined in ASTM A 615. The simulated concrete solutions were 0.6 M KOH, O.2M NaOH and 0.001 M $Ca(OH)_2$ [3]. This composition is also known as simulated pore solution (SPS). Reference electrodes were either saturated calomel (SCE) or Hg/HgO.

Several different types of reinforcing steel surfaces were prepared from the same lot of steel. Flat polished coupons, approximately 5/8 in. (15.9 mm) in diameter and 1/8 in. (3.2 mm) thick, were ground to 600 grit with SiC paper, ultrasonically cleaned in methanol, and air dried. The sample holder exposed a surface area of 1.00 cm^2. As-received specimens were approximately 5/8 in. (15.9 mm) in diameter and about 3 in. (76 mm) long. They were wire brushed before being ultrasonically cleaned in methanol. A small piece of copper contact wire was connected to one end of the specimens and both cut faces of the specimens and the wire were coated with epoxy and cured for 2 days.

Specimens about 5/8 in. (15.9 mm) in diameter and 6 in. (152 mm) long were machined to provide a smooth surface and mounted in a test fixture that was used in the modified diffusion cell in Phase 3. All samples (at least 6 replicates) were ultrasonically cleaned in methanol and air dried prior to testing.

In Phase 1, tests were performed on steel samples in simulated pore solutions containing the following fixed amounts of NaCl: 0.0035 % NaCl, 0.035 % NaCl, 0.35 % NaCl. and 3.5 % NaCl by weight of SPS solution. In Phase 2, gradual introduction of chloride into simulated pore solutions was studied. Samples were placed in the solutions, corrosion potentials were measured over a 12 hour period, and polarization resistance measurements were made. This procedure was repeated and then enough NaCl was added in order to bring the concentration to 0.0035 % NaCl. The solution was gently stirred with a magnetic stirrer, and the measurements described above were repeated. Incremental additions were added until a concentration of 3.5 % NaCl was reached. The procedure was performed on at least 6 replicates.

Phase 3 utilized a two compartment diffusion cell similar to those used to study the diffusion of chlorides into cement pastes [20]. The steel bars were placed in a two compartment cell separated by a concrete disk made from ordinary Portland cement and having a water/cement ratio of about 0.55. Compartment 1 contained the simulated pore solution plus 3.5 % NaCl and Compartment 2 contained the simulated pore solution plus the steel bars. In this way, chlorides could be transported through the concrete to the initially salt-free side, and an increase in chloride concentration would occur due to transport of chlorides through the concrete. The accumulation of chlorides was determined by measuring the solution concentration with a chloride electrode. A double three-electrode test fixture (Figure 2a) was used to hold the steel specimens. One electrode served as a reference electrode by using the unpolarized corrosion potential for the polarization of the other two electrodes. The polarization resistance of the working electrodes (in the test fixture) was measured periodically. These measurements were also compared with those made using a single compartment cell into which incremental additions of NaCl were added. A schematic of the diffusion cell containing a double three-electrode test fixture is shown in Figure 2b. The diffusion cell was made of acrylic and sealed with silicone caulk.

In addition to specimens which were immersed for the entire duration of the experiments, some specimens were removed at various times of exposure in order to observe changes in surface morphology and composition using (SEM/EDS) and Auger Electron Spectroscopy in some cases.

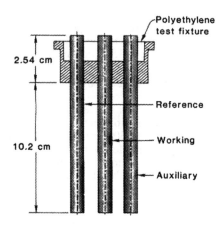

Figure 2a. Three electrode set-up made up of working, reference, and auxilliary electrodes.

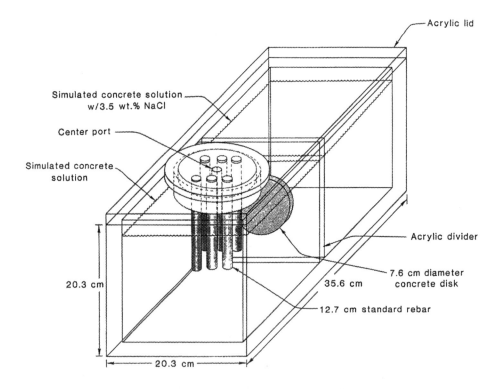

Figure 2b. Schematic of diffusion cell containing a double three-electrode configuration.

3 Results and Discussion

Representative results of Ecorr vs time for steel immersed in simulated pore solutions containing fixed initial amounts of NaCl are shown in Figures 3a and b . In general, Ecorr values became more negative as the NaCl concentration was increased. In the case of the polished flat coupons, this was evident at a concentration of approximately 0.35 % NaCl, while in the case of the as-received bars Ecorr values became more negative at a concentration of approximately 3.5 % NaCl. The trend toward higher current densities as a function of NaCl concentration was also observed in the anodic polarization curves taken once specimens had reached stable Ecorr values.

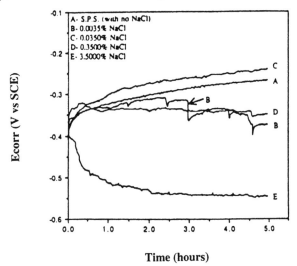

Figure 3a. Ecorr vs time for polished flat rebar coupons in SPS with and without NaCl.

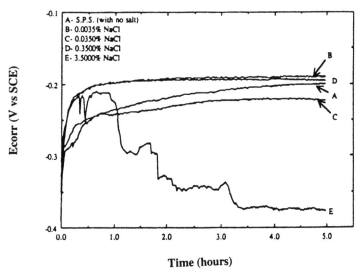

Figure 3b. Ecorr vs time for as-received bars in SPS with and without NaCl.

In general, the results of Ecorr and 1/Rp data for the bars subjected to gradual NaCl additions were similar to those described above. For the polished flat specimens, there was a sudden depression in Ecorr accompanied by an increase in 1/Rp that occurred at about 0.35 % NaCl. In the case of the as-received bars, this condition occurred at a higher concentration of about 3.5 % NaCl. In fact, for one as-received bar, Ecorr and 1/Rp remained low even at a concentration of 3.5 % NaCl. It was only after the bar was allowed to remain in the solution for 72 hours that the depression in Ecorr and the increase in 1/Rp occurred. This seems to suggest that merely achieving a "critical" chloride concentration may not be sufficient to achieve depassivation. Certain factors may increase the required concentration and time dependence may be involved.

A representative plot showing the effect of chloride transport on Ecorr by means of the modified diffusion cells is shown in Figure 4. No polished cylindrical bars experienced depassivation (according to our criteria) after 150 days even though chloride levels were in excess of 4000 ppm. This was still true after one year of exposure and it again emphasizes the fact that factors other than chloride concentration play a significant role in depassivation.

One of the most interesting observations was found on the micrographs taken at the presumed depassivation concentrations (when Ecorr exhibited a sharp decrease and 1/Rp exhibited a sharp increase) in Phase 2. There were scale-like deposits on the bars and chloride-rich regions as determined from SEM/EDS. In addition, selected Auger data showing Fe/O concentrations that levelled off (as a result of argon sputtering) at a ratio of between 0.5 and 0.7 for as-received specimens suggested the presence of iron oxides having those ratios. The Fe/O ratio for Fe_2O_3 is 0.667. Selected Auger data are shown in Figure 5.

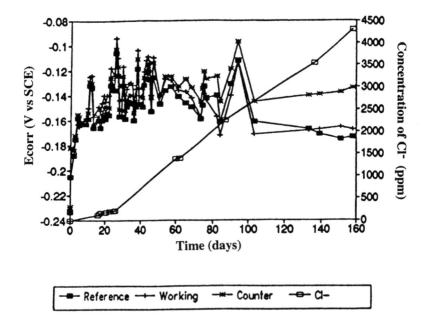

Figure 4. Ecorr and concentration vs time for polished rebars in modified diffusion cell.

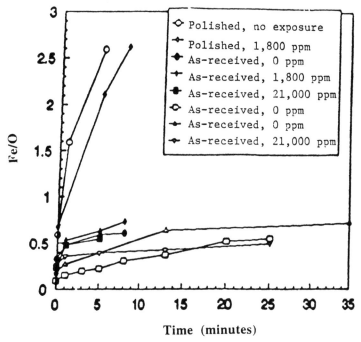

Figure 5. Fe/O vs argon sputtering time for selected samples exposed to SPS in the presence or absence of NaCl. (Chloride concentration is given in ppm).

4 Conclusions

a. The chloride threshold for corrosion initiation is not a fixed value but a function of steel surface condition and the rate of chloride addition.

b. Specimens removed from exposure at the time of presumed depassivation showed chloride-rich regions.

c. Long periods of protection in the presence of large amounts of chloride were observed using the modified diffusion cell, suggesting that factors other than chloride concentration play a significant role in depassivation.

d. Auger results indicated that the protective film was much thinner for the polished specimens than the as-received specimens. For the latter specimens, the outer layer was composed of mainly Ca and K (and sometimes Fe) while the inner layer was composed of one or more iron oxides having an Fe/O ratio of between 0.5 and 0.7.

5 Acknowledgments

The authors are very grateful for the financial support provided by the National Science Foundation (Grant No. MSS-8921091).

6 References

1. Pourbaix, M. (1974) *Atlas of Electrochemical Equilibria in Aqueous Solutions,* National Association of Corrosion Engineers, Houston, TX.
2. Clear, K.C. and Hay, R.E. (1973) Time-to-corrosion of reinforcing steel in concrete slabs, Vol. 1: Effect of Mix Design and Construction Parameters. Federal Highway Administration, Report FHWA-RD-73-32.
3. Barneyback, R.S., Jr. and Diamond, S. (1981) Expression and analysis of pore fluids from hardened cement pastes. *Cement and Concrete Research.* Vol. 11, pp 279-285.
4. Page, C.L. and Treadaway, K.W. J. (1982) Aspects of the electrochemistry of steel in concrete. *Nature,* Vol 297, pp 109-114.
5. O'Grady,W.E.(1980) Mossbauer study of the passive oxide film on iron. *J. Electrochemical Society,* Vol 127, p 555.
6. Sagoe-Crentsil, K.K. and Glasser, F.P. (1990) Analysis of the steel:concrete interface in *Corrosion of Reinforcement in Concrete,* (ed. C.L. Page. K.W.J., Treadaway, and P.B. Bamforth, Elsevier Applied Science Publishers, London, pp 74-86.
7. Page, C.L. (1975) Mechanism of corrosion protection in reinforced concrete marine structures. *Nature,* Vol. 258, pp 514-515.
8. Leek, D.S. and Poole, A.B., (1990) The breakdown of the passive film on high yield mild steel by chloride ions in *Corrosion of Reinforcement in Concrete* (ed. C.L. Page, K.W. J. Treadaway, and P.B. Bamforth Elsevier Applied Science Publishers, London, pp 65-73.
9. Hoar, T.P.(1967) The anodic behavior of metals. *Corrosion Science.,* Vol. 7, pp 341-355.
10. Foley, R.T. (1970) Role of the chloride ion in iron corrosion. *Corrosion,* Vol. 26, No. 2, pp 58-70.
11. Alvarez, M.G. and Galvele, J.R. (1984) The mechanism of pitting of high purity iron in NaCl solutions. *Corrosion Science,* Vol 24, pp 27-48.
12. Zhang. H. and Wheat, H.G. (1994) Electrochemical testing of reinforced concrete in *Corrosion and Corrosion Protection of Steel in Concrete* (ed. R.N. Swamy) Sheffield Academic Press Ltd. , Sheffield.
13. Van Daveer, J.R. (1975), Techniques for evaluating reinforced concrete bridge decks. *American Concrete Institute Journal,* Vol. 72, p 697, 1975.
14. Berke, N.S., and Hicks, M. (1993) "Predicting chloride profiles in concrete," Corrosion '93, Paper No. 341, National Association of Corrosion Engineers, Houston,TX.
15. Hausmann, D.A. (1967) Steel corrosion in concrete. *Materials Protection,* Vol. 6, pp 19-23.
16. Wheat, H.G. (1985) *An Electrochemical Investigation of the Corrosion of Steel in Concrete,* Ph.D. Dissertation, The University of Texas at Austin, Austin, TX.
17. Ramirez, C.W., Borgard, B., Jones, D. and Heidersbach, R. (1990) Laboratory simulation of corrosion in reinforced concrete. *Materials Performance,* Vol. 29, pp 33-39.
18. Gouda, V.K. (1970) Corrosion and corrosion inhibition of reinforcing steel. I. Immersed in alkaline solutions. *British Corrosion Journal,* pp 199-203.
19. Williamson, T.J. , Kasthurirangan, J. and Wheat, H.G. (1992) "The evolution of the degradation of steel in salt-contaminated concrete," Corrosion '92, Paper 207, National Association of Corrosion Engineers, Houston, TX.
20. Page, C.L., Short, N.R., and El Tarras, A. (1981) Diffusion of chloride ions in hardened cement pastes. *Cement and Concrete Research,* No. 11, p 396.

18 THE EFFECT OF FLOURIDE ON THE CORROSION RESISTANCE OF STEEL IN CONCRETE

A. MACIAS
Instituto de Ciencias de la Construcción "Eduardo Torroja", CSIC, Madrid, Spain
M.L. ESCUDERO
Centro Nacional de Investigaciones Metalúrgicas, CSIC, Madrid, Spain

Abstract
Fluoride can be present as a minor component in low-energy cements. The influence of fluoride content of cement on the corrosion of reinforcements has been studied in alkaline solutions used to simulate the aqueous phase present in concrete pores and in mortar of cement manufactured using CaF_2 as a mineraliser and flux agent. Results show that fluoride anions are able to produce pitting of reinforcing steel in alkaline media of high pH but in $Ca(OH)_2$ saturated media, the precipitation of CaF_2 lowers F^- concentration below the minimum that promotes a pitting corrosion process. This results were confirmed in the tests on mortar with CaF_2 as a minor component in which corrosion rates of steel were similar to those measured in a traditional cement even in presence of corrosive agents such as chlorides or carbonation of cement.
Keywords: Fluoride, steel, corrosion, rebars, concrete.

1 Introduction

Fluxes and mineralizers are introduced in cement manufacture to decrease the consumption of energy (1). Fluorine compounds act both as a mineralizer and as a flux in promoting the formation of alite (2). Consequently, fluoride can be present as a minor component in low-energy cements.

Corrosion of steel in concrete has become over recent years an increasingly important factor when considering the maintenance of buildings and structures and the effects produced by fluoride presence in modern cements have to be analysed in order to make correct predictions of service life.

In contrast to corrosion research in chloride-containing solutions, there have been

Mechanisms of Chemical Degradation of Cement-based Systems. Edited by K.L. Scrivener and J.F. Young.
Published in 1997 by E & FN Spon, 2–6 Boundary Row, London SE1 8HN. ISBN: 0419215700.

few investigations on the effect of fluoride ions in the passivity breakdown of iron and the   results obtained by different authors do not lead to the same conclusions. Thus, some authors reported a passivation or found no evidence of localized attack of steel in fluoride containing solutions (3) (4) (5), whereas others researchers obtain results that show that fluoride anions are able to produce pitting of mild steel (6) (7).

Due to this lack of basic knowledge, the modification that the presence of fluoride has on the corrosion behaviour of reinforcing corrugated steel bars was first studied in alkaline solutions which simulated the aqueous phase present in concrete pores (8). The solutions used were a saturated solution of $Ca(OH)_2$ and a 0.04 M NaOH solution with addition of NaF in different proportions. Then, the corrosion behaviour of reinforced steel in mortar made with a cement manufactured using CaF_2 in its clinkerization was analysed (9). Corrosion rates were compared with those obtained in a cement of the same characteristic but made traditionally in the same factory.

2 Experimental

2.1 Materials and Procedure

Reinforcing corrugated 6 mm nominal diameter and 8 cm long steel bars with an exposed area of 5.6 cm², were used.

As in previous work (10), cells of polyethylene were used. The saturated solutions of $Ca(OH)_2$ were prepared by adding 1.5 g of calcium hydroxide to 250 cc of solutions of the different compositions tested.   The pH values and the F⁻ concentrations of the solutions used measured at the end of the experiment with a combined electrode for the pH range of 0-14 and a specific fluoride ion, respectively, are presented in Table 1.

Table 1. pH values and F⁻ concentrations of the samples tested

Sample	$pH_{(exp.)}$	$F^-_{(exp.)}$ (ppm)	$pH_{(theor.)}$	$F^-_{(theor.)}$ (ppm)
$Ca(OH)_2$ sat.	12.58	-	-	-
$Ca(OH)_2$ sat. + 0.1 M NaF	12.93	11.7	13.2	12.6
$Ca(OH)_2$ sat. + 0.6 M NaF	13.06	8720	13.2	8350
0.04 M NaOH	12.56	-	-	-
0.04 M NaOH + 0.01 M NaF	12.49	232	12.6	188
0.04 M NaOH + 0.1 M NaF	12.55	2032	12.6	1880
CaF_2 sat.	8.52	6.7	-	12.6

Cements used were: cement A, a normal white cement and Cement B, a white cement manufactured with CaF_2.   The chemical analysis of these cements is presented in Table 2. The fluoride content of cement A is due to fluoride content of raw materials.

Mortars specimens of $8 \times 5.5 \times 2$ cm with w/c = 0.5 and c/s = 1/3 were made. Two mild steel rods and a central carbon bar - counter electrode - were cast in the mortar prisms. A saturated calomel electrode was used as reference. One third of these specimens were made with 2% of NaCl by weight of cement.

Table 2. Chemical analysis of cements

Sample	SiO_2	Al_2O_3	CaO	Fe_2O_3	MgO	Na_2O	K_2O	SO_3	CaF_2
A	21.2	3.92	66.05	0.29	0.83	0.05	0.16	3.2	0.59
B	24.3	1.98	67.60	0.37	0.55	0.05	0.10	2.7	1.65

The specimens were cured during 24 hours in humidity saturated atmosphere and afterwards kept for 83 days in a chamber at 50% or 100% R.H. Then the samples at 50% R.H. were partially immersed (P.I.) in distilled water in individual polyethylene bottles. Samples kept at 100% were totally immersed (T.I.). In the case of specimens with 2% NaCl addition, the total or partial immersion was in natural sea water.

Once again the steel has get a passivated state (after 13 days), one series of the specimens without any addition was taken out of the chambers at 50% or 100% R.H. and rapidly carbonated in CO_2 atmosphere during 11 days. Then the specimens were put again in the previous R.H. conditions.

2.2 Techniques used

The corrosion intensity i_{corr} was estimated from the polarization resistance using the Stern-Geary equation (11). The polarization resistance measurements were carried out by means of an AMEL potentiostat. Rp was estimated by the ratio between applied polarization (10 mV when the steel is actively corroding and 30 mV when is passivated) and the current response measured after a time stabilization of 30 seconds.

3 Experimental Results

3.1 Test in solution

Fig. 1 shows the electrochemical results obtained for 0.04 M NaOH solutions plus 0.01 M and 0.1 M NaF additions. The addition of 0.01 M NaF hardly affects the passivation behaviour pattern of steel in 0.04 M NaOH solution. Increasing the fluoride concentration above 0.1 M, causes the corrosion rate to increase to values about 0.3 μA cm^{-2} , a value which remains constant from 5 days until the end of the test. This higher fluoride concentration initiates a pitting corrosion process as visual observation of steel bars corroborated.

Fig. 2 shows the i_{corr} values versus the time of immersion in the $Ca(OH)_2$ saturated solution with addition of 0.1 and 0.6 M of NaF. As would be expected for calcium hydroxide solution without aggressive anions steel becomes passivated and

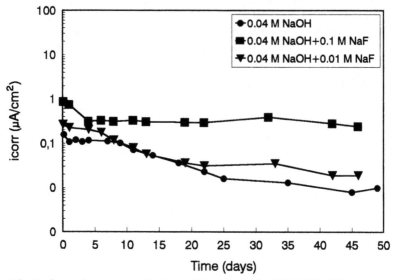

Fig. 1. i_corr values versus the time of immersion in 0.04 M NaOH solutions plus 0.01 and 0.1 M NaF.

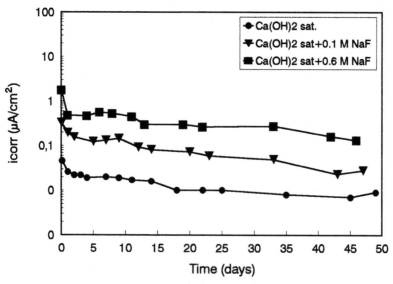

Fig. 2. i_corr values versus the time of immersion in the Ca(OH)$_2$ saturated solution with addition of 0.1 and 0.6 M NaF.

corrosion rates about 0.01 μA cm^{-2} are measured. In the presence of fluoride anions once again after a short period of initial dissolution, corrosion rates measured gradually decrease to values of 0.1 μA cm^{-2} at 45 days, equivalent to a material loss of 1.16 μm year $^{-1}$. In presence of 0.6 M NaF addition corrosion rates are 10 times higher than those in $Ca(OH)_2$ saturated solution. Nevertheless, it should be pointed out that the corrosion rates measured in $Ca(OH)_2$ saturated solution are 10 times lower than those obtained for NaF addition in the same concentration to a 0.04 M NaOH solution.

3.2. Test in mortar

The i$_{corr}$ values of steel bars embedded in mortar A and B without any addition, carbonated and mixed with 2% NaCl are presented versus time in Fig. 3. In this figure it could be seen that initially i$_{corr}$ values are around 0.1 μA cm^{-2} for all samples.

For mortars A and B without any addition, after one week, i$_{corr}$ values are below 0.02 μA cm^{-2}. These values indicate the passivation of steel. The change in the environmental conditions to a higher relative humidity, influences directly the resistivity of the mortar which decreases drastically. However, there is only a light increase in corrosion rates; i$_{corr}$ values keep under 0.2 μA cm^{-2} until the end of the test. This corrosion rates lead to a material loss of 116 μm in 50 years.

The passivation state is destroyed by the decrease in pH of the pore phase when the mortars are carbonated. Accordingly, steel stars to corrode with a corrosion rate higher than 0.1 μA cm^{-2}. On the other hand, the calcium carbonate precipitated in the mortar pores as a consequence of the reaction between the calcium hydroxide and the CO_2, decreases the porosity of the material and increases its resistivity. In these conditions, the corrosion process is under resistivity control. This fact justifies the gradual decrease of corrosion rate registered after carbonation of samples and the lowers i$_{corr}$ for samples at 50% R. H. compared with those at 100% R. H.. When the environmental conditions change to partial or total immersion, the resistivity of the mortars diminish considerably and i$_{corr}$ increases to values around to 0.5 μA cm^{-2} that remain constant until the end of the test. Those corrosion rates mean a material loss of 5.8 μm year^{-1}.

The addition of 2% of NaCl to the mix water is not enough to promote a pitting corrosion of steel embedded in mortar A or B as can be deduced from the results presented in Fig. 3. The corrosion rates measured are only slightly higher than those measured in mortars without any addition in the same condition. To increase the aggressiveness of the conditions tested in order to magnified the possible different behaviour of mortars A and B, the total or partial immersion of these samples was in natural sea water. The present of higher chloride concentration leads to corrosion rates about 0.5 μA cm^{-2}. Again corrosion behaviour of steel embedded in both cements is similar.

4 Discussion

The results obtained in 0.04 M NaOH solution tests show that fluoride anions are

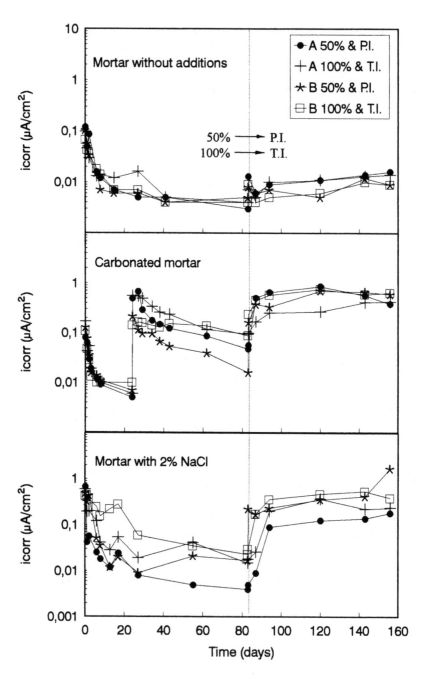

Fig. 3. i_corr values versus time for steel bars embedded in mortars A and B

able to produce pitting of reinforced steel in alkaline media of a pH similar to the concrete pore fluid. However, corrosion rates are lower than those found in the presence of chloride anions under comparable concentrations and experimental conditions. In consequence, fluoride can be considered less aggressive than chloride to steel corrosion in concrete.

In calcium containing media, F^- precipitates along with Ca^{2+} in the form of CaF_2, one of the sparingly insoluble fluorides ($K_{sp} = 1.46 \times 10^{-10}$), resulting in an increase in pH. As the $Ca(OH)_2$ concentration of saturated solutions used in the present tests is about 5 g l^{-1}, the maximum concentration of F^- that can be precipitated in form of CaF_2 is about 0.15 M. Approximated theoretical results of pH and $[F^-]$ obtained from the chemical balances, along with the experimental results are given in Table 1.

According to these results, in $Ca(OH)_2$ saturated solutions with enough calcium content to precipitate all F^- added in form of NaF, the fluoride concentration in solution keeps below 0.1 M for NaF additions lower than 0.3 M approximately. In these conditions the F^- concentration in solution is not high enough to promote a pitting process but the presence of fluorides in this media causes a modification of the pH and of the composition and characteristics of the passive film so that overall quality or protective ability is affected; in consequence the corrosion rates measured are higher than that in saturated $Ca(OH)_2$ solutions.

The liquid phase in concrete made with normal Portland cement is essentially a solution of calcium and alkaline hydroxides with a pH between 12 and 13. The solid phase contains an important amount of $Ca(OH)_2$ coming from cement hydrated reactions. Therefore, the corrosion behaviour of steel reinforcements in concrete made with cement manufactured employing fluorides as fluxes and mineralizers cannot be very different from that in concrete made with traditional cement because of the high $Ca(OH)_2/F^-$ relation and the low solubility of fluoride anion with Ca^{2+} present. The behaviour of cement B corroborates these conclusions.

The results of the present work indicate that corrosion rates of reinforcing corrugated steel bars embedded in mortar of the new cement B manufactured adding CaF_2 are similar to corrosion rates measured for a traditional made cement A. The small differences detected in the corrosion behaviour of both cement can be explain in base to factors other than F^- content. Thus, the lowest corrosion rates measured in cement A for samples keep at 50% R. H. are due to its higher resistivity. And the highest corrosion rates for mortar B in presence of chlorides are related to the different bond chloride capacity of both cements (Cement A has a C_3A content three times higher than cement B).

5 Conclusions

- Fluoride anions are able to produce pitting of reinforced steel in alkaline media of high pH.
- In $Ca(OH)_2$ saturated solution, the precipitation of CaF_2 lowers F^- concentration below the minimum that promotes a pitting corrosion process. The presence of fluoride in these media causes an increase in corrosion rates of reinforced steel

with concentration.

- Corrosion behaviour of reinforced steel embedded in mortar of cement manufactured employing fluorides to lower the clinkering temperature, is similar to that of cement traditionally made.
- Fluoride concentration of this cement manufactured employing fluorides is not high enough to promote a pitting corrosion or to increase significantly the corrosion rate of steel.

6 References

1. Klemm, W.A. and Skalny, J. (1977) Mineralizers and fluxes in the clinkering process. *Cement Research Progress 1976*, Am. Ceram. Soc., Columbus, Ohio. pp. 259-291.
2. Palomo, A. (1986) *Doctoral Thesis*, Universidad Complutense de Madrid.
3. Thiongo, K. and Cottis, R. A. (1992) The passivation of steel in neutral sodium fluoride solutions, *Advances in Corrosion and Protection*, paper 310. Manchester.
4. Carroll, W.M. (1990) The influence of temperature, applied potential, buffer and inhibitior addition on the passivation behaviour of a commercial grade 316L steel in aqueous solutions. *Corros. Sci.*,Vol. 30, No. 6/7, pp. 643-655.
5. Tachibana, K., Mizushiro, M. and Kumagai, Y. (1992) Passivation behaviour of 304 stainles steel in fluoride solutions. *Advances in Corrosion and Protection*, paper 299. Manchester.
6. Vásquez Moll, V.D., Acosta, C.A., Salvarezza, R. C., Videla, H.A. and Arvía, A.J. (1985) The kinetics and mechanism of the localized corrosion of mild steel in neutral phosphate-borate buffer containing sodium fluoride. *Corros. Sci.*, Vol. 25, No. 4, pp. 239-252.
7. Mayer, P., Manolescu, A.V. aand Rasile, E.M. (1984) Corrosion of medium carbon steel in aqueous solutions containing fluoride ions at 300 C. *Corrosion.* Vol. 40, No. 4, pp. 186-189.
8. Macías, A. and Escudero, M.L. (1994) The effect of fluoride on corrosion of reinforcing steel in alkaline solutions. *Corros. Sci.*, Vol. 36, No. 12, pp. 2169-2180.
9. Escudero, M.L. and Macías, A. (1995) Corrosion of reinforcing steel in mortar of cement with CaF_2 as a minor component. *Cem. Concr. Res.*, Vol. 25, No. 2, pp. 376-386.
10. Macías, A. and Andrade, C. (1990) The behaviour of galvanized steel in chloride-containing alkaline solutions. *Corros. Sci.*, Vol. 30, No. 4/5, pp. 393-407.
11. Stern, M. and Geary, A.L. (1957) Electrochemical polarization. *J. Electrochem. Soc.*, Vol. 104, No. 1, pp. 56-63.

19 EARLY DETECTION OF CORROSION OF REINFORCING STEEL IN STRUCTURAL CONCRETE

F.M. LI and Z.J. LI
Department of Civil and Structural Engineering, The Hong Kong University of Science and Technology, Kowloon, Hong Kong

Abstract

The rebar corrosion in structural concrete is mainly caused by the presence of chloride ions in the matrix. When the concentration of chloride ions in concrete reaches a level of 0.5–1 Kg/m^3 of concrete, the chloride ions will dissolve the protective oxidized film thus allowing the underlying steel to freely corrode. The rebar corrosion is held responsible for most deterioration of infrastructure. Obviously, the early corrosion detection is critical for long term pro-active infrastructure management. Along this line, acoustic emission (AE) technique is a strong candidate due to its capability of detecting a weak stress wave. When corrosion products are formed on a corroding rebar which push out to the surrounding concrete, microcracks are also formed and stress waves are generated. The growth of the microcracks is directly related to the amount of corrosion product. Thus, by detecting the AE event rate and their amplitude, the degree of the corrosion can be interpreted.

This paper clearly establish the correlation between the characteristics of the acoustic emission event and the behaviour of the rebar corrosion through an accelerated corrosion experimental investigation. The correlation obtained in this investigation can be used for a field application to characterize the change of a corroded rebar and to help making decisions on maintenance and repair.

Keywords: Acoustic emission, concrete, infrastructure, microcrack, rebar corrosion.

1 Introduction

Corrosion of reinforcing steel (rebar) in concrete is responsible for most deterioration of structural concrete. There are two kinds of rebar corrosion, chloride–

Mechanisms of Chemical Degradation of Cement-based Systems. Edited by K.L. Scrivener and J.F. Young.
Published in 1997 by E & FN Spon, 2–6 Boundary Row, London SE1 8HN. ISBN: 0419215700.

induced corrosion and carbonation–induced corrosion. Generally speaking, the chloride–induced corrosion is more serious than the carbonation–induced corrosion. Chloride can get into the concrete at the time of mixing or can penetrate into the hardened concrete later on. For the case of chloride in the surrounding environmental penetrating into the concrete, a sufficient quantities of chloride ions have to be accumulated first. Next, a localized breakdown of the passive film on the steel is formed by the action of these accumulated chloride ions and thus a galvanic cell is created. The local active areas behave as anodes, while the remaining passive areas become cathodes where reduction of dissolved oxygen takes place. As the steel increases its state of oxidation, the volume of the corrosion products expands. The unit volume of Fe can be doubled if FeO formed. The unit volume of the final corrosion product, $Fe(OH)_3 \cdot 3H_2O$, is as large as six and a half times of the original Fe volume. This expansion creates cracking and spalling inside concrete, and finally destroy the integrity of the structural concrete and cause a failure of buildings and infrastructure. To prevent the serious failure caused by corrosion, it is essential to detect the corrosion rate in existing buildings and infrastructures. Unfortunately however, current commonly used inspection methods for corrosion lack accuracy and early detection capacity. For example, the two commonly used methods for determining concrete rebar corrosion are visual observation and half–cell potential measurements. Visual observation uses direct or remote inspection to detect obvious signs of corrosion, such as physical damage in the form of spalling or cracking. This method, however, is subjective and provides corrosion detection only after significant corrosion has occurred. Half–cell potential measurements can be used to determine the probability of corrosion activity taking place at the time of the reading. But this method is often inconclusive because the measurements depend on the condition of the concrete. Moisture level, the amount of carbonation, and salt concentration can affect the reading and lead to erroneous judgement.

Since corrosion detection at an early stage is critical for long term pro–active infrastructure management and cost–effective scheduling of structural concrete repair, it is an urgent task to develop an advanced technology for infrastructure maintenance operations in general. The capability of acoustic emission (AE) technique in detecting a weak stress wave makes it a strong candidate for early detection of rebar corrosion in concrete. The primary advantage of AE over other conventional nondestructive evaluation (NDE) techniques is that it can directly detect the process of a flaw growth. When corrosion products are formed on a corroding rebar, they swell and apply pressure to the surrounding concrete. Microcracks will be formed and stress waves will be generated during the expansion process. The formation and propagation of the microcracks is directly related to the amount of corrosion product of a corroding rebar. Thus, by detecting the AE event rate and their amplitude, the degree of the corrosion can be interpreted.

A few papers has been published that report the use of AE for monitoring the metallic corrosion process [1]–[4]. Rettig and Felson [1] reported a linear relationship between the AE count rate and the rate of hydrogen evolution during the corrosion of iron wire in HCl solution and aluminum wire in salt water. Mansfeld and Stocker [2] correlated qualitatively the rate of pitting corrosion of aluminum alloys in 3.5% NaCl (by weight loss measurements) to the rate of AE (count rate).

In a recent study on the pitting corrosion of AISI 316L austenitic stainless steel in acidified 3% NaCl solution, Mazille [4] reported "a good correlation between AE activity (number of events) and the pitting corrosion damage". Concerning the corrosion of mild steel, Seah *et al.* [3] reported a "quite clear correlation between the AE activities and corrosion rate" during the corrosion of AISI 1020 mild steel immersed in HCl solution.

In civil engineering, AE technique has been utilized to study damage and microcrack nucleation and localization in a stressed materials such as a tensioned concrete specimen by Li and Shah [5]. A method for locating AE source and for calculating the wave velocity in a damaged material has also been developed by Li and Shah [5]. It has been found that the source location of microcracks recovered by using this technique has a very good correlation with the macroscopic crack propagation. Recently, the corrosion of reinforcing steel in concrete was also studied by applying AE technique [6].

In this research, a systematic study was performed on the applicability of AE technique to corrosion detection. The correlation between AE rate and corrosion rate was carefully established. And the capability of AE in early corrosion detection was further verified by an accelerated corrosion tests.

2 Experimental study of corrosion in HCl solution

2.1 Experiment method

The schematic experimental setup for the rebar corrosion in HCl solution is shown in Fig. 1. The specimen is held vertically by a clamp stand. The bottom portion of the specimen is immersed into a solution with a length of 100 mm. A piezoelectric transducer is attached to the top of the sample to detect the AE signals caused by corrosion. The signal detected by the transducer was amplified and filtered by a preamplifier first and then transferred into digital signals by a A–D module. The digital signals were fed into a computer via shielded coaxial cables.

The specimen material was mild steel rebar with a diameter of 20 mm. Each specimen was 400 mm in length. Its surface was polished using 150# sandpaper first and then cleaned carefully using acetone before experiment. The immersed end surface of the rebar was coated with glue to avoid corrosion on the bottom surface which would produce big bubbles. For test coupled with copper, a 40 mm copper bar with the same diameter was connected to the end of rebar. The corrosive reagent used was diluted HCl solution of various concentrations. For each test, new rebar and new solution are used. The solution has a fixed volume of 2000 ml with the same beaker. For the anodic current polarization test, two new rebars are used for anode and cathode respectively. The polarization was applied at a current density of 2 mA/cm² with a power supply system.

The acoustic emission was monitored using an AMS3 system. It is a 12 channel instrumentation consisted of transducers, preamplifiers, filters and amplifiers. The acquisition system was completely computer controlled. The AE information was transferred and stored on hard disk as soon as detected. The recorded signals were processed to obtain the characteristic parameters such as amplitude, risetime and energy. The results were displayed on the computer screen for monitoring.

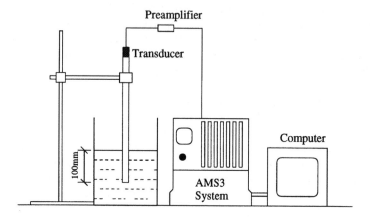

Fig. 1.   Experimental setup of rebar corrosion in HCl solution

2.2   Experimental results

There are three basic groups of tests: rebar alone; rebar coupled with copper; rebar with anodic current polarization. Each group consists of four different tests in four different HCl solutions. The program is shown in table 1.

Typical AE curves of event numbers versus time are shown in Fig. 2–4. Fig. 2 is for the case of rebar alone. It can be seen that the AE rate increases with the increase of HCl concentration. Fig. 3 shows the response of the specimen group of rebar coupled with copper. Fig. 4 shows the number of AE events recorded of the corrosion of rebar with anodic current polarization ($I = 2$ mA/cm^2) in different HCl solutions. Similarly, it can be seen from the figure that the higher the concentration of the solution, the larger the AE event numbers are obtained. Fig. 2–4 clearly indicate that the AE activities follows the sequence of $a_{15\%} > a_{10\%} > a_{5\%} > a_{1\%}$. It is thus verified that AE rate is closely related to the corrosion process.

Fig. 5–8 compare the responses of the three group specimens in HCl solutions with concentration of 1%, 5%, 10%, and 15% respectively. It is clearly demonstrated that the AE rate of polarized specimen is the highest and that of pure rebar specimen is the lowest. This phenomenon corresponds to the order of the corrosion rate for these three group of specimens ($a_{polar.} > a_{Fe+Cu} > a_{Fe+HCl}$). Once more, it is proved that the AE rate is proportional to the corrosion of rebar. Thus, rebar corrosion can be detected by detecting AE activities.

Table 1.   Experimental program for rebar corrosion in HCl solution

Group Number	Description	HCl concentration				Remarks
		1%	5%	10%	15%	
(1)	Fe+HCl	3	3	3	3	Freely corroding
(2)	Fe+Cu+HCl	2	2	2	2	Coupled with copper
(3)	Anodic Curr. polar.	3	3	3	3	$I = 2$ mA/cm^2

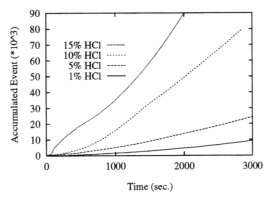

Fig. 2. Event number vs. time (rebar corrosion in HCl solution: Fe+HCl)

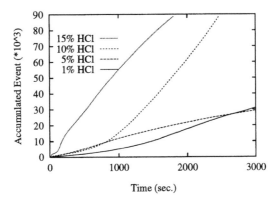

Fig. 3. Event number vs. time (rebar corrosion in HCl solution: Fe+Cu+HCl)

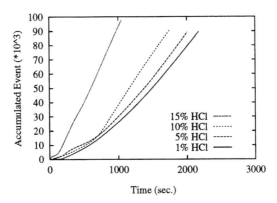

Fig. 4. Event number vs. time (rebar corrosion in HCl solution: with curr. polarization)

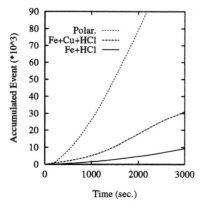

Fig. 5. Event number vs. time
(rebar corrosion in 1% HCl solution)

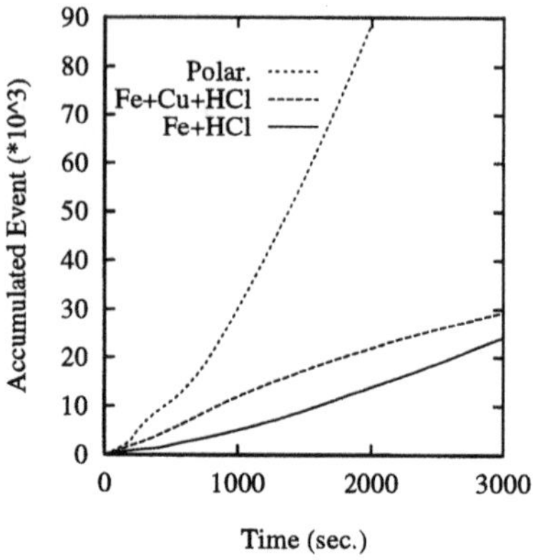

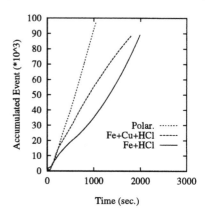

Fig. 6. Event number vs. time
(rebar corrosion in 5% HCl solution)

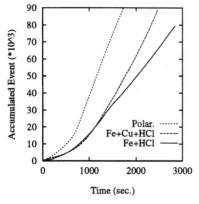

Fig. 7. Event number vs. time
(rebar corrosion in 10% HCl solution)

Fig. 8. Event number vs. time
(rebar corrosion in 15% HCl solution)

3 Experimental study of corrosion in concrete

To study the applicability of AE technique to the detection of rebar corrosion
in concrete, the accelerated corrosion tests were carried out. The concrete rebar
specimens were exposed to a cyclic salt electrolyte exposure, through an acrylic
tank which was attached to the top surface of the concrete specimens, for a period
of three months to initiate corrosion of the steel reinforcement. The specimens were
subjected to a 15% NaCl electrolyte with a cycle of 3 days wet and 4 days dry.
This simulates the condition of wet/dry cycling of salt water on concrete bridge
decks and substructures. Three deformed rebars, 1-inch in diameter, were placed
in a 1-cubic-foot concrete block. One was positioned at approximately 1-inch from
the top surface of concrete and the other two 10 inches below the surface. The
steel rebars were coated with an epoxy resin, one inch in length, at the surfaces
where they entered into concrete to prevent edge effect. The two rebars placed at

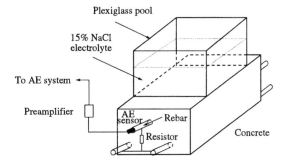

Fig. 9. Experimental setup of rebar corrosion in concrete

the bottom of the concrete specimens were electrically connected to the top rebar by a shunt resistor. This allowed the galvanic current due to corrosion of the top rebar to be measured with exposure time. Three AE transducers were used in the experiment. Two of them were mounted on one side of the top rebar with one inch apart and another was located on the opposite end of the top rebar. The output of the sensors was amplified and filtered by preamplifiers and transferred into digital signals by a Locoy A–D module. These digital signals were fed into a computer via shielded coaxial cables under the control of LabView program. The schematic of the test setup is shown in Fig. 9.

A set of AE signal recorded during the corrosion test is given in Fig. 10. It can be seen that the first arrival time is different for different transducer and this difference can be used to locate the position of the corrosion. The calculated corrosion location was further verified by breaking the concrete covers to expose the embedded rebar and a good correlation was found by present study. The AE signal is also of a high frequency, as expected for a rapid crack growth.

The AE data indicate that numerous AE events are obtained when corrosion of the rebar is initiated. Fig. 11 shows the accumulated number of AE events recorded during the beginning period of experiment as a function of cyclic exposure time. It can be seen that there is a big increase in AE events at about 20 days into the exposure which is most likely due to microcracking caused by building up of corrosion product on the rebar. On the other hand, the measurement of galvanic current and the half–cell potential did not show any obvious change at same time. It is thus verified that AE monitoring can detect a corrosion earlier than these two detecting methods.

4 Conclusions

Acoustic emission monitoring was performed on rebar corrosion in HCl solution and in steel reinforced concrete specimens by using the steel rebar as a waveguide. The experimental results indicated that there is a clear relationship between the AE rate and the rebar corrosion rate. Also, AE monitoring can detect the onset of rebar corrosion earlier than other methods such as galvanic and half–cell potential

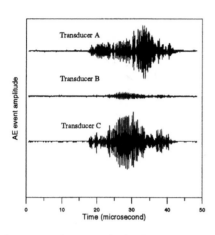

Fig. 10. AE signal of rebar corrosion in concrete

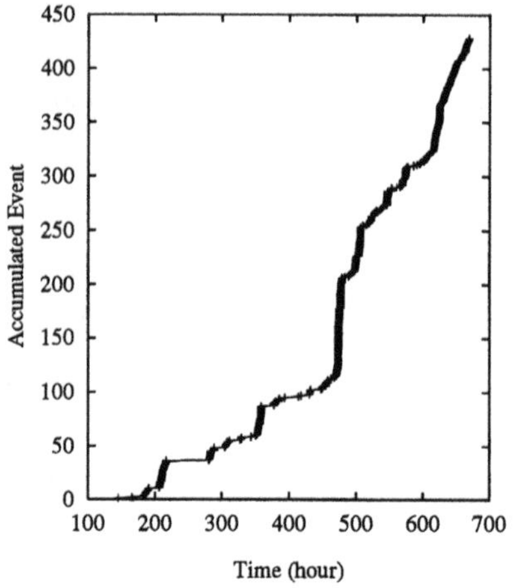

Fig. 11. Event number vs. time (rebar corrosion in concrete)

measurements. It was verified by experiments that AE technique can provide the true locations of rebar corrosion in concrete. However, further research is needed to quantify and calibrate the relationship between the AE signal and rebar corrosion in concrete before applying this technique to field practice.

5 References

1. Rettig, T.W. and Felsen, M.J. (1976) Acoustic emission method for monitoring corrosion reactions. *Corrosion–NACE*, Vol. 32, No. 4, pp. 121–126.
2. Mansfeld, F. and Stocker, P.J. (1979) Acoustic emission from corroding electrodes. *Corrosion–NACE*, Vol. 35, No. 12, pp. 541–544.
3. Seah, K.H.W., Lim, K.B., Chew, C.H. and Teoh, S.H. (1993) The correlation of acoustic emission with rate of corrosion. *Corrosion Science*, Vol. 34, No. 10, pp. 1707–1713.
4. Mazille, H., Rothea, R. and Tronel, C. (1995) An acoustic emission technique for monitoring pitting corrosion of austenitic stainless steels. *Corrosion Science*, Vol. 37, No. 9, pp. 1365–1375.
5. Li, Zongjin and Shah, S.P. (1994) Localization of microcracking in concrete under uniaxial tension. *ACI Material Journal*, Vol. 91, No. 4, pp. 372–381.
6. Zdunek, Alan D., Li, Zongjin, Prine, David, Landis, Eric, and Shah, S.P. (1995) Early detection of steel rebar corrosion by acoustic emission monitoring. *Corrosion/95, The NACE Annual Conference and Corrosion Show*, Orlando, U.S.A., March, 1995.

20 INFLUENCE OF THE SERVICE LOAD ON DURABILITY OF REINFORCED HIGH PERFORMANCE CONCRETE IN PRESENCE OF CHLORIDE

A. KONIN, R. FRANÇOIS and G. ARLIGUIE
Laboratoire Matériaux et Durabilité des Constructions, INSA-UPS
Génie Civil, Toulouse, France

Abstract

This paper deals with the effect of the loading on the durability of reinforced high performance concrete. In the case of reinforced concrete structures, the cracking is only the visible sign of the mechanical degradation and is accompanied by damage to the concrete (microcracks).

Furthermore, this damage occurs before the appearance of visible crack. In order to take into account this damage, we have performed experiments on two models. The first one when it is loaded, leads to a transversal cracking which corresponds to the case of the main reinforcement. The second one when it is loaded leads to a longitudinal cracking which corresponds to the case of secondary reinforcement. Both models were stored under different loading levels into an aggressive environment (cyclic salt fog). Three different concretes were used, their compressive strength are 45 MPa, 80 MPa, 100 MPa.

The loadless initial state of concrete and the damage state after loading were described by the use of the single replica technique observed on SEM. We notice an increase of the specific area of microcracks with the increase of the load.

This study shows that the loading which leads to damage to the concrete increases the chlorides penetration in the concrete, even when this damage is not accompanied by a visible crack.

Keywords: Chloride, cracking, Interface damage, corrosion, reinforced concrete, Tensile strength, Concrete cover, absorption

1 Introduction

The corrosion of re-bars is the major cause of reinforced concrete structure deterioration. During the corrosion process, the increase of volume of rust products creates tensile stresses causing secondary cracking and spalling of concrete. This can

Mechanisms of Chemical Degradation of Cement-based Systems. Edited by K.L. Scrivener and J.F. Young.
Published in 1997 by E & FN Spon, 2–6 Boundary Row, London SE1 8HN. ISBN: 0419215700.

result in a reduced load-bearing capacity and then considerably reduce the service life of concrete structures.

Among many factors affecting reinforced concrete durability, the chloride penetration remains one of the major causes of embedded steel corrosion.

It is well-known, that high strength concrete (H.S.C.) offers better performances in term of durability.

But in the case of reinforced concrete structures, the penetration of chlorides does not depend only on concrete transfer properties but also on the loading applied, on the state of strains, mainly characterized by the presence of cracking, and on the exposure to the aggressive environment.

In the development of this phenomenon, the cracks which for the most part are clearly visible, have been quickly implicated. With this mind, the rule books have emphasized the importance of the crack widths as a durability criterion. These rules have lead to an increase in the quantity of steel reinforcement (50 kg/m^3 in 1950 to 160 kg/m^3 in 1990) in concrete in order to control this cracking. The overcost due to these measures has lead to a development of research on this topic. The most recent developments show that the cracks, as long as their width doesn't exceed 0.5 mm, are not the essential factor in the corrosion process. The concrete cover quality and the cover width seem to play the most significant role. Nevertheless, there is a relation between the quality of concrete and the cracking because a crack is only the visible part of the concrete damage. So in the area of the concrete structure where the cracks have occurred, the quality of concrete has decreased due to the damage.

In order to take into account these different parameters, we have performed experiments on reinforced concrete elements, over a long period. These investigations compares ordinary and high strength concretes in order to quantify the durability properties in presence of cracks.

The results presented in this paper deals with the study of both initial damage and damage after loading in relation to the penetration of chloride ions.

2 Experimental program

2.1 Reinforced concrete specimens
In each series, all samples are made with the same concrete composition. Three concrete composition are used. These concrete compositions (Table 1) were made to obtain concrete of 45, 80 and 100 MPa average compressive stress measured on cylinders specimens. The cement used is a CPA CEM I 52.5R. The first composition is used to make control specimens. The second composition allows to obtain a high strength concrete (HSC). The third composition allows to obtain a very high strength concrete (VHSC). In the second and third composition a part of cement is substituted by silica fume.

Table 1: Concrete composition

Concrete composition	control group	HSC	VHSC
Gravel (kg/m^3)	1220	1166	1214
sand (kg/m^3)	820	727	650
cement (kg/m^3)	400	405	540
silica fume (kg/m^3)		45	50
superplasticizer (kg/m^3)		13.5 - (3%)	23.6 - (4%)
water (kg/m^3)	200 - (0.50)	157.5 - (0.35)	147.5 - (0.25)
Average compressive stress (MPa)	45	80	100

2.2 Test methods

Two different models have been performed. The first one uses cylinders of concrete 11x22 cm reinforced with a centered re-bar which get out from one side of the sample.

The reinforcing steel is loaded by a tensile strength. Because of the pull out of the steel, the loading lead to the formation of a longitudinal cracking. In the following comments, we called this model 'longitudinal model'.

The second one uses prismatic samples 10x10x50 cm also reinforced with a centered re-bar which get out from each side of the concrete. The steel bar is solicited by a tensile strength between 0 and 50 kN at the both sides. This loading leads to the formation of one or many transverse cracks in relation of the loading level. In the following comments, we called this model 'transverse model'.

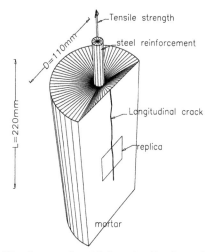

Fig. 1: sample with longitudinal crack

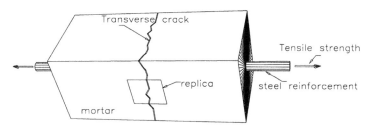

Fig. 2: sample with transverse crack

Specimens are stored in loading state in a confined salt fog. At different terms, electrochemical measurements are performed on both models. The evolution of the state of damage is assessed by the use of the replica technique [1]. In particular, the microcracking formed in concrete by the expansive rust will be characterized. The alteration of transfer properties due to the rust formed in concrete by the entering agents will also be investigated on both samples.

The two models lead to two different types of cracking but probably also to two different types of micro-cracking because both stresses fields in the concrete due to the strength applied on the reinforcement, are very different. In the case of the transverse model, the concrete surrounding the reinforcement is submitted to a tensile stress field, whereas in the case of the longitudinal model, the stress field in the concrete is more complex and a large part of the concrete is not in a tension state.

In this paper, only the first results obtained with the transverse model are presented.

2.3 Conservation mode

The salt fog (35 g/l of NaCl) is generated by means of 4 sprays located in each upper corner of the confined chamber. The exposure is a sequence of wetting and drying by periods of 15 days.

2.4 Experimental method

Some non destructive testings are made on the specimens. The state of steel is monitored periodically (every 2 months) by means of electrode potential measurements. Macroscopic and Microscopic damages are monitored on the outside surface of the samples.

At different terms, specimens were used to performed destructive testing. First, samplings are performed by drilling per sequence of 5 mm. At every depth, the powder obtained from this dry process is used to measure out the chloride content in concrete. Then, some specimens are sawed to allow microscopic observations around the reinforcement.

3 Concrete Damage

3.1 Macroscopic Damage

3.1.1 Initial state (prior to loading)

We notice that cracks are not visible on the surface of the samples.

3.1.2 Damage after loading (prior to aggressive environment action)

The cracks due to the tensile strength applied to the reinforcement are basically located near the central part of the sample.

A load increase results in a greater crack width and in an increase in cracking density. The width of visible cracks is between 0.05 mm and 0.1 mm.

3.2 Microscopic Damage.

The progression of the microcracking was studied by the replica technique and quantified by total projections [2]. The second process can be resumed as follow. The SEM images are digitized and the microcracking thus extracted appears on a map. Two essential parameters characterize the microcracking: firstly, the specific area which quantifies the importance of the microcracking network and secondly, the degree of orientation which quantifies the anisotropy.

3.2.1 Initial state (prior to loading)

For ordinary concrete, the results are the same as those reported in previous works [3][4] and allow to conclude that initial microcracking in the paste (at the resolution of 0.1 µm) and at the paste-aggregate interface is absent from the reinforced concrete samples.

For HSC, the results show the presence of microcracks due to the self-desiccation. The specific area of microcracking is about 0.5 mm^{-1}. Nevertheless, the density of this microcracks network is quite important (Fig. 3)

For VHSC, the results show the absence of microcracks in the paste due to self-desiccation (Fig. 4). Other random studies on the same replica have shown some microcracks which specific area can varied until 0.3 mm^{-1} . So, microcracks due to self-desiccation exist but are not present on the area studied.

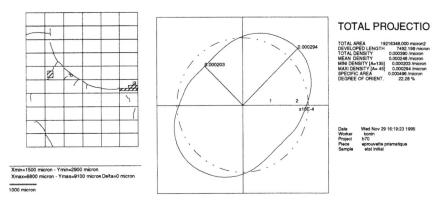

Xmin=1500 micron - Ymin=2900 micron
Xmax=6800 micron - Ymax=9100 micron Delta=0 micron

1000 micron

Fig. 3: microcracking map and analysis for HSC sample

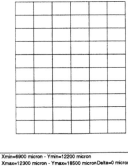

Xmin=6900 micron - Ymin=12200 micron
Xmax=12300 micron - Ymax=18500 micron Delta=0 micron

1000 micron

Fig. 4: microcracking map for VHSC sample

3.2.2 Damage after loading (prior to aggressive environment action)

For ordinary concrete, the results allow to conclude that the tensile strength applied to the reinforced samples leads to the absence of microcrackings wider than 0.1 μm in cement paste between aggregates but the paste-aggregate interface is damaged.

For HSC, the results show that the tensile strength applied to the reinforced samples leads to the increase of the density of microcracks which are basically located near the macro-crack. The specific area of microcracking is about 1.1 mm^{-1} after cracking.

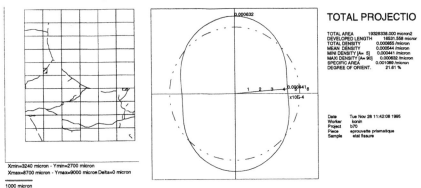

Xmin=3240 micron - Ymin=2700 micron
Xmax=8700 micron - Ymax=9000 micron Delta=0 micron

1000 micron

Fig. 5: loaded state after cracking

For VHSC: the result show that the tensile strength applied to the reinforced samples leads to the increase of the density of microcracks which are basically located near the macro-crack (Fig. 6). The specific area of microcracking is about 2 mm^{-1}.

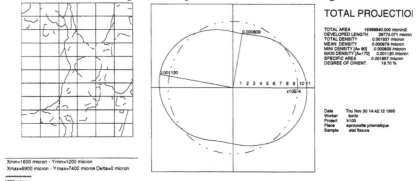

TOTAL PROJECTIO

TOTAL AREA	19368840.000 micron2
DEVELOPED LENGTH	29774.071 micron
TOTAL DENSITY	0.001537 /micron
MEAN DENSITY	0.000679 /micron
MINI DENSITY [A=80]	0.000809 /micron
MAXI DENSITY [A=170]	0.001120 /micron
SPECIFIC AREA	0.001957 /micron
DEGREE OF ORIENT.	19.70 %

Date	Thu Nov 30 14:42:12 1995
Worker	konin
Project	b100
Piece	eprouvette prismatique
Sample	etat fissure

Xmin=1600 micron - Ymin=1200 micron
Xmax=6900 micron - Ymax=7400 micron Delta=0 micron

1000 micron

Fig. 6: loaded state after cracking

The applied of the loading leads to a damage of the both HSC and VHSC concretes especially at the interface paste-aggregate. Furthermore, this damage occurs before the formation of the first macro-crack (Fig. 7).

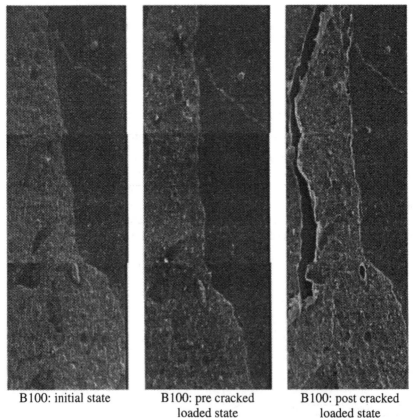

B100: initial state	B100: pre cracked loaded state	B100: post cracked loaded state

Fig. 7: Evolution of damage in relation to the loading state

4 Penetration of chlorides ions into the concrete

In this paper, only the experimental result concerning the HSC (80 MPa) are presented. Four samples have been studied to compare the penetration of chlorides according to the loading state of the sample area and the intensity of the mechanical stress applied.

EP1 is an unloaded sample, EP2 is loaded until to reach the cracking level; EP3 is loaded before the cracking level (there is no visual damage) and EP4 is loaded (as EP2) to reach the cracking level but the visible crack is coated to avoid the direct penetration of chloride by the crack.

The total chloride content after 12 and 17 months of storage in the salt fog chamber is presented respectively on Fig. 8 and Fig. 9. In the surface layer of the concrete (between 0 and 10 mm), the total chloride content is greater than 0.1 % whatever the sample type and the time of the sampling. This high content means that the penetration of the chloride ions is not a diffusion process but a suction process linked to the wet and dry cycles. Moreover the chloride content found in a macrocrack is in the same scale (only twice). So, the chloride ions penetrate very quickly in the surface layer of concrete even if the concrete is a High Performance Concrete.

In the bulk of the concrete (over 10 mm), the total chloride content is about 0.03 % and increase in the relation to the time of exposure to chloride environment. This low level means that the penetration of the chloride ions is a diffusion process which could be influenced by the loading level of the sample. Moreover the chloride content found in a macrocrack is not in the same scale (five time over). It seems that the unloaded sample (EP1) has the lower content of total chloride than the loaded samples (EP2, EP3 and EP4).

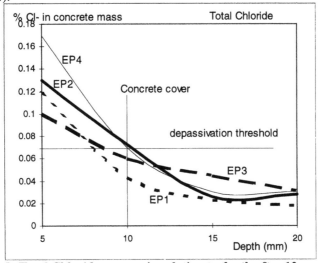

Fig. 8: Total Chloride content in relation to depth after 12 months

According to the electrode potential measurement, all the loaded specimens with a concrete cover of 10 mm, are corroded. These results are in good agreement with the chloride rate measured at a depth of 10 mm. According to the profile of chloride content, the rate of total chloride is about 0.07% in relation to the concrete mass, which correspond to the depassivation threshold generally presumed [5][6]. Moreover, it is generally presumed that the depassivation threshold decrease with the admixture of silica fume [7].

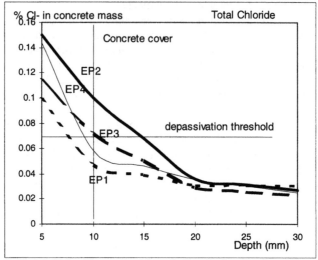

Fig. 9: Total Chloride content in relation to depth after 17 months

6 Conclusion

It seems that the increase of the specific area of microcracking with the loading and the damage of concrete even in the case of HSC or VHSC, lead to an increase of the chloride penetration. Moreover, this study shows that the loading leads to damage to the concrete before the formation of a visible crack. This damage occurs preferentially at the interface paste-aggregate even if it is generally presumed that the use of silica fume to realize HSC leads to a resorption of ITZ (Interfacial Transition Zone).

Even with the use of HSC, a concrete cover of 10 mm is not enough to avoid the risk of corrosion.

7 References

1.   J.P. Ollivier (1985) A non destructive procedure to observe the microcracks of concrete by scanning electron microscope, *Cement and Concrete Research*, vol.15.
2.   E. Ringot (1988) Automatic quantification of microcracks network by stereological method of total projections in mortars and concretes, *Cement and Concrete Research*, Vol. 18. pp. 35-43.
3.   R. François, J.C. Maso (1987) Microfissuration initiale d'un béton de structure, *Annales de l'ITBTP* n°457.
4.   R. François, G. Arliguie (1991) Reinforced concrete: Correlation between cracking and corrosion - Second CANMET/ACI International Conference on Durability of Concrete - SP126 - vol.II - Montréal 08/1991.
5.   "Durability of Concrete Bridge Decks", National Cooperative Highway Research Program, Synthesis of Highway Practice 57, Transportation Research Board, National Reasearch Council, Washington,DC, 1979.
6.   G. de Wind, P. Stroeven (1987) Chloride penetration into offshore concrete and corrosion risks, ACI SP-100, vol.2, pp.1679-1690.
7 .   M.D.A. Thomas (1995) Chloride penetration and reinforcement corrosion in marine-exposed fly ash concretes, RILEM International Workshop on Chloride Penetration into Concrete, St Rémy les Chevreuse, France, October 15-18 1995.

PART THREE

MICROSTRUCTURE AND MECHANISMS OF CHEMICAL DEGRADATION II

21 SOME CHEMICAL AND MICROSTRUCTURAL ASPECTS OF CONCRETE DURABILITY

H.F.W. TAYLOR and R.S. GOLLOP
Blue Circle Industries PLC, Greenhithe, Kent, UK

Abstract
Several processes that can damage concrete take place through the inward movement of a reaction front. Studies on external sulfate attack are used to illustrate aspects of experimental methods for studying microstructure and microchemistry and their interpretation. Significant findings obtained from microstructural and other studies in this field include the following. (1) C-S-H is attacked not only by $MgSO_4$ but also by Na_2SO_4 solutions, though to a lesser extent. (2) Al present in C-S-H or hydrotalcite is not available for reaction with sulfate solutions. (3) The amount of sulfur entering the hydration products of a slag blend as sulfide can be as great as that entering them as sulfate, and this has implications for the interpretation of microanalytical data, including those pertaining to sulfate attack. (4) Expansion and cracking are probably caused, directly or indirectly, by ettringite formation, while softening and disintegration are caused by destruction of C-S-H. These latter processes appear to be at least as important in practice as expansion and cracking. Destruction of C-S-H is probably of greater relative importance with slag blends than with plain Portland cements; this could explain observations that slag blends often fail in laboratory tests by weakening or disintegration before much expansion has occurred.
Keywords: Sulfate attack, decalcification, hydrotalcite, ettringite, expansion, softening, sulfide, microanalysis.

1 Introduction

A number of processes that can damage concrete take place through the inward movement of a reaction front. Carbonation, acid attack and possibly delayed ettringite

Mechanisms of Chemical Degradation of Cement-based Systems. Edited by K.L. Scrivener and J.F. Young. Published in 1997 by E & FN Spon, 2–6 Boundary Row, London SE1 8HN. ISBN: 0419215700.

formation are examples. Another example is external sulfate attack. In this paper, points from some investigations on this last topic are summarised to illustrate features of experimental procedure or interpretation, some of which may be of more general relevance. Much of the material is reported more fully elsewhere [1 - 4].

If the part of a cube or other sample of a paste or mortar that has been attacked is only a layer extending inwards a short way from the surface, there is not much hope of obtaining significant information by grinding up the entire sample and examining it by some standard procedure, such as XRD or thermal analysis. This nevertheless appears to have been done in at least one classic and otherwise well planned investigation. Similarly, while expansion measurements on samples of a single size may be useful as a comparative procedure, they cannot provide information of a fundamental nature.

To obtain information about phase assemblages and microstructures, two approaches have proved particularly effective.

2   Experimental approaches to the study of phase changes and microstructures in sulfate attack

2.1   Removal of surface layers of known thickness
In one approach, surface layers of various known thicknesses are removed and the resulting surfaces examined by XRD or other methods. This method has been used to study attack on pastes of Portland cements [5,6] and fly ash blends [5].

In these investigations, the maximum depth at which any changes were observed was typically in the order of a millimetre. In typical cases with Portland cement, the phases present in the unattacked core of the sample included monosulfate and calcium hydroxide. On moving outwards towards the surface, monosulfate was replaced by ettringite; further out, there was a zone containing gypsum. Alongside these changes, there was a gradual depletion of the calcium hydroxide.

Clearly, a reaction front advances into the material. The first substantial phase change is that observed at the greatest depth below the surface, which is that producing ettringite. The results do not support some earlier views [7,8] according to which gypsum is the first main product.

This approach has the merit of showing phase compositions directly and reasonably unequivocally, but it does have some limitations. It does not provide a detailed picture of the microstructure. It does not show whether any changes occur in the C-S-H composition. Also, because of the relatively deep penetration of X-rays, it is rather imprecise as regards the depths to which the data relate.

2.2   SEM examination of polished sections

2.2.1   General
In another approach, a cube or other specimen that has been attacked is sawn across the middle. The cut surface is polished and examined in the SEM by backscattered electron imaging and X-ray microanalysis. The first application of this method to the study of sulfate attack was that of Crumbie, Scrivener and Pratt [9]. It was later used by Bonen and Cohen [10,11] and Bonen [12,13], who examined pastes and mortars of a Type I cement with and without addition of microsilica, and by Gollop and Taylor

[1–4], who examined pastes of normal Portland cement (Type I; PC), sulfate resisting Portland cement (Type V; SRPC) and various slag blends.

2.2.2 Microstructure of a Portland cement paste after storage in Na_2SO_4 solution

Examination by this method of a PC paste that had been stored in Na_2SO_4 solution [1] gave results consistent with those of the XRD investigations mentioned above [4,5] and additional information. The ettringite, which was the first product of attack, was mixed with the C-S-H at or below a micrometre level, and its presence was inferred only from microanalyses. Some of the gypsum was also similarly mixed with C-S-H, but some had separated out into veins. As the surface of the cube specimen was approached, the average Si/Ca ratio of the C-S-H increased. Cracks, often associated with the gypsum veins, were observed sub parallel to the cube faces and there was more extensive cracking or loss of material at the cube edges.

2.2.3 Decalcification of C-S-H

Crumbie, Scrivener and Pratt [9] showed that the Si/Ca ratio of the C-S-H in a PC paste tended to increase as a result of attack by Na_2SO_4. This observation, which has been confirmed in studies on pastes of PC, SRPC and slag blends [1–4], shows that the widely held view that attack on the C-S-H occurs with $MgSO_4$ but not with Na_2SO_4 is incorrect, though the extent of decalcification is certainly much greater with $MgSO_4$. Decalcification begins when the CH has been depleted, or, perhaps, when CH from this source is no longer readily available. In decalcification, the Si/Ca ratio of the C-S-H increases due to removal of Ca^{2+} and OH^-, and this causes the amount of C-S-H to decrease. This contrasts with the increase in Si/Ca that occurs through pozzolanic action, which occurs through addition of SiO_2 and thus causes the quantity of C-S-H to increase.

2.2.4 Expansion and cracking

The crack patterns observed in cement pastes that have been attacked by Na_2SO_4 solutions suggest that expansion has taken place in a surface layer, which is tending to become detached. The cracking occurs in the zone of gypsum formation and thus nearer the surface than the zone in which the ettringite is formed [1,3]. Some workers (e.g., Wang [6]) have considered that the expansion is due to formation of gypsum rather than to that of ettringite, and the present observation is certainly consistent with that hypothesis. However, the extensive evidence that expansion is related to the content of available Al_2O_3 from C_3A or other sources does not support it. More probably, the damage is associated with ettringite formation, but it may be an indirect rather than a direct result of it. The immediate cause of the expansion is possibly water uptake, by a mechanism broadly of the type suggested by Thorvaldson [14] and subsequently by Mehta [15].

3 Graphical presentation of microanalysis results

Individual X-ray microanalyses relate to regions some 1 –2 micrometres across in each direction. Because of the fine scale of the microstructure, it is often not practicable to distinguish between microstructurally different regions, such as inner product of clinker grains, inner product of slag grains or outer product. However, by excluding analyses that gave totals above a certain value after including contents of oxygen calculated by stoichiometry, one can be reasonably certain that all those considered relate to hydrated material. For this purpose, a value of 85 % is generally appropriate, but gypsum undergoes partial dehydration during specimen preparation or in the SEM, and usually gives higher totals [1–4].

In general, for cement hydration products, the absolute contents of elements obtained by X-ray microanalysis, whether by wavelength or energy dispersion, are not meaningful, and it is normally safer to consider atom ratios. In order to present the data in an intelligible form, it is almost essential to use graphical methods. These include triangular diagrams, used by Bonen and Diamond [16] in a study on normal hydration, and Cartesian plots of one atom ratio against another or against depths below a face or edge of a cube. Five types of Cartesian plot have proved especially useful [1,3,4], viz:

- *Si/Ca against depth:* this shows the Si/Ca ratio of the C-S-H in the unattacked core material and the extent of decalcification at different depths.
- *S/Ca against depth:* this shows the depth of sulfate penetration and gives an indication of the ranges of depth at which ettringite and gypsum are formed.
- *Mg/Ca against Al/Ca [17]:* for slag cements, this shows the Mg/Al ratio of the hydrotalcite-type phase and the Al/Ca ratio of the C-S-H. Neither of these quantities appears to vary with depth, indicating that neither the Al in the hydrotalcite nor that in the C-S-H is available for reaction with sulfate solutions.
- *Al/Ca against Si/Ca:* This plot, which is made for varying ranges of depth, shows the distributions of Si/Ca and Al/Ca in the C-S-H and thus the extent of any decalcification. It also shows the nature of the hydrated aluminate or sulfo-aluminate phases, and gives an indication of the closeness of mixing of the phases. For slag cements, the quantity Mg/(p·Ca), where p is the Mg/Al ratio of the hydrotalcite, is subtracted from Al/Ca to eliminate Al present in hydrotalcite [18].
- *S/Ca against Al/Ca:* This gives further information for given ranges of depth about the C-S-H, AFm and AFt compositions and also about the content of gypsum and the closeness of mixing of this phase with C-S-H and ettringite. For slag cements, the correction to Al/Ca mentioned above is applied.

4   Problems arising from the presence of sulfide ion in slags

4.1   Quantity of sulfide in the hydration products

The sulfur in granulated or pelletised blastfurnace slags occurs almost entirely as sulfide, present in the glass, and is released on hydration at the same relative rate as the other constituent ions of that phase [19]. Typically, the slag contains about 1 % of sulfide. Based on determination of the fraction of slag that has reacted [20] and other relevant data, one may estimate how much sulfide has been released at any given age. Such calculations show that, for a blend containing 60–70 % of slag at an age of 6 months, comparable amounts of sulfur will have been released into the hydration products as sulfate and as sulfide. One cannot, therefore, afford to ignore the sulfide in any discussion of the chemistry of hydration or of sulfate attack.

4.2 Hydrated phases containing sulfide

AFm phases containing sulfide have been described, and form extensive solid solutions with tetracalcium aluminate hydrate [19,21]. The extent to which sulfide can substitute in ettringite appears to be much more restricted [19]. There is little direct evidence on what happens to the sulfide on hydration of slag blends; a small proportion goes into the pore solution, but most probably enters an AFm phase [19].

The sulfide in AFm phases is easily oxidised, but in a paste, the extent of atmospheric oxidation appears to be small [19]. A study on an unusual type of Portland cement relatively high in sulfide confirmed that atmospheric oxidation of the sulfide in the paste was indeed quite restricted, but some of the sulfide was oxidised by the Fe^{3+} and Mn^{3+} in the cement [22]. If this is also the case with blends high in slag, there is unlikely to be much Fe^{3+} in the hydration products after a few months, as all will have been reduced by the sulfide to the +2 state.

4.3 Possible oxidation in surface layers

Even though atmospheric oxidation may be slight in the bulk of a well made paste, it could still be significant in the very thin surface region to which microanalyses relate. This introduces some uncertainties into the interpretation of microanalyses for slag cement pastes, including those attacked by sulfate solutions.

Microanalyses, as normally carried out, do not show the oxidation state of the sulfur. We have to ask, how much, if any, is present as sulfide, and whether the results truly reflect what is happening in the body of the material and not merely in the surface layer that has been sawn and polished and thus exposed to the atmosphere. Experimental data on this subject are lacking.

4.4 Implications for interpretation of microanalysis results

The most serious problem concerns the ettringite. If it is true that ettringite cannot accommodate much sulfide, the possibility arises that any ettringite detected by X-ray microanalysis of a paste of a slag blend that has undergone sulfate attack has resulted partly or wholly from atmospheric oxidation, which occurred to a significant extent only in the surface layer. It is unlikely that the ettringite formation could be wholly due to this cause, because sulfate is being supplied in substantial proportion from the external solution. One may also ask whether that part of the aluminium present in the AFm phase that is balanced by sulfide is available for reaction with a sulfate solution, and if it is, what happens to the sulfide. Again, data on these matters are lacking.

5 Physical effects of sulfate attack

5.1 Expansion and cracking versus softening and disintegration

Much attention has been paid to expansion and cracking as the damaging effects of sulfate attack, but Mehta [23] concluded that field experience shows that loss of adhesion and strength are usually more important. In laboratory studies, mortars of slag blends reportedly often fail by weakening or disintegration before much expansion has occurred [24 - 26]. These latter effects may thus be of greater relative importance in the case of slag cements. We have to ask what microstructural changes produce expansion and cracking, and which cause weakening or disintegration.

The following discussion is based very largely on the conclusions of Rasheeduzzafar et. al. [24,27] and Al-Amoudi et al. [28] in conjunction with the microstructural and microchemical evidence summarised in this paper. The essential points are that expansion and cracking are caused, directly or indirectly, by the formation of ettringite, and softening and disintegration by decomposition of C-S-H. Slag blends tend to fail by softening and disintegration because, compared with plain PC, decomposition of C-S-H is more important relative to ettringite formation.

5.2   Ettringite formation

This requires, besides sulfate, sources of water and of Ca^{2+} , OH^- and $Al(OH)_4^-$ ions. The only significant direct source of $Al(OH)_4^-$ is the AFm phase; as seen earlier, Al^{3+} substituted in C-S-H, or present in hydrotalcite, is not available. This is also true of Al^{3+} present in ettringite already formed independently of sulfate attack. Unreacted slag or clinker phases may be considered an indirect source. The microstructural evidence shows that the Ca^{2+} and OH^- used in forming either ettringite or gypsum are obtained preferentially from CH and that decalcification of C-S-H begins only when that phase has been used up, or perhaps, when what remains is not easily accessible.

We therefore expect ettringite formation to be the dominant effect if available Al_2O_3 and CH are abundant. Cases of this include attack by sodium sulfate solutions on pastes of Type I PC high in C_3A or of blends containing fairly small proportions of slags high in Al_2O_3. Failure in these cases is therefore likely to occur through expansion and cracking. Ettringite formation resulting from sulfate attack is minimal with SRPC (Type V cements), because there is relatively little Al_2O_3 in the hydration products, and much of what there is is not available [2,3]. Similarly, it is minimal with blends high in slag, because although there is much Al_2O_3 in the system, not much of it is in an available form [4]. The resistance of mixer blends lower in slag can be much increased by adding supplementary calcium sulfate, thereby causing normal hydration to produce ettringite which persists indefinitely [25].

5.3   Destruction of C-S-H

Attack on the C-S-H of a PC or SRPC paste by Na_2SO_4 solution is limited because of the presence of CH. In the case of attack by $MgSO_4$ solution, it is more serious, and can be the dominant effect [12,27]. It yields gypsum, hydrous silica, brucite and a magnesium silicate hydrate, shown [1,3,4] to be poorly crystalline serpentine ($M_3S_2H_2$). The reactions involving Mg^{2+} greatly lower the pH and therefore destroy the C-S-H completely in the regions attacked. Destruction of C-S-H is thus favoured by low or zero contents of CH and, very strongly, by the presence of Mg^{2+} ions.

CH contents are low or even zero with blended cements of all kinds, and this may be why slag cement pastes or mortars tend to fail by weakening and disintegration. There may be other relevant effects. Pastes of slag blends are less permeable than those of plain Portland cements, and this may restrict penetration by SO_4^{2-}. They also contain much fine porosity, and this possibly enhances the ability of the microstructure to accommodate an increase in solid volume due to ettringite formation without undergoing damage. On the other hand, if plain PC or SRPC pastes are attacked by $MgSO_4$, a composite layer of gypsum and brucite is formed on the surface [1,3,10 - 13], and may have a protective effect [11,13,27]. Such layers are not formed in blends containing slag [4], and this may lower their resistance to attack by $MgSO_4$ solutions.

6 Acknowledgement

The authors wish to thank Blue Circle Industries PLC for permission to publish this work and for sponsoring RSG for an External Ph.D. course at Imperial College, University of London. The work forms part of a programme of research being carried out by RSG under the Public Institutions and Industrial Laboratories scheme at Imperial College, where HFWT is a Visiting Professor. The authors are indebted to Dr G.K. Moir (Blue Circle) and to Dr K.L. Scrivener, lately of Imperial College, and the late Professor P.L. Pratt for encouragement and advice.

7 References

1. Gollop, R.S. and Taylor, H.F.W. (1992) Microstructural and microanalytical studies of sulfate attack. I. Ordinary Portland cement paste. *Cement and Concrete Research*, Vol. 22, No. 6. pp. 1027 - 38.
2. Gollop, R.S. and Taylor, H.F.W. (1994) Microstructural and microanalytical studies of sulfate attack. II. Sulfate-resisting Portland cement: ferrite composition and hydration chemistry. *Cement and Concrete Research*, Vol. 24, No.7. pp. 1347 - 58.
3. Gollop, R.S. and Taylor, H.F.W. (1995) Microstructural and microanalytical studies of sulfate attack. III. Sulfate-resisting Portland cement: reactions with sodium and magnesium sulfate solutions. *Cement and Concrete Research*, Vol. 25, No.7. pp. 1581 - 90.
4. Gollop, R.S. and Taylor, H.F.W., in preparation.
5. Cabrera, J.G. and Plowman, C. (1988) The mechanism and rate of attack of sodium sulfate solution on cement and cement/pfa pastes. *Advances in Cement Research*, Vol. 1, No. 3. pp. 171 - 79.
6. Wang, J.G. (1994) Sulfate attack on hardened cement paste. *Cement and Concrete Research*, Vol. 24, No. 4. pp. 735 - 42.
7. Calleja, J. (1980) Durability, in *7th International Congress on the Chemistry of Cement, Vol. 1*, Editions Septima, Paris, pp. VII-2/1 - 2/48.
8. Bensted, J. (1981) Chemical considerations of sulfate attack. *World Cement Technology*, Vol. 12, No. 4. pp. 178 - 84.
9. Crumbie, A.K., Scrivener, K.L. and Pratt, P.L. (1989) The relationship between the porosity and permeability of the surface layer of concrete and the ingress of aggressive ions, in *Materials Research Society Symposium Proceedings, Vol. 137*, (ed. L.R. Roberts and J.P. Skalny), Materials Research Society, Pittsburgh, pp. 279 - 84.
10. Bonen, D. and Cohen, M.D. (1992) Magnesium sulfate attack on portland cement paste. I. Microstructural analysis. *Cement and Concrete Research*, Vol. 22, No. 1. pp. 169 - 80.
11. Bonen, D. and Cohen, M.D. (1992) Magnesium sulfate attack on portland cement paste. II. Chemical and mineralogical analyses. *Cement and Concrete Research*, Vol. 22, No. 4. pp. 707 - 18.
12. Bonen, D. (1992) Composition and appearance of magnesium silicate hydrate and its relation to deterioration of cement-based materials. *Journal of the American Ceramic Society*, Vol. 75, No. 10. pp. 2904 - 06.

13. Bonen, D. (1993) A microstructural study of the effect produced by magnesium sulfate on plain and silica-fume-bearing portland cement mortars. *Cement and Concrete Research*, Vol. 23, No. 3. pp. 541 - 53.

14. Thorvaldson, T. (1954) Chemical aspects of the durability of cement products, in *Proceedings of the 3rd International Symposium on the Chemistry of Cement*, Cement and Concrete Association, London, pp.436 - 66

15. Mehta, P.K. (1973) Mechanism of expansion associated with ettringite formation. *Cement and Concrete Research*, Vol. 3, No.1. pp. 1 - 6.

16. Bonen, D. and Diamond, S. (1994) Interpretation of compositional patterns found by quantitative energy dispersive X-ray analysis for cement paste constituents. *Journal of the American Ceramic Society*, Vol. 77, No. 7. pp. 1875 - 82.

17. Harrisson, A.M., Winter, N.B. and Taylor, H.F.W. (1987) Microstructure and microchemistry of slag cement pastes, in *Materials Research Society Symposium Proceedings, Vol. 85* (ed. L.J. Struble and P.W. Brown), Materials Research Society, Pittsburgh, pp. 213 - 222.

18. Wang, S.-D. and Scrivener, K.L. (1995) Hydration products of alkali activated slag cement. *Cement and Concrete Research*, Vol. 25, No. 3. pp. 561 - 71.

19. Vernet, C. (1982) Comportement de l'ion S⁻⁻ au cours de l'hydratation des ciments riches en laitier (CLK). Formation de solutions solides de S⁻⁻ dans les aluminates hydrates hexagonaux. *Silicates Industriels*, Vol. 47, No. 3. pp. 85 - 89.

20. Lumley, J.S., Gollop, R.S., Moir, G.K. and Taylor, H.F.W. (1996) Degrees of reaction of the slag in some blends with Portland cements. *Cement and Concrete Research*, Vol. 26, No. 1, in press.

21. Dosch, W. and Keller, H. (1976) On the crystal chemistry of tetracalcium aluminate hydrate, in *The 6th International Congress on the Chemistry of Cement, Vol. 3,* Stroyizdat, Moscow, pp. 141 - 46 [Russ. with Engl. preprint].

22. Brückner, A., Lück, R., Wieker, W., Andreae, C. and Mehner, H. (1992) Invesigation of redox reactions occurring during the hardening process of sulfide containing cement. *Cement and Concrete Research*, Vol. 22, No. 6. pp. 1161 - 69.

23. Mehta, P.K. (1992) Sulfate attack on concrete – a critical review, in *Materials Science of Concrete III* (ed. J. Skalny), American Ceramic Society, Westerville, Ohio, pp. 105 - 30.

24. Rasheeduzzafar, Dakhil, F.H., Al-Gahtani, A.S., Al-Saadoun, S.S. and Bader, M.A. (1990) Influence of cement composition on the corrosion of reinforcement and sulfate resistance of concrete. *ACI Materials Journal*, Vol. 87, No. 2. pp. 114 - 22.

25. Kollek, J.J. and Lumley, J.S. (1991) Comparative sulfate resistance of SRPC and Portland slag cements, in *Proceedings of the 5th International Conference on the Durability of Building Materials and Components* (ed. J.M. Baker, P.J. Nixon, A.J. Majumdar and H. Davies), E & FN Spon, London, pp. 409 - 20.

26. Lawrence, C.D. (1992) The influence of binder type on sulfate resistance. *Cement and Concrete Research*, Vol. 22, No. 6. pp. 1047 - 58.

27. Rasheeduzzafar, Al-Amoudi, O.S.B., Abduljanwad, S.N. and Maslehuddin, M. (1994) Magnesium-sodium sulfate attack in plain and blended cements. *Journal of Materials in Civil Engineering*, Vol. 6, No. 2. pp. 201 - 22.

28. Al-Amoudi, O.S.B., Maslehuddin, M. and Saadi, M.M. (1995) Effect of magnesium sulfate and sodium sulfate on the durability performance of plain and blended cements. *ACI Materials Journal*, Vol. 92, No. 1. pp. 15 - 24.

22 MECHANISMS OF DEGRADATION OF PORTLAND CEMENT-BASED SYSTEMS BY SULFATE ATTACK

C.F. FERRARIS, J.R. CLIFTON, P.E. STUTZMAN and E.J. GARBOCZI
National Institute of Standards and Technology, Gaithersburg, MD, USA

Abstract
Although the chemical mechanisms of sulfate attack have been studied by many researchers, considerable disagreement exists over the mechanics of the associated expansion and cracking processes. Studies were carried out by exposing mortar and concrete specimens of different geometries (prisms, cylinders, and spheres) to sodium sulfate solutions with controlled pH (7,9,11). Relationships between the formation of reaction products and the expansion of portland cement mortars were investigated by identifying the reaction products present as a function of attack depth and amount of expansion. Simulation modeling based on a finite element approach is being performed to elucidate the mechanics of the cracking induced by sulfate attack.
Keywords: Sulfate attack, microstructure, mortar, environment influence

1. Introduction

A potentially significant degradation process for underground concrete vaults made for the disposal of low-level radioactive waste is sulfate attack [1]. In a previous paper [2], a model of the kinetics of sulfate attack of cement-based materials was described. Further studies have been carried out for the purpose of developing relationships among the formation of reaction products, the depth of sulfate penetration, and the amount of expansion. Simulation modeling, based on a finite element approach, is also being performed to elucidate the mechanics of expansion and cracking. In the present paper, results of experimental studies and simulation modeling of the mechanics of expansion and cracking during sulfate attack are given.

Mechanisms of Chemical Degradation of Cement-based Systems. Edited by K.L. Scrivener and J.F. Young.
Published in 1997 by E & FN Spon, 2–6 Boundary Row, London SE1 8HN. ISBN: 0419215700.

2.   Experimental Studies and Results

2.1.   Influence of sodium sulfate concentration and pH

The focus of NIST sulfate attack experimental is the determination of the influence of the environment and the mechanisms of the deterioration.   The variables in the experimental design were:

- pH of the controlled-pH sulfate solutions: 7, 9 and 11
- sulfate solution concentrations: 0 to 10 % sodium sulfate solution (by weight)
- C_3A content of the cements: 2.8, 4, 6, 8, 11.1 and 12.8 % (ASTM C150 [3])

All the tests were performed using mortar prism specimens (25 mm x 25 mm x 279 mm).   The mortar mixtures were prepared according to ASTM C109 [4] specifications.   Three cements were blended to vary the C_3A content (ASTM C150 [3]) of the cement used in the sample preparation:

- Cement 1: 2.7 % C_3A
- Cement 2: 11.1 % C_3A
- Cement 3: 12.8 % C_3A

Cement 1 was mixed with either Cement 2 or Cement 3 to prepare "mixed" cements with 4, 6, and 8% C_3A content.

The pH of the solution in which the samples were placed was controlled using an automatic burette that added sulfuric acid (1 N) when the pH rose above a preset value [5].   In a separate experiment, the uncontrolled pH of the solution rose to 12 in a few days [5].   The changes in length of each specimen was monitored until disintegration of the specimen occurred. Each value reported is the average of 3-7 bars. Figure 1 shows the results obtained.

All the factors examined seem to have an effect on the rate of deterioration of the samples as measured by the time to reach 0.1% elongation.   This criterion was selected, instead of 0.5% as reported elsewhere [5], because most of the samples had already disintegrated before they had expanded by even 0.3%. All specimens were mortar prisms 25 mm x 25 mm x 276 mm. The mix design was as described in ASTM C109 [4] . In Figure 1, a shorter time indicates a faster expansion of the specimen. The C_3A content is the main factor affecting the rate of expansion (Figure 1b).   When the solution pH was not controlled, but instead the solution was changed each time a measurement was made according to ASTM C1012 [6], a longer time to reach 0.1% expansion was required, i.e., the deterioration was slower (Figure 2).   Therefore, a controlled pH environment allows a faster assessment of the sulfate resistance as was also determined previously [5].   As would be predicted, a higher concentration of sodium sulfate in the solution led to faster expansion of the specimens.   This series of tests confirms the importance of sodium sulfate concentration and pH on service life.

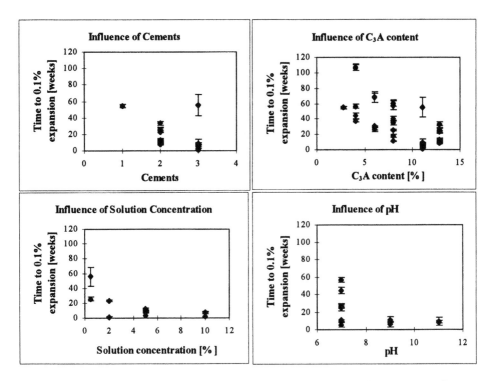

Figure 1: Influence of various factors on the time in weeks to reach 0.1% expansion. In the lower graphs, the data shown are with all three cements. (error bars are ± σ)

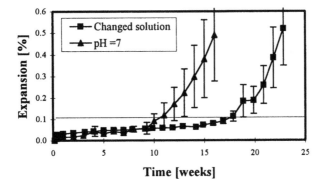

Figure 2: Expansion versus time in solutions in which the pH was a) controlled or b) uncontrolled. The initial concentration of sodium sulfate solution was 5%.(Cement 3 used) (error bars are ± σ)

2.2. Microstructure

The deterioration mechanism was studied by analyzing the microstructures of the specimens, the composition of the reaction products, and observing the form of distress. Samples were taken for X-ray powder diffraction (XRD) analysis at selected times during the tests shown in Figure 2, to provide information on the changes in crystalline products as a function of depth and time. A sliced cross section was sampled as follows: the outer 3 mm, labeled as "outer", the next 3 mm, labeled as "inner", and the center of the section, labeled as "core". Each sample was gently crushed and sieved through a 100 mesh (150 μm, ASTM E 11 [7]) screen to remove the bulk of the quartz sand. The hardened cement paste powder was stored in a desiccator under a constant relative humidity of 33% to prevent dehydration of the calcium aluminosulfates. The powdered samples were analyzed within 24 hours to minimize influence of carbonation.

Scanning electron microscopy using backscattered electron and X-ray imaging of resin-impregnated, polished sections provide information on: phases, phase composition, textural and morphological relationships, and porosity. The resin impregnation was carried out as follows. Sections cut at selected time intervals throughout the experiment were placed in ethanol to stop hydration, and then placed in a low-viscosity resin to impregnate the microstructure without drying and to minimize the possibility of inducing drying shrinkage cracks. The resin was cured at 60 °C for 24 hours, and the samples polished with a 0.25μm diamond paste on a lap wheel. The surface was coated with a thin layer of carbon to provide the conductive surface necessary for SEM imaging.

The following observations were made. First, the sulfur distribution determined by X-ray imaging indicated a high concentration to about 2 mm from the surface relative to the control samples, cured in limewater. This corresponds to a gypsum-rich zone in the outer microstructure. The transition between gypsum-rich paste and paste beyond this zone is quite abrupt, and in severely-damaged mortar bars, the interface separates an outer region of coarse cracking typically parallel to the mortar bar surface, and the interior region which contains inter-aggregate cracking. The core at later ages is not devoid of gypsum or ettringite, however, as will be noted in the discussion of the X-ray diffraction (XRD) results. Quantitative energy dispersive X-ray spectrometry (EDS) analysis of sodium levels in the calcium silicate hydrate (C-S-H) showed high concentration in the outer 200 μm while sodium was not detected in C-S-H at greater depths. Calcium depletion was observed in the outer C-S-H as calcium/silicon ratios are markedly lower in the sulfate-exposed mortars relative to the control mortar. This ratio increases to the value of the control mortar around one millimeter of depth. The outer (less than 250 μm) microstructure of the sulfate-exposed specimen relative to the control (equal age in their respective solutions) exhibits scaling and appears to be almost completely devoid of unhydrated cement particles and overall darker (Figure 3). This darkening of the paste may originate from two sources: compositional changes (as in the leaching of Ca), or an overall increase in porosity. These features were also observed by Taylor [8].

Second, after three days, the outer layer contains ettringite, gypsum, monosulfate, and gypsum. Void space appears to be "filled" with reaction product. In some cases replacement of calcium hydroxide with gypsum and large deposits of ettringite are

Outer layer   **Core**

Figure 3: Microstructure by SEM; 15 weeks of exposure in uncontrolled pH sodium sulfate (5% by weight) solution.

common. In contrast, the inner layer and core sampled at the same time show the presence of calcium hydroxide, monosulfate and traces of ettringite. Massive deposits of ettringite are not found in the core until late stages of testing, after the rapid increase in expansion.

Third, at 28 days, gypsum and ettringite were detected throughout the specimen, while the monosulfate appears to decrease. Gypsum and ettringite increase with age at all depths while calcium hydroxide decreases, though the changes in phase proportions do not appear to be as great as in the outer few millimeters of the specimen.

2.3. Specimen shape and spalling

Three specimen geometries were used: cylinders (various diameters), spheres and prisms. The cylinders had diameters of 25, 50 and 75 mm. To determine if one of the leading mechanisms of deterioration by the sulfate is governed by the transport of the sulfate ions into the specimen, the cylinders were capped at the ends so that the penetration of the sodium sulfate was only perpendicular to the cylinders main axis. It was observed that the larger diameter cylinders expanded more slowly than the smaller diameter cylinders (Figure 4). This observation suggests that the transport rate of sulfate ions into the material governs the deterioration. Companion prisms (25mm x25 mm x 279mm) tested at the same time expanded like the 25mm diameter cylinders, confirming that the size and not the shape was the important factor in the time scale of the expansion due to sulfate attack, although the cracking pattern was different. During this test, it was observed that the outer layer of the 75-mm-diameter specimen

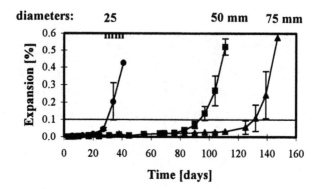

Figure 4: Influence of the diameter on the expansion of the cylinders. Type I Portland cement was used.

was peeling like an onion (several layers). To determine the influence of aggregate size on the depth of spalling and simultaneously avoid end effects, spheres were cast of mortar (2.36 mm maximum diameter sand), cement paste, and concretes (12-19 mm or 4-8 mm diameter coarse aggregates). It was observed that the samples peeled like an onion (successive layers) for the mortar and the finer concrete after 3 and 5 months, respectively, while the cement paste and the coarser aggregate concretes did not show cracks after 9 months. The thickness of the skin did not reflect the different size of the aggregates used, probably because it depends more on the paste-rich outer layer that forms on any sample cast in a mold.

3.   Discussion of mechanics of expansion and cracking

The kind of cracking seen in the sulfate-attacked specimens was observed to depend on the specimen geometry. For example, the stress concentrations introduced by the sharp edges on the sides and ends of the prisms and the ends of the cylinders led to cracking along these edges first, which tended to dominate the subsequent random surface cracking and spalling. The spheres that showed cracking had random surface cracking only, followed by spalling. This has led to an attempt to use elasticity theory and finite element modeling to try to understand the mechanics of the degradation process as modified by sample geometry.

The conceptual model is the following. Sulfate ions enter the material and react with various phases of the mortar, which then grow in size, inducing stresses throughout the specimen. The specimen is modeled as a composite of expansive inclusions in an elastic matrix. The inclusions are modeled as having an intrinsic strain ε^{o}_{j}, so that the stress σ_i in the inclusions is given by [9]

$$\sigma_i = C_{ij} (\varepsilon_j - \varepsilon^{o}_{j}). \qquad (1)$$

where C_{ij} is the elastic tensor and ε_j is the strain. In this case, the diagonal components of the ε^{o}_{j} matrix are positive and identical, allowing for a pure expansion only.

A recent effort at analytical modeling [10] of alkali-silica reaction, which in terms of elasticity theory is similar mathematically, gave a uniform value of ε^o_j for a given surface layer. If this idea is applied to sulfate attack in a macroscopic way, the entire surface of the sample would have, down to the depth of the sulfate attack, to have a positive intrinsic strain and thus tend to expand. This however results in compressive-only stresses in the outer layers, since the expansive layer remains in compression, due to the unreacted layers underneath that restrain the expansion so that $\varepsilon < \varepsilon^o_j$, resulting in a compressive stress according to eq. (1). Uniform compressive stress in the layer will not result in the kinds of surface cracking seen experimentally. This analysis seems to rule out any exact mathematical analysis of the mechanics, and thus suggests that a random model must be used.

It is necessary to go back to the expansive inclusion model mentioned above. In this model, any surface layer will contain both matrix and inclusions. The inclusions will still be in compression, due to the restraining matrix, but the matrix will be in tension, in order to balance the expansive inclusions and result in an overall average stress of zero, since the samples can freely expand. These random tension fields can result in the kind of random cracking seen experimentally. The causes of the delamination seen are more subtle, and are clearly related to the cement paste "skin" effect arising from the mold. It would be interesting to redo the cylinder experiments with cores, so there is no possible effect from the mold.

A numerical solution of the random inclusion model can be attempted by building up samples of the appropriate geometry out of cubic pixels, and then applying a 3-D finite element model [11]. A cube of material with a volume fraction of about 5% expansive material was generated, and then the appropriate sample was "cut out", exposing some of the expansive sites on the surface. Only these sites are allowed to grow, simulating a fairly early stage of sulfate attack where the sulfate ions have not penetrated all the way through the sample. Solving the elastic equations will then induce stresses mainly in the sample surface. The principal stresses for each pixel were calculated by diagonalizing the average stress matrix per pixel. Only preliminary results are available at present, for the case of a cylinder whose entire surface was open to sulfate penetration. In this case, random tensile stress fields were seen over the sample surface, and somewhat higher principal tensile stresses were seen at the cylinder ends, a result of the stress concentration from the cylinder edges. These preliminary results seem to confirm the ideas discussed above, although the role of creep in response to these localized stresses needs also to be considered.

4.   Conclusions

The following conclusions can be drawn:

- Environmental conditions such as pH of the solution and the sulfate concentration played an important role in the rate of expansion of the cement-based specimens
- The C_3A content of cements also influenced the rate of deterioration of the specimen
- Massive deposits of ettringite appeared in the core only at late stages of testing, after a large amount of expansion.

- The outer paste (near the mold surface) present in a molded specimen plays an important role in the deterioration due to sulfate attack
- The feasibility of finite element modeling of the expansion and cracking caused by sulfate attack was demonstrated

5.   Acknowledgments

This work was sponsored by the "High Performance Construction Materials" program at NIST, by the Nuclear Regulatory Commission and the National Science Foundation Center for Science and Technology of Advanced Cement-Based Materials (ACBM). We would like to thank M. Yang for producing the finite element results, John Winpigler and Frank Davis for performing some of the experimental measurements.

6.   References

1 .   Clifton, J.R., Knab, L.I., (1989), *Service Life of Concrete*, NIST report #89-4086, National Institute of Standards and Technology, Gaithersburg MD 20899

2 .   Pommersheim, J.M., Clifton, J.R., (1994), Expansion of Cementitious Materials Exposed to Sulfate Solutions, *Mat. Res. Soc. Symp. Proc.* Vol 333, pp 363-368.

3.   ASTM (1994), *Standard Specification for Portland Cement*, ASTM designation C150-94, Annual Book of ASTM Standards Vol. 04.02.

4.   ASTM (1994), *Standard Test Method for Compressive strength of Hydraulic Cement Mortars (using 2-in. or 50-mm cube Specimens*, ASTM designation C109-93, Annual Book of ASTM Standards Vol. 04.02.ASTM C109

5.   Brown, P.W., (1981), An Evaluation of the Sulfate Resistance of Cements in a controlled Environment. *Cement and Concrete Research*, Vol. 11 pp. 719-727

6.   ASTM (1994), *Standard Test Method for Length Change of Hydraulic-Cement Mortars Exposed to a Sulfate Solution*, ASTM designation C1012-89, Annual Book of ASTM Standards Vol. 04.02.

7   ASTM (1995), *Standard Specification for Wire Cloth and Sieves for Testing Purposes*, ASTM designation E 11-95, Annual Book of ASTM Standards Vol. 04.02

8.   Taylor, H.F.W., (1993) Sulfate Reactions in Concrete-Microstructural and Chemical Aspects, *Cement and Technology*, ed. by Gartner, E. M. and Uchikawa H., Ceramic Transactions vol. 40

9.   Cook, R.D., Malkus, D.S., Plesha M.E., (1989) *Concepts and Applications of Finite Element Analysis*, J, Wiley and Sons, New York 3rd ed

10.   Goltermann,P., (1994) Mechanical predictions on concrete deterioration. Part I: Eignestresses in concrete, *Amer. Conc. Inst. Mat. Journal* vol. 91, pp. 543-550

11   Garbozci, E.J., Day, A.R., (1995), An algorithm for computing the effective linear elastic properties of heterogeneous materials: Three-dimensional results for composites with equal phase Poisson ratios, *J. Mech. Phys. Sol.*, vol. 43, 1349-1362

23 ASSESSMENT OF THE CONDITIONS REQUIRED FOR THE THAUMASITE FORM OF SULPHATE ATTACK

N.J. CRAMMOND and M.A. HALLIWELL
Building Research Establishment, Watford, UK

Abstract

In the thaumasite form of sulphate attack, the main mechanism of deterioration is the breakdown of the calcium silicate hydrate phases in hardened cement pastes in the presence of sulphate and carbonate ions, followed by the formation of the mineral thaumasite. The following paper describes two large programmes of work in which this form of deterioration was reproduced in laboratory-prepared concretes and mortars containing two different sources of limestone. The effect of mix composition on the degree of reactivity is discussed along with the effect of environmental conditions such as temperature and different sulphate sources.

Keywords: Blended cements, concrete, limestone aggregates, limestone filler cements, Portland cements, sulphate attack, thaumasite

1 Introduction

In the last few years, the Building Research Establishment (BRE) has been involved with three case studies where concrete foundations have deteriorated as a result of the thaumasite form of sulphate attack [1] [2] [3]. The conditions favoured for thaumasite ($CaSiO_3.CaCO_3.CaSO_4.15H_2O$) formation are cold wet environments, a source of calcium silicate, and a readily available supply of carbonate and sulphate ions.

This form of sulphate attack is different from the classic form where ettringite ($3CaO.Al_2O_3.3CaSO_4.31H_2O$) is produced because it is the calcium silicate hydrates (CSH), in the hardened Portland cements which are targeted in the reaction and not the calcium aluminates. CSH is the main binding agent produced in all Portland cements including sulphate resisting Portland cements (SRPC). Thaumasite formation is therefore accompanied by a reduction in the binding ability of the cement in the

Mechanisms of Chemical Degradation of Cement-based Systems. Edited by K.L. Scrivener and J.F. Young. Published in 1997 by E & FN Spon, 2–6 Boundary Row, London SE1 8HN. ISBN: 0419215700.

hardened concrete resulting in loss of strength and transformation into a mushy incohesive mass as shown in Fig. 1.

As already mentioned, thaumasite formation requires sources of sulphate and carbonate ions and whereas the sulphate ions originated in the surrounding groundwaters of the three sites, close examination of the deteriorated concretes showed that the carbonate ions were derived from finely divided particles of limestone within the concretes themselves.

Although the field evidence is insufficient to prove that this form of sulphate attack constitutes a major durability problem, the rate of attack observed was extremely rapid and even occurred in foundations where good quality SRPC concretes were used. Such evidence therefore warranted further laboratory-based research and two programmes of work were initiated at BRE in order to investigate the 'finely divided limestone' and 'poor performance of SRPC' issues more closely.

At about the same time as this field evidence was discovered, a new British Standard [4] was introduced for Portland limestone cement, permitted to contain up to 20% of finely divided limestone of defined purity. The first programme of work therefore involved monitoring the sulphate resistance of a wide range of mortars made with such limestone filler cements. This work followed on from a much larger durability study carried out by a joint working party comprising BRE and the British Cement Association [5].

Limestone dust present in concretes can also be derived from incorporated aggregates. As the rate of thaumasite formation was thought to be greater the dustier the aggregate, it was believed that concretes made with good quality limestone aggregates may be immune to this form of deterioration. The second programme of work was initiated in order to develop the basis for guidance on the safe use of limestone aggregates by carrying out sulphate resistance tests on concretes containing commonly used limestone aggregates.

Fig. 1 The thaumasite form of sulphate attack can reduce hardened concrete to mush

Detailed accounts of both programmes of work including composition of materials and mixes, experimental conditions and discussion of results are presented in two separate BRE reports [6] [7]. The following paper contains abridged versions of both of these and highlights the conditions necessary for the thaumasite form of sulphate attack to occur.

2 Assessment of limestone filled cements

2.1 Experimental details
In the first laboratory experiment, five series of limestone-filled mortars were subjected to a range of different sulphate environments in order to monitor their performance. Each series included a parent Portland cement (of varying tricalcium aluminate (C_3A) contents :- 0%, 5%, 9%, 10% and 13-14%) control mix and up to four additional blended mixes containing the same Portland cement and percentages of limestone filler in the range 5-35%. A list of the compositions of the cements used and the mix formulations have previously been given [5] [6].

The test samples comprised small truncated mortar prisms (dimensions 20 x 20mm tapering over 13mm to 11 x 11mm) prepared from a 1:3 cement:quartz sand mix with a w/c ratio of 0.6. Freshly cast mortar specimens were cured for 24 hours at 20°C in moist air. After demoulding, they were stored in distilled water for 20 days at 20°C before being placed in sulphate solutions namely three $MgSO_4$ (0.14%, 0.42% and 1.8% by weight SO_4), one saturated with respect to $CaSO_4$, one Na_2SO_4 (1.8% SO_4), one mixture containing $CaSO_4$ (0.14% SO_4), $MgSO_4$ (0.07% SO_4) and Na_2SO_4 (0.05% SO_4) and an artificial seawater. The solutions were changed every two weeks. The tests were carried out at four temperatures, 5, 10, 15 and 20°C. The specimens were inspected visually at regular intervals up to a period of two years. X-ray diffraction (XRD), petrography and scanning electron microscopy were carried out on a limited number of deteriorated samples.

2.2 Results
The main results from this laboratory study are listed in Table 1. The comments refer to those samples stored under cold conditions as faster rates of deterioration were observed in samples stored at 5°, 10° and 15°C compared with 20°C. When deterioration occurred, it occured in all the different sulphate solutions to varying degrees. A fixed pattern did not emerge for the performance of mortars in one solution compared with other solutions except in the case of the three MgSO4 solutions which proved to be roughly the same. Therefore the comments in Table 1 only refer to the condition of samples stored in the magnesium sulphate solutions.

3 Assessment of sulphate resisting concretes containing commonly used limestone aggregates

3.1 Experimental details
An assessment of sulphate resistance was also made on 100mm concrete cubes made with commonly used limestone aggregates including Jurassic limestone gravel, crushed

Table 1.   Results of sulphate resistance tests on mortars made with limestone filler cement stored in $MgSO_4$ solution (0.42% as SO_4) at 5°C

Series ID*	% C_3A of parent cement	Limestone filler contents (%) of cements	Comments
G	13.0-14.0	0, 5, 20, 25	Rates of deterioration were rapid for all the mortars irrespective of limestone filler content. Reaction products were mostly ettringite and gypsum
H	10.2	0, 5, 20, 35	Positive correlation was found between the amount of limestone filler and the degree of deterioration. Deteriorated samples contained thaumasite, ettringite and gypsum
F	9.0	0, 5, 20, 25, 35	Positive correlation was found between the amount of limestone filler and the degree of deterioration. Gypsum was the main reaction product detected along with lesser amounts of thaumasite and ettringite
D	5.3	0, 5, 25	Rate of deterioration was much slower in this series. There was a positive correlation between limestone filler content and amount of attack. Substantial amounts of thaumasite were detected in mortars containing higher levels of limestone filler
S	0 (SRPC)	0, 5, 10, 35	Rate of deterioration was even slower in this series. Deterioration only occurred in mortars with higher limestone filler contents. Reaction products included thaumasite, ettringite and gypsum.

* Full details of series given in references [5] and [6]

Carboniferous limestone and crushed magnesian limestone in various combinations. Siliceous Thames Valley gravel was used as a control aggregate. All the mixes studied in this programme were designed to be sulphate resistant as recommended in BRE Digest 250/363 [8]. The cements used were sulphate resisting Portland cement (SRPC) containing almost zero C_3A, pulverised fuel ash (pfa) blended with ordinary Portland cement (OPC with 6.4% C_3A) at 25% and 40% levels of replacement, and ground granulated blastfurnace slag (ggbfs) blended with the same OPC at 70% and 90% levels of replacement. Three different cement contents were also used; $290kg/m^3$, $330kg/m^3$ and $370kg/m^3$ for the SRPC mixes and 300, 340 and 380 kg/m^3 for the blended cements.

The concrete mixes were designed to constant workability and full compaction and a fairly narrow range of workabilities was achieved (compacting factor: 0.93 to 0.98). Free and total water contents were adjusted for different aggregates to give constant workability and similarly water reductions were made for pfa mixes (in comparison with SRPC and slag mixes made with the same aggregates) to keep workability constant.

All the100mm concrete cubes were cured for 24 hours under damp hessian and polythene. After demoulding, the cubes were stored in water for 28 days at 20°C and then transferred to four different sulphate solutions at 5°C and 20°C. The solutions used were magnesium sulphate and sodium sulphate in strong (1.8% SO_4) and weak (0.42% SO_4) concentrations. The solutions were prepared using distilled water and those maintained at 20°C were changed every three months in order to be consistent with previous sulphate resistance concrete tests carried out at BRE. The solutions stored at 5°C were not replenished during the course of the experiment in an attempt to simulate the more 'static' conditions characteristic of a clay soil on site.

The conditions of the cubes were assessed at one and two years using a 'wear rating' measurement similar to the one employed by Harrison [9] and these data and photographs are presented elsewhere [7]. Wear rating values were used to categorise the sulphate resistance of cubes into good, satisfactory or poor. XRD and microscopic techniques were used to study products of deterioration. The current paper concentrates on a smaller-scale, purely visual assessment which was carried out after three years.

3.2 Results

The results from this study are encapsulated in Figs. 2-4. Overall, they have shown that after one year of exposure to some of the sulphate solutions at 5°C, the SRPC and OPC/pfa (25% and 40% replacement) concretes containing limestone aggregate started to deteriorate significantly and that thaumasite was the main reaction product formed. The deterioration took the form of gradual erosion of the faces, corners and edges of the cubes and became steadily worse after two and three years exposure. In contrast, all the control specimens containing siliceous aggregate showed good sulphate resistance up to the age of three years. The condition of a selection of these cubes after three years exposure to cold $MgSO_4$ solution (1.8% SO_4) are shown in Figs. 2 and 3.

Concretes made with Carboniferous limestone are not represented in these photographs. Like the cubes containing Jurassic limestone gravel, their performance was poor and both sets of cubes were similar in appearance. After one year exposure to strong $MgSO_4$ solution at 5°C, the performance of cubes containing the three different types of limestone was comparable but after three years, the performance of cubes containing magnesian (dolomitic - $CaMg.2CO_3$) limestone was significantly worse.

The effect of increasing the cement content of the SRPC concretes was to provide a better resistance to sulphate attack as shown in Fig. 2. A similar effect was noted in concretes containing OPC/pfa.

Temperature plays a very important role as all the SRPC and OPC/pfa concretes performed well at 20°C, independent of aggregate type. This is consistent with previous sulphate resistance cube test results carried out at BRE and elsewhere.

Fig. 2 Condition of SRPC cubes; Left-magnesian limestone, Centre-Jurassic limestone, Right-flint gravel

Fig. 3 Condition of OPC/pfa cubes; Left-Jurassic limestone, Right-flint gravel

Fig. 4 Condition of OPC/ggbfs cubes; Left-Jurassic limestone, Right-flint gravel

In general, most deterioration occured in the cubes stored in the two magnesium sulphate solutions even under 'static' conditions when the solutions were not replenished. Less attack was found in the strong sodium sulphate solution and none at all in the cubes stored in weak Na_2SO_4 solution.

Fig. 4 shows that the thaumasite form of sulphate attack had not occured in concretes containing OPC/ggbfs after three years exposure to cold $MgSO_4$ solution. Similar results were also observed in the cubes stored at 20°C and in cubes with higher cement contents.

4 Discussion on the conditions necessary for the thaumasite form of sulphate attack

The work described in this paper has shown that the thaumasite form of sulphate attack can be reproduced in laboratory concretes and mortars subjected to wet sulphate environments. This occurred in those concretes which contained a source of limestone either in the form of aggregate particles or as a filler added to the cement but did not occur in the limestone-free control specimens. This form of deterioration has been named the thaumasite form of sulphate attack although mineralogical analyses have revealed that thaumasite is not always the main reaction product formed. It is however a very distinct type of deterioration characterised by the breakdown of CSH in hardened cement pastes in the presence of sulphate and carbonate ions. The current study has identified some of the main conditions needed for the thaumasite form of sulphate attack as follows:

- *Temperature:* As expected from previous published papers, cold temperatures in the range 5° to 15°C are favoured. However, new information has revealed that the rate of attack drops off markedly somewhere between 15° and 20°C.
- *Sulphates:* The degree of sulphate attack in the limestone filled mortars did not depend on the concentration of sulphate ions in the three magnesium sulphate solutions. This was not the case in the concrete cube experiment in which the sulphate solutions were not replenished.
- *Limestone source:* Results from this study have shown that there was a positive correlation between the quantity of limestone dust present in a mortar and the degree of deterioration obtained, provided the C_3A content of the cement was 10% or less. The marked deterioration of concretes containing good quality limestone aggregates, where the amount of dusty material was minimal, was unexpected. This finding, however, dismisses the theory that only finely divided limestone dust can supply the carbonate ions required for the thaumasite form of sulphate attack. In the longer term, it appears that the use of aggregates containing dolomite, as opposed to calcium carbonate, may prove to be more detrimental.
- *Cement types:* All mortars or concretes made with Portland cement (OPC and SRPC) are susceptible to the thaumasite form of sulphate attack, with faster rates of deterioration coinciding with increased C_3A contents. Attack was also observed in concretes made with OPC/pfa cements with 25% and 40% replacement levels. Increasing the cement content had the effect of reducing the degree of deterioration but did not eliminate it. One of the most important observations from the study was that OPC/ggbfs (70% and 90% slag replacement levels) concretes have not

undergone the thaumasite form of sulphate attack after three years exposure to cold sulphate solutions.

5 Further work

The relative ease with which the thaumasite form of sulphate attack can be reproduced in the laboratory does not explain why the field evidence for its occurrence is so limited. Either the problem exists and has gone undetected or there are other conditions not yet discovered which are preventing or slowing up attack in the field. Further work is under way at BRE to study the effects of pH, curing and the type of cations associated with the sulphate solutions. BRE are also investigating, in greater detail, the potential benefits of using slag cements in concretes containing limestone and intend to carry out a field study on the performance of laboratory-prepared concretes in a sulphate-bearing clay site. BRE Digest 363 [8] incorporates a short paragraph warning that concretes buried in sulphate soils which contain. a finely divided source of calcium carbonate can be susceptible to the thaumasite form of sulphate attack. Publication of further advice has been planned in order to take into account the findings described in this conference paper.

6 References

1. Crammond N.J. and Nixon P.J. (1993) Deterioration of concrete foundation piles as a result of thaumasite formation. *6th Conference on the Durability of Building Materials, Japan*, vol 1, pp295-305.
2. Bickley J.A., Hemmings R.T., Hooton R.D. and Balinski J. (1994) Thaumasite related deterioration of concrete structures. *Proc. Concrete Technology: Past, Present and Future*, ACI SP: 144-8, pp159-175.
3. Crammond N.J. and Halliwell M.A. (1995) The thaumasite form of sulphate attack in concretes containing a source of carbonate ions. *2nd Symp. Advances in Conc. Tech*, ACI SP154-19, pp357-380.
4. British Standards Institution (1992) *Specification for Portland limestone cement.* BSI, London. BS 7583
5. Matthews J.D. (1994) Performance of limestone filler cement concrete. *Eurocements: Impact on concrete construction*, (ed R.K. Dhir and M.R. Jones), E & FN Spon, London, pp113-147.
6. Halliwell M.A., Crammond N.J. and Barker A.P. (1996) *The thaumasite form of sulphate attack in limestone-filled cement mortars,* Building Research Establishment Laboratory Report, Garston, Construction Research Communications Ltd (in press)
7. Crammond N.J. and Halliwell M.A. (1996) *The thaumasite form of sulphate attack in laboratory-prepared concretes,* Building Research Establishment Laboratory Report, Garston, Construction Research Communications Ltd (in press).
8. Building Research Establishment (1991) *Sulphate and acid resistance of concrete in the ground,* BRE Digest 363 (originally Digest 250).
9. Harrison W H. (1992) *Sulphate resistance of buried concrete. The third report on a long-term investigation at Northwick Park,* BRE Report 164.

24 DETERIORATION OF CONCRETE BRIDGE PIERS IN ICELAND

G. GUDMUNDSSON
The Building Research Institute, Reykjavik, Iceland

Abstract
Damaged piers in a bridge in western Iceland is the focus of this research. The deterioration was most severe in the lowest part of the tidal zone, decreased upward, and was non existent in the splash zone. The damage is best described as the cement paste is peeling off the surface, leaving the aggregates exposed to the air.

At the surface of the damaged samples the Ca content of the cement paste is depleted and the Mg content is enriched, relative to undamaged samples. In the damaged samples the air voids and cracks that lie parallel to the surface are filled with secondary ettringite. As the air voids are filled the resistance to freezing and thawing is lowered. It is concluded that freeze/thaw cycles are more significant than expansion due to secondary ettringite formation in deterioration process.

The Cl content of both damaged and sound concrete was relatively high, with a diffusion coefficient between $1 \cdot 10^{-11}$ and $1 \cdot 10^{-12}$ m^2/sec, but formation of Cl-rich phases in the cement paste reduce the risk of corrosion of the steel reinforcements.

Keywords: Concrete, sea water, secondary ettringite formation, freeze/thaw cycles

1 Introduction

The mild but unstable climatic conditions in Iceland impose some problems for concrete structures. In the Capital there are about 80 annual freeze/thaw cycles, which is relatively high, in Copenhagen (Denmark) for instance the annual cycles are about 60 [1]. In order to prevent frost damage the concrete must be designed to withstand this attack. Normally, the preventive measures are to use air entrained concrete and to lower the w/c ratio.

The most common damage to concrete structures is due to relative high humidity and many freeze/thaw cycles. Alkali-aggregate reactions (AAR) caused serious damages in Iceland in the sixties and the seventies. Preventative actions, taken in 1979, when mandatory use of silica fume and the ban of using unwashed sea-dredged material, have eliminated all AAR in concrete structures build after 1979.

AAR is a classical example of secondary reactions that take place within the concrete, reactions that involve expansions and the build up of pressure and the eventual break down of the concrete [2].

Another type of harmful reactions is secondary ettringite ($C_3A \cdot 3CaSO_4 \cdot 32H_2O$) formation. Primary ettringite is a common mineral in concrete, as such it does not impose any problems. Secondary ettringite is formed after the concrete is fully cured, usually due to changes in the activities of some of the environmental components. Secondary ettringite formation may involve expansion and may lead to the break down of the concrete structure. Transformation of monosulphate ($C_3A \cdot CaSO_4 \cdot 12H_2O$) to ettringite due to the ingress of external sulphate is well documented [3]. This reaction is

Mechanisms of Chemical Degradation of Cement-based Systems. Edited by K.L. Scrivener and J.F. Young. Published in 1997 by E & FN Spon, 2–6 Boundary Row, London SE1 8HN. ISBN: 0419215700.

usually referred to as delayed ettringite formation. Another example is the formation of ettringite from the break down of Al-rich paste and uptake of sulfur, for instance from atmospheric pollution [4]. This type of reaction is usually called sulphur attack.

The purpose of this paper is to describe and analyze the damage that has taken place in concrete piers of a bridge from a shore environment in Iceland.

2 The Bridge

Between 1976 and 1980 a bridge was build across the bay of Borgarfjörður, in western part of Iceland. The west side of the bay is open to the Atlantic Ocean and on the east side is the river Hvítá, a glacial river. The piers were constructed in the first two years, a total of 12. The deck was laid in the years 1978 to 1979 and the bridge was formally open in 1980. The total length of the bridge is 520 m. The piers were cast on site with a Danish rapid hardening Portland cement with low alkali content ($Na_2O_{eq.}$ = 0.86). The cement content was about 400 kg/m^3, with a w/c-ratio ranging from 0.5 to 0.6. The compressive strength (cylinder - 28 day) was between 37 and 51 MPa. The cover over the steel rebars was about 50 mm. Finally the entrained air content was between 4 and 6 %.

In Table 1 is shown the concentration of selected ions in the bay water, at high and low tides, for comparison an average composition of sea water is given [5]. At high tides the sea water is diluted by about 50 % and at low tides the dilution is about 85 %.

Table 1. Composition of the bay water, from [5].

	Average Sea Water	High Tides	Low Tides
pH		7.8	7.6
Ca^{2+}	400 ppm	206 ppm	53 ppm
Mg^{2+}	1272 ppm	593 ppm	157 ppm
SO_4^{2-}	2650 ppm	1310 ppm	353 ppm
Cl^-	18980 ppm	9250 ppm	2460 ppm
Dilution	*0*	*50 %*	*85 %*

On Figure 1 is shown a schematic drawing of one of the piers. The locations of cores taken in this study are also shown.

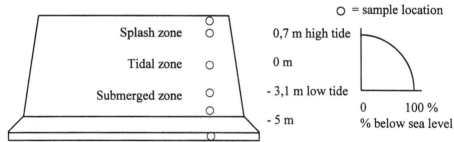

Figure 1. Schematic drawing of one pier and the relations to high and low tides. The sample locations are shown.

A few years after the bridge was completed, surface scaling was observed in the piers. The scaling was most prominent in the area that is exposed in the lowest part of the tidal zone, but gradually it is working its way up through the tidal zone. The current status is that about 30-40 mm have been eroded, where the scaling is most intense [5]. Leaving only about 10 to 20 mm to cover the rebars. The degree of scaling in the submerge zone is unknown.

3 Research procedure and analysis

Cores were taken from two of the piers (shown on Figure 1); from the top of the piers, where no damage has occurred; in the middle (little damaged); in the lowest part of the tidal zone (most damaged area) and at three depth locations in the submerged zone, shown on Figure 1. These samples were initially observed in an optical microscope and then selected samples from the tidal zone were investigated further with the aid of an electron microscope. The water soluble chloride content was measured in samples from the top and the bottom part of the tidal zone.

The electron microscopical investigation was carried out by a Cambridge 240 Scanning Electron Microscope equipped with backscatter and energy dispersive x-ray detector. The analysis were performed at an accelerating voltage of 15 kv, reference standard and matrix corrections were applied by the ZAF procedure. All analysis of hydration products were normalised to a water free basis.

The water soluble chloride was analysed with an Volhard titration.

4 Result of analysis

4.1 Optical microscopy
The samples from the top part of the piers show no sign of degradation. There is some carbonation and very little if any secondary ettringite formation. Primary Portlandite $(Ca(OH)_2)$ is unstable in the carbonated part, but stable below it and very common in the interfacial transition zone (ITZ), shown on Figure 2. There is no sign of secondary ettringite or monosulphate in these samples.

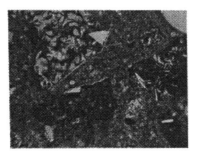

0,5 mm

Figure 2. Microphotograph of an undamaged sample. Portlandite precipitation is observed in the ITZ.

In the samples from the middle of the piers have developed some cracks. The cracks lie parallel to the surface. Secondary ettringite formation has taken place in the cracks and in voids. The cracks penetrate about 1 mm into the sample and in this layer Portlandite is not found. Aggregates are usually not penetrated by the cracks. Below this layer there is very little secondary ettringite formation, the sample is uncracked and air voids are empty. Portlandite is commonly found in the cement paste and the ITZ.

In the lowest part of the tidal zone there is a heavy crack formation, cracks that lie parallel to the surface are filled with secondary ettringite and Portlandite is not observed. All air voids are also filled with secondary ettringite, see Figure 3. This layer extends from the surface some 10 mm into the sample. Below it there is some ettringite formation in air voids, but the formation decreases further away from the surface. Portlandite is observed. Observation with the aid of fluorescence microscopy revel that the cement paste is more porous in the surface layer than in the bulk of the sample.

0,5 mm

Figure 3. Microphotograph of a damaged sample from the tidal zone. The crack at the bottom of the figure is parallel to the surface. Secondary ettringite fills air voids and cracks. In this sample about 40 mm have been scaled off.

In many ways the submerged zone is similar to the middle part of the tidal zone. The crack formations and the scaling seem to be similar although the actual extend of the surface scaling in the submerged zone is not known. The ettringite formation specially in air voids is more intense than in the sample from the tidal zone and extends further into the samples.

4.2 SEM results

Observation with SEM reveal the existence of a few micron thick layer of calcium carbonate at the surface of the undamaged samples. Just below it there is another thin layer, relatively Mg-rich. The CaO/SiO_2 ratio of the undamaged cement paste below these two thin surface layers is about 1,7 and with a Al_2O_3 content of 2,4 wt %. Although areas with much higher CaO/SiO_2 ratio of 4,4 and an Al_2O_3 about 15 wt % are found.

At the surface of the damaged samples leaching of calcium ions has taken place and the cement paste has taken magnesium ions up from the sea water. This exchange is shown in Figure 4 where MgO is plotted vs. CaO content of the cement paste. The

MgO content is as high as 32,5 wt %. In these samples the CaO contents is down to almost 3,5 wt % and the MgO/SiO_2 ratio is about 0,75. Magnesium was only found in the cement paste, other magnesium bearing phases like $Mg(OH)_2$ (brucite) were not found.

On Figure 5 is shown the SO_3/CaO ratio plotted vs. Al_2O_3/CaO ratio. Analysis of crystalline phases correspond to secondary ettringite, found in cracks and voids, it is also apparent from the figure that no crystalline monosulphate is present in the samples. The cement paste can have relatively high Al_2O_3/CaO ratio, but in most cases the ratio is less than one.

The chloride concentration of the paste is usually very low, average content about 3 wt %, but in the paste chloride rich phases have formed. The Ca and Al content of these phases resemble that of Fridel salt ($C_3A \cdot CaCl_2 \cdot 10H_2O$). These phases are generally very small and were hard to find with the microscope, but appear to be present in all the specimens.

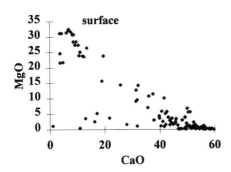

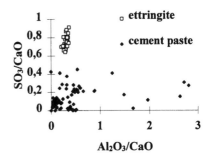

Figure 4. MgO vs. CaO content (wt %) of the cement paste in the damaged samples. At the surface the paste is enriched in MgO and depleted in CaO, going into the sample away from the surface the MgO content decreases and the CaO content increases.

Figure 5. The SO_3/CaO ratio against the Al_2O_3/CaO ratio. Crystalline ettringite is shown as open boxes and the cement paste is shown as filled diamonds.

4.3 Water soluble chloride content

The water soluble chloride content of the concrete was analysed as function of depth from the surface at about 10 mm interval. Samples from the top and the bottom of the tidal zone were analysed. Duplicate analyses were carried out. The results, expressed as percentages relative to cement, of the analysis is shown on Figure 6. The chloride content of the specimens is similar. It turned out to be impossible to measure the concentration in the surface layers, but extrapolation of the data suggest a value between 3 and 4 %. The chloride content decreases with depth from the surface, and at about a depth of 80 mm the content is less than 0,5 %. On Figure 6 is also shown the results calculated diffusions profiles for Cl (Flicks second law of diffusion), with the coefficient of diffusion being 10^{-11} and 10^{-12} m²/sec, respectively. In the calculations the time was restrained to be 18 years, the same as the age of the piers.

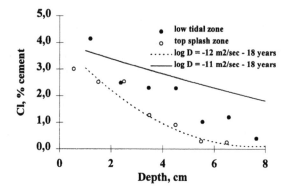

Figure 6.   Chloride depth profile of damaged and undamaged samples.   Two theoretical diffusion-based profiles are shown.

5 Discussions

The result of this research shows that the secondary ettringite formation is a result of uptake of sulphur from the sea water and the break down of Al-rich paste, similar to a model calculations of Reardon [4], in this case the sulphur is in solution. The interaction with sea water results in lowering the pH value of the pore water. The reduction in pH results in dissolution of Portlandite. In undamaged concrete Portlandite is commonly found on ITZ as well as the cement paste. The dissolution of Portlandite leaves open space in the cement paste, where the secondary ettringite formation takes place. When the space is full, pressure is build up and eventually the concrete can crack. The fact that the cracks are parallel to the surface must imply that its formation is controlled by diffusion of elements from the sea water into the cement paste. The aggregates do not take place in the reactions and are rarely penetrated by the cracks.

The fact that secondary ettringite has also filled all voids in the damaged concrete, suggest that the air voids do not serve as a protection against freeze/thaw damages. Alternate wetting and drying, specially in the lowest part of the tidal zone affects the concrete in such way that the piers are in a great danger of frost damages, due to the high number of annual freeze/thaw cycles in Iceland. Since the scaling does not seems to be as serious in the submerged zone as the lowest part of the tidal zone suggest that freeze/thaw actions are very important in this process and must be considered. Certainly, the secondary ettringite formation lowers the freeze/thaw resistance of the structures, allowing the freeze/thaw actions to take part in this process. It is not clear if the ettringite formation will result in much expansion.

The key element in the formation of secondary ettringite is aluminium. At high pH Al^{3+} hydrolysis to produce $Al(OH)_4^-$ therefore it can be carried in solution at relatively high pH. It complexes easily with sulphate. Progressive additions of sulphur dissolves more aluminium, the pH is lowered and Portlandite becomes unstable and it dissolves. Eventually the solution becomes saturated with respect to ettringite and it precipitates.

At the surface of the concrete leaching of calcium from the cement paste has occurred. Simultaneously the cement paste has taken up magnesium from sea water. In this process the CSH gels are more or less converted to MSH gels, other magnesium bearing phases were not found. The effect of this process on the durability of the structure is not known. It is apparent that the MSH gel is less dense than the CSH gel, which will in the long run cause problems for the structure like increased diffusion of elements like chlorides, etc.

In general, the fact that some of the chlorides are bound in solid phases must reduce the corrosion risk of the reinforcing steel. In this case the chloride content of the pore water is relatively high, the danger zone for chloride content is generally considered to lie between 0,4 and 1 % of cement [6]. This suggest that for most part, the situation is adequate, but where the scaling has penetrated about 30-40 mm in to the structure the situations is severe and must be attended to.

6 Conclusions

Due to secondary ettringite formations most of the air voids in the deteriorated part of the concrete are filled with ettringite. Cracks that penetrate the samples are also filled. The formation of ettringite is due to uptake of sulphur from sea water. In the process aluminium and Portlandite are dissolved, and subsequently ettringite is precipitated. This process has a net volume increase and can therefore possibly lead to expansion and crack formation. As the air voids are filled with ettringite the concrete will become more vulnerable to freeze/thaw actions which may also lead to cracking and scaling. Scaling does not seem to be as prominent in submerged zone, where no freezing and thawing cycles occur, as in the lowest part of the tidal zone. This suggest that the freeze/thaw actions are more responsible for the break down of the piers than expansion caused by secondary ettringite. The scaling is gradually working its way up the piers, at relatively slow rate.

The effect of the decalcification of the cement paste and the subsequent conversion to magnesium rich paste on the durability of the structure is not fully understood.

The Cl content of both damaged and sound concrete is relatively high. The diffusion coefficient between lies somewhere between $1 \cdot 10^{-11}$ and $1 \cdot 10^{-12}$ m^2/sec. The formation of Cl-rich phases in the cement paste have probably delayed the risk of corrosion of the steel reinforcements.

In order to avoid this problem from repeating itself in future bridges it is necessary to take some preventive measures, like using denser concrete, with relatively lower w/c-ratio and perhaps thicker cover over the steel rebars. Those are relatively simple actions that will certainly increase the lifetime of marine structures like this one.

7 Acknowledgements

Thank are due to Hákon Ólafsson, Einar Hafliðason, Rögnvaldur Gunnarsson and Pétur Ingvarsson for discussions and suggestions at various stages of the investigation. The author is also grateful to Halldór Guðmundsson for his aid with the SEM analysis and to dr. Karl Grönvold for carbon coating the samples. Appreciation is also extended to Jón Baldvinsson and Nikulás Magnússon. The author is also indebted to Dr. Karen L. Scrivener for constructive comments on the manuscript.

8 References

1. Gudmundsson, G. (1971) *Alkali-silica reactions in concrete.* The Icelandic Building Research Institute Report, Rb-12 (in Icelandic).
2. Stanton, T.E. (1940) *Expansion of concrete through reaction between cement and aggregate.* Proceedings of American Society of Civil Engineers, 66, pp. 1781-1811.
3. Scrivener, K.L., Taylor, H.F.W. (1993) *Delayed ettringite formation: a microstructural and microanalytical study.* Advances in Cement Research, Vol. 5, pp. 139-146.
4. Reardon, E.J. (1990) *An ion interaction model for the determination of chemical equilibria in cement/water systems.* Concrete and Cement Research, Vol. 20, pp. 175-192.
5. Haflidason, E. (1995) *Deterioration of piers in the bridge across the bay of Borgarfjördur, research and restoration.* The Icelandic Concrete Association, 20 pages.
6. Bamforth, P., (1994) *Admitting that chlorides are admitted.* Concrete, November/December, pp. 18-21.

PART FOUR

IMPLICATIONS OF CURING TEMPERATURES FOR DURABILITY

25 IMPLICATIONS OF CURING TEMPERATURES FOR DURABILITY OF CEMENT-BASED SYSTEMS

M. MORANVILLE-REGOURD
LMT, ENS Cachan, France

Abstract
Elevated temperature treatment used for high early strengths of cement based systems modify the microstructure of cement pastes i.e. composition and structure of hydrates, ionic concentrations in pore solution, pore size distribution, microcracking. These changes can favour the ingress of aggressive agents as chloride or the formation of expansive delayed ettringite or alkali-silica gels. New cement-based products have been optimized for high strength and durability. Adapted thermal treatments involve characteristic temperature level, duration and applied pressure.
Keywords : Temperature curing, microstructure, chemical durability, expansion, cracking.

I Introduction

Elevated temperature treatments are implicated for durability of cement based materials. During the accelerating process of cement hydration, they modify the nature of hydrates and microstructure of the matrix. Changes are function of the level of temperature and duration of the heat treatment. Related to the formation of delayed ettringite or alkali-silica gels, they are complex. Numerous parameters like the percentage of C_3A, SO_3, alkalies, MgO in cement and precracking of the matrix have been considered. New systems have been optimized for high strength and durability. This paper will report recent progress in this field.

Mechanisms of Chemical Degradation of Cement-based Systems. Edited by K.L. Scrivener and J.F. Young.
Published in 1997 by E & FN Spon, 2–6 Boundary Row, London SE1 8HN. ISBN: 0419215700.

2 Microstructure of heat treated cement-based systems

2.1 Cement hydrates
2.1.1 Silicates
Following Arrhenius' law, elevated temperatures accelerate the kinetics of silicate hydration. As an example, the hydration of β C_2S is complete in 1 day at 80°C, 3 days at 60°C, 14 days at 40°C, 27 days at 25°C [1]. After complete hydration, [29] Si NMR spectra show Q^1 and Q^2 silicate ions, for C_3S and C_2S. But, the relative intensity of the Q^2 peak increases from 25° to 200°C while that of Q^1 decreases in C-S-H samples (Fig.1) [2]. If Si-OH is considered in Q^1, Si-O-Si appears in Q^2 sites newly formed. For every Q^2 sites, two Q^1 sites disappear and one water molecule too. This is in agreement with the amount of bound water during the hydration of C_3S at temperatures between 50 and 75°C [3]. The C/S ratio of C-S-H remains unaltered up to 75°C but increases at higher temperatures. Already at room temperature, Al ions are in intermixed layers of AFm [4] coexisting with Mg-Al hydroxyde-type phases [5]. Al can also substitute for Si in tetrahedral coordination. These configurations are enhanced at high temperature.

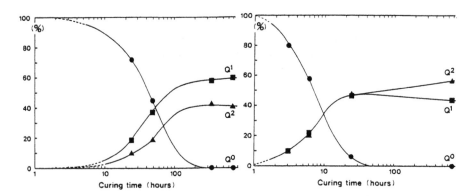

Fig.1 Variation of Q^0 , Q^1 , Q^2 given by [29] Si NMR in function of curing time at 40 and 80°C [2].

Autoclaved mixtures of 20% OPC, 15% CaO and 65% quartz treated at 180°C under saturated steam pressures an investigated by Analytical Electron Microscopy, show that chemical reactions follow the sequence dicalcium silicate hydrate, Ca-rich C-S-H, C-S-H, poorly crystalline 1.1 nm tobermorite, highly crystalline 1.1 nm tobermorite [6]. The reaction rate forming tobermorite depends on the specific surface area of quartz. Minor elements Mg, Fe, K and S are in the range of 0.00 to 0.03 based on Si. K and S could be balanced by substitution of Al for Si. The Al/(Al+Si) ratio is around 0.03-0.04. Highly crystalline tobermorite has a Ca/(Al+Si) ratio of O.8. Long autoclaving results in the decomposition of tobermorite into xonotlite.
2.1.2 Sulfoaluminates

AFt is not normally observed just after the heat treatment, in cement pastes, mortars and concretes, at temperatures higher than 70°C. The only sulfoaluminate is AFm.

In autoclaved systems, the aluminate phase is considered to be hydrogarnet containing Ca, Si and Al together with Fe, Mg, S, and K [6].

2.2 Microtexture

Elevated temperature treatments result in non uniformly distributed hydration products and coarse interconnected pores in OPC pastes [7]. For a same water/cement ratio, C-S-H crystallises in fibers longer than at room temperature [8] (Fig.2). AFm appears in thin plates, leaving spaces like Hadley grains. Crystals of $Ca(OH)_2$ are smaller, in relation with the decrease of CH solubility with increasing temperature [9]. The pore size distribution is refined in smaller pores, from 75 to 40 nm [3]. The decreased water demand during the polymerisation of silicate ions leads to an increased porosity of the cement paste at a given degree of hydration [7].

In autoclaved systems, tobermorite appears as platy crystals and xonotlite in fibers. The overall porosity remains the same during the sequence of silicate hydrates but the pore size is shifted to 50-100 nm when tobermorite is formed [6].

Fig. 2. ISO mortar 20 days after 4h at 80°C. x AFm in Hadley grain, o C-S-H..

2.3 Pore solution

The pore solution of cement pastes treated up to 190°C is higher in SO_4^{2-} ions and lower in OH^- ions than at 20°C [10]. These changes are more pronounced with high alkali contents or with SO_3/Al_2O_3 molar ratios of the cement.

In mixtures cement+quartz, heat cured in the same conditions, alkalis and sulfate ions are preferably bound in hydrates: 96.7% of SO_3 in hydrates within the 60% cement+40% quartz sample treated 4h at 90°C compared to 67.4% in absence of quartz [10].

2.4 Microcracking

Elevated temperature curing, high temperature drying, shrinkage can result in microcracking. The post-cracking behavior of concrete steam cured ties showed that microcracks that already deteriorated the material [11], appeared during the heat treatment [12]. The free energy analysis [13] showed that three days after 4h at 80°C, the structure was damaged in its thicker part [14]. The damage modified the elastic characteristics and more particularly the elastic modulus, and created mechanical stresses

3. Durability of heat treated cement-based systems

3.1 Delayed ettringite formation

All the primary AFt disappears in cement-based systems cured at temperatures higher than 70°C. Some SO_4^{2-} are in the pore solution. Others are dispersed in the C-S-H, together with aluminate species [15]. Just after the treatment, AFm crystals are observed in Hadley grains (Fig.2). At longer term, several months or years, AFt reappears in the matrix and at the cement paste-aggregate interface. This secondary ettringite crystallises in Hadley grains (Fig.3) or in cracks. In the pore solution, SO_4^{2-} ions decrease and OH^- ions increase, in a reverse way as during the heat treatment [10]

Fig. 3. ISO mortar 6 months after 4h at 80°C. Recrystallisation of AFt in a Hadley grain.

The late expansion and cracking of heat cured concretes have been attributed to the delayed ettringite formation. Numerous studies have considered different parameters as temperature, SO_3 content in the cement, water/cement ratio[16], Na_2O and MgO in cement [17], length of curing [18], humidity of the environment after curing [19].

The mechanism of expansion has been largely discussed . It seems now admitted that the ettringite observed at aggregate interfaces recrystallises in open spaces created by the paste expansion [15,17,20]. First ettringite appears as very small crystals in C-S-H and then grows in cavities with no disruption of the surrounding system [15]. The paste expansion has been related to the existence of pre-cracks due to the thermal effects like shrinkage during drying [21]. Ettringite nucleation occurs in the crack tip-zone and precipitated nucleus on the crack wall lowers the free energy. The crystal pressure of further ettringite crystals will increase the stress intensity at the crack tip. The crack will extend and new nucleation sites will be created. Continuous crack growth is possible if the supersaturation of the pore solution is maintained [22]. This is confirmed by the fact that at the end of the expansion process, the sulfoaluminate is AFm, intermixed with C-S-H [23].

3.2 Alkali-silica reaction
Steam cured concretes have also exhibited alkali-silica gels. The expansion was attributed to both alkali-silica gels and ettringite (Fig.4) but in many cases ASR appeared as the first cause of expansion [24-27]. A probabilistic model of alkali-silica reaction applied to steam cured concrete ties showed that a temperature of 70°C induced the onset of expansion ten times faster than at 20°C [13]. Moreover, the chemical reaction is favoured by the alkalinity of the pore solution after the heat treatment [10].

Fig. 4. Concrete 100 d at 38°C in an AAR reactor after 4h at 80°. Coexistence of ASR gel (+) and secondary ettringite (o).

4. Optimisation of heat treated cement-based systems

Studies on the durability of heat treated cement-based systems have highlighted the role of C_3A, SO_3, alkalis in cement, water/cement ratio and level of temperature. In order to avoid the delayed ettringite and alkali-silica gel formation and their resulting expansion, and to keep a high strength at long term, it is necessary to optimise the formulation and treatment of these materials. Reactive powder concretes are an example [28].

4.1 Reactive powder concrete
Reactive powder concretes [28] contain OPC with 3.8% C_3A, silica fume and sand (average grain size .250 μm), crushed quartz (average grain size 10 μm). Their basic formulation includes:

OPC: 1, Silica Fume:O.25, Water/ (Cement + Silica Fume): 0.12.
When quartz is used, it is at 40% by weight of cement.

Some powder concretes were treated at 90, 200, 250 and 400°C. Some were moreover pressed under 625 atm. Thermal treatments enhanced the amount of bound water and pozzolanic reaction. Silicate hydrates were in function of temperature C-S-H, C_2SH, tobermorite and xonotlite. The total porosity as measured by mercury intrusion did not exceed 9%. The porosity threshold is the smallest for samples treated between 90 and 250°C (Fig.5). No ettringite was detected after the heat treatment. Compressive strengths were 230 MPa for RPC treated at 90°C and 680 MPa for samples moreover pressed under 500 atm and containing quartz

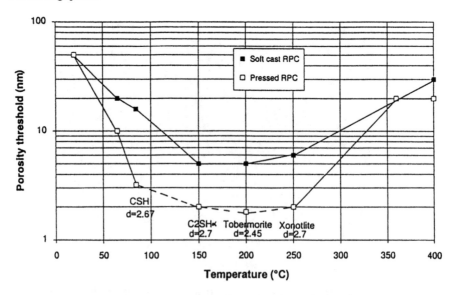

Fig. 5. Porosity threshold versus heat treatment temperature [29]

4.2 Durability of reactive powder concrete

Two industrial RPC 200 (200 MPa) treated at 90°C during 3 days after a 7-day water curing at 20°C,were tested in the laboratory [30]. The durability factor is equal to 100 for different mixes performing in freezing and thawing cycles. The chloride ion permeability measured by the ASHTO test (10 coulombs) translates a quasi impermeability of the material.

5. Conclusion

Elevated temperatures modify the microstructure of cement-based systems. The stability of cement hydrates, silicates and sulfoaluminates, results in a sequence of minerals of different porosity, bound water and ionic pore solution.

Delayed ettringite responsible for late expansions occurs first in pre-existing microcracks, in cavities like Hadley grains or spaces at the aggregate-matrix interface. The precracking of the paste is due to the thermal treatment involving drying shrinkage and mechanical stresses. When delayed ettringite coexists with alkali-silica gel, it appears that alkali-silica reaction accelerated by high temperature also induces cracking and favours the growth of secondary expansive ettringite

Characteristic parameters like the alumina and sulfate content in cement, water/cement ratio, alkalinity of the pore solution have to be taken into account for high strength and durability. New systems have been optimised in that way. Powder Reactive Concretes, containing OPC, silica fume, fine sand, crushed quartz, treated at high temperature (90-400°C) and pressed under 600 atm., are an example.

6. Acknowledgements

The author would like to thank Dr Sasaki and Prof. Cheyrezy for providing Figures 1 and 5 respectively.

7. References

1. Ishida, H., Sasaki, K., Okada, Y. and Mitsuda, T. (1992) *Journal of the American Ceramic Society*, Vol. 75, No. 9, pp. 2541-46
2. Cong, X. and Kirpatrick, R.J. (1995) *Cement and Concrete Research*, Vol. 25, pp.. 1237-46.
3. Odler, I., Abdul-Maula, S. and Lu Zhongya .(1987) *Materials Research Society*, Vol. 85, pp. 139-44.
4. Taylor, H.F.W. (1993) *Advances in Cement Based Materials*, Vol. 1, pp. 38-46.
5. Richardson, I.G. and Groves, G.W. (1993) *Cement and Concrete Research* Vol. 23, No.1, pp. 131-8.
6. Mitsuda, T., Sasaki, K. and Ishida, H. (1992) *Journal of the American Society*, Vol. 75,No. 7, pp. 1858-63.
7. Cao, Y. and Detwiler, R.J. (1995) *Cement and Conrete Research*, Vol. 25, No.

3, pp. 627-38.

8.. Regourd, M. and Gautier, E. (1980) *Annales ITBTP, Série Bétons,* Vol. 387, No. 198, pp. 83-96.

9. Patel, H.H., Bland, C.H. and Poole, A.B.   (1995) *Cement and Concrete Research,* Vol. 25, No. 3, pp. 485-90.

10. Wieker, W., Bade,Th., Winkler, A. and Herr, R (1992) in *Hydration of Cements,* E & F.N. Spon, London, pp. 125-36.

11. Tepponen, P. and Erickson, E.B. (1987) *Nordic Concrete Research,* No. 6, pp. 199-209.

12. Sylla, H.M. (1988) *Beton,* Vol. 38, Part. II, pp. 44-54

13. Bournazel, J.P. and Moranville-Regourd, M. (1994) *Concrete Technology, P Past, Present and Future,* ACI SP-144, pp. 233-250

14. Bournazel, J.P. and Moranville-Regourd, M.. (1995) *Materials Research Society,* Vol. 370, pp. 57-66

15. Scrivener, K.L. and Taylor, H.F.W. (1993) *Advances in Cement Research,* Vol. 5, No. 20, pp. 139-46.

16. Heinz, D. and Ludwig U. (1987) *American Concrete Institute,* SP 200, Vol. 2, pp. 2059-69

17. Lawrence, C.D. (1995) *Cement and Concrete Research,* Vol. 25, No. 4, pp. 903-14

18. Crumbie, A.K., Pratt, P.L. and Taylor, H.F.W. (1992) *9th International Congress on the Chemistry of Cement,* Vol. IV, p. 131

19. Odler, I. and Chen, Y. (1995) *Cement and Concrete Research,* Vol. 25, No. 4, pp. 853-62

20. Johansen, V., Thaulow, N. and Idorn, G.M. (1994) *Zement-Kalk-Gips,* Vol. 3, p. 150.

21. Fu, Y. Xie, P. and Beaudoin, J.J. (1994) *Cement and Concrete Research,* Vol. 24, No. 6, pp. 1015-24.

22. Xie, P. and Beaudoin, JJ. (1992) .*Cement and Concrete Research,* Vol. 22, No. 4, pp. 597-604.

23. Lewis, M.C. Scrivener, K.L. and Kelham, S. (1995) *Materials Research Society* Vol. 370, pp. 67-76.

24. Shayan, A. and Wick, G.W. (1992) *American Concrete Institute Materials Journal,* Vol; 89, No. 4, pp. 339-43.

25. Taylor, H.F.W. (1994) *Advances in Cement and Concret,* ASCE, pp. 122-31.

26. Diamond, S. and Ong, S. (1993) *Ceramics Transactions, Cement Technology* Vol. 40, pp. 79-90.

27. Johansen, V., Thaulow, N. and Skalny, J.   (1993) *Advances in Cement Research,* Vol. 5, No. 17, pp. 23-9.

28. Richard, P. and Cheyrezy, M. (1995) *Cement and Concrete Research,* Vol. 25, No. 7, pp. 1501-11.

29. Cheyrezy, M., Maret, V. and Frouin,L. (1995) *Cement and Concrete Research,* Vol. 25, No. 7, pp. 1491-1500.

30. Bonneau, O., Dugat, J. and Aitcin, P.C. (1995) *American Concrete Insttitute, Materials Journal,* to appear.

26 WHAT CAUSES DELAYED ETTRINGITE FORMATION?

N. THAULOW, V. JOHANSEN and U.H. JAKOBSEN
G.M. Idorn Consult A/S, Bredevej, Denmark

Abstract
Hardened concrete that has been subjected to high-temperature heat curing may suffer from expansion and cracking during subsequent exposure to moisture. The deterioration begins after months or years and continues for several years. This type of concrete deterioration is known as "Delayed Ettringite Formation" DEF.

Three groups of parameters influence the potential for DEF, namely parameters relating to:

- concrete composition
- curing conditions
- exposure conditions

These parameters are discussed based on research and field experience. Furthermore some thoughts on the reaction mechanism that leads to expansion of the cement paste and cracking of the concrete are presented.
Keywords: Concrete composition, curing conditions, DEF, exposure conditions, gaps around aggregates, paste expansion.

1 Introduction

This paper discusses the mechanism of Delayed Ettringite Formation (DEF).

Hardened concrete that has been subjected to high-temperature heat curing may suffer from expansion and cracking during subsequent exposure to moisture. The deterioration begins after months or years and may continue for several years. This type of concrete deterioration is known as "Delayed Ettringite Formation" (DEF).

Mechanisms of Chemical Degradation of Cement-based Systems. Edited by K.L. Scrivener and J.F. Young.
Published in 1997 by E & FN Spon, 2–6 Boundary Row, London SE1 8HN. ISBN: 0419215700.

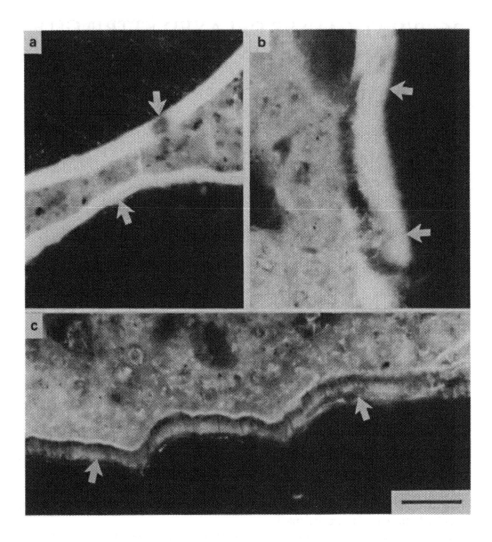

Fig. 1. Micrographs showing the microstructure of DEF. (a) Empty gaps around aggregates, (b) Needle shaped ettringite in gap around aggregate, and (c) Massive ettringite in gap around aggregate. The images are taken in fluorescent light. Scale bar measures 0.1 mm.

A characteristic feature of concrete suffering from DEF is the development of gaps completely surrounding aggregate particles. The gaps may be empty (Figure 1a) or more or less filled with needle shaped ettringite crystals (Figure 1b). Sometimes massive bands of densely compacted ettringite are found in the gaps (Figure 1c).

2 Delayed Ettringite Formation, DEF

Secondary (recrystallized) ettringite is commonly found in voids and cracks in concrete exposed to moisture and is not a diagnostic feature of DEF. DEF is diagnosed based on the typical crack pattern which indicates expansion of the cement paste causing it to break the bonds to the aggregate and pull away from the aggregate surfaces leaving a gap [1][2][3]. The width of the gaps created is roughly proportional to the diameter of the aggregate [1]. Another school of thought is that the gaps are created by growth of ettringite crystals [4][5]. This theory, however, fails to explain the crack pattern and the correlation between aggregate size and gap width [1].

The fundamental reaction mechanism is still debated amongst researchers [6]. Generally DEF is seen as a form of internal sulphate attack. The lack of a clear definition of DEF has caused some confusion in the literature. Some researchers claim that DEF may occur in concrete never exposed to high temperature [7]. However, no documented examples of this exists according to our knowledge. Other researchers, as e.g. Lawrence [6], include high temperature in their definition. We propose that Lawrence's [6] definition is used for DEF and that other sulphate related expansion phenomena are classified according to Figure 2.

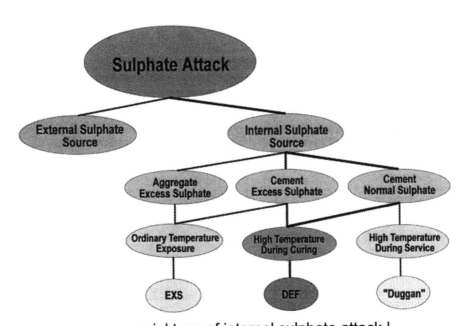

- a special type of internal sulphate attack !

Fig. 2. Sketch showing the various types of sulphate attack occurring in concrete. Sulphate attack with an external sulphate source is not treated in this paper. "EXS" = EXcess Sulphate

In this classification the primary distinction is between internal and external sulphate sources. With an external sulphate source we are dealing with ordinary sulphate attack which may be further classified according to the type of sulphate. DEF on the other hand is a type of sulphate attack with an interior sulphate source, <u>and</u> where the fresh concrete is cured at high temperatures. Another type of interior sulphate attack may occur when <u>hardened</u> concrete is cyclic treated at high and low temperatures. This scenario is known from the Duggan test [8].

Interior, low temperature sulphate attack may occur if an excessive amount of gypsum is present in the aggregate or in the cement. It is proposed that this type of reaction is <u>not</u> designated delayed ettringite formation.

The common factor in all types of sulphate attack is a high concentration of sulphate (SO_4^{2-}) in the pore liquid of the hardened concrete.

A number of parameters influence the potential for DEF. The parameters can be divided into three main groups as shown in Figure 3. The three main groups cover parameters relating to :

- Concrete Composition
- Curing Conditions
- Exposure Conditions

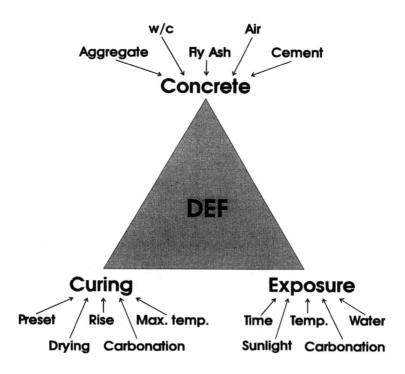

Fig. 3. Factors influencing Delayed Ettringite Formation, DEF

3 DEF Parameters

3.1 Concrete Composition
Parameters related to concrete composition which may influence DEF include, but are not limited to,:

- Aggregate type and grading
- Cement content and cement chemistry
- Fly ash or filler type and content
- Water to cement ratio (w/c)
- Air content and porosity of the concrete.

The mineralogy of the *aggregate* is important as some minerals like limestone form an improved interfacial bond with the cement paste. This may prevent breakage of the bond and limit expansion and cracking.

The *chemistry of the cement* is the parameter that has attracted most of the research over the years. Lawrence has discussed the literature and published his own data [6][9] showing that % MgO and % SO_3 correlate with expansion. From the literature it appears that most cements produced in the world have potential for DEF if the concrete is cured at high temperature as they contain more than 2.5% SO_3 or more than 1.0% MgO, which seem to be the lower threshold values. Furthermore analyses of Lawrences data [6] using an expert system seem to indicate that very low sodium oxide (Na_2O) and low iron oxide (Fe_2O_3) reduces or eliminates DEF.

Fly ash or filler may mitigate or even prevent DEF. Fly ash is known to mitigate external sulphate attack. This is an area for further research.

Low *water to cement ratio* is generally better than higher w/c as the internal ion transport is slowed down. Low absorption has a similar effect.

3.2 Curing
European experience with steam-cured concrete railroad ties over the last 25 years has proved that in most cases curing parameters are the key to DEF [10]. Recently, the authors have found this to be true for US railroad ties as well.

The curing conditions which influence DEF are:

- Preset time
- Temperature rise
- Maximum temperature
- Drying
- Carbonation

If the fresh concrete is exposed to heat before it sets, it shows lower late strength and a greater risk of DEF. Sufficiently long *preset time* allows ettringite to form in the cement paste in a non-harmful way. Recommended preset time before heating is 2-4 hours. The rate of *temperature rise* may also be critical as a too rapid temperature rise may cause microcracking of the concrete due to differences in thermal coefficient of expansion of air, water, cement and aggregate, thus, weakening the bond between cement paste and aggregate. Recommended rate of heating is approximately 20°C/hour (~40°F/hour).

The *maximum temperature* of concrete during curing is the single most important parameter controlling DEF. In this respect it is the temperature in the interior of the concrete that counts and not the ambient temperature. Concrete generates a lot of heat in the interior due to the chemical reaction between cement and water (hydration). The total heat generation is in the order of 320 kJ/kg cement, depending on the cement type. Often the peak temperature of concrete is substantially higher than the ambient temperature.

In laboratory tests at 100% relative humidity curing temperatures higher than about 60°C (140°F) may cause DEF. Under production conditions where temperature gradients in the concrete may cause stress, the critical temperature level may be lower. Furthermore, if premature drying of concrete occurs due to the application of dry heat the critical temperature level is expected to be lower because drying slows down the hydration of cement resulting in more cement minerals and sulphate available for later reaction during the service life of the concrete.

Carbonation during curing due to the presence of carbon dioxide from fuel combustion may exacerbate the situation because ettringite is destroyed by carbonation and sulphate is liberated for later reaction.

3.3   Exposure Conditions
Potential DEF in concrete introduced by the concrete composition and the curing conditions cannot develop unless the concrete is exposed to moisture for extended periods of time during service.

The following exposure conditions control DEF:

- Water
- Time
- Temperature
- Sunlight
- Carbonation

Liquid *water* is needed inside the concrete to allow transport of ions to the reaction sites. Furthermore, water participates in the reaction. About half the volume of solid ettringite consists of water fixed in the crystal lattice. A relative humidity higher than 80-90% is needed to fuel the reaction.

If concrete is wet on one side and dry on the other side, the moisture gradient may cause accumulation of alkali sulphates where water evaporates and aggravates deterioration.

The dimensions of concrete are also important in controlling the rate of expansion because the ingress of water may be diffusion controlled. This means that a 35 mm thick concrete slab may expand much earlier and faster than a 350 mm thick railroad tie.

Time is an important factor as DEF is known [5] to follow an S-shaped curve with a dormant period followed by a period of accelerating rate of expansion, and finally a period of decelerating rate of expansion. In principle this means that concrete which seems sound today may expand and crack tomorrow. Normally expansion due to DEF in field concrete occurs after 2 to 6 years.

The rate of cracking and expansion is to a large extent controlled by the average ambient *temperature*. DEF is a chemical reaction which is accelerated by temperature. A $10°$ C increase in temperature ($18°$ F) may double the rate of reaction. The exact value of the activation energy for the reaction is however, not known to the authors. A warm climate is worse than a cold climate (not considering freeze/thaw) and a wet climate is worse than a dry climate.

Concrete surfaces exposed to *sunlight* may reach temperatures at more than $30°$ C ($\sim54°$ F) higher than ambient temperatures. It is known from laboratory studies [8] that hardened concrete subjected to high temperature excursions may suffer from internal sulphate attack ("Duggan test"). This applies to non-heat cured as well as to heat cured concrete. Three cycles from $21°$ C to $82°$ C ($\sim180°$ F) is sufficient to cause expansion due to internal sulphate attack.

However, it has not been investigated whether a higher number of heating cycles having a lower maximum temperature also causes internal sulphate attack. It is conceivable that temperature cycles induced by very intense sun radiation in a hot climate may exacerbate DEF. Field observation on concrete elements show that elements exposed to sun on the south side of a building cracked and expanded while elements in the shade on the north side of the same building were undamaged [11].

Carbonation of the concrete causes already formed ettringite to break down to form insoluble calcium carboaluminate and soluble sulphate [12]. The sulphate can migrate to non-carbonated cement paste where new reaction takes place followed by expansion and cracking.

4 Other deterioration mechanisms parallel to DEF

Other deterioration mechanisms may run parallel to DEF and to a certain extent interact with DEF. This includes alkali silica reaction, ASR, if reactive aggregate is present and the alkali concentration of the pore liquid is sufficiently high. Initial confusion between the durability problem arising from DEF and those from ASR may partly have been the result of frequent simultaneous appearances of these expansive reactions in field concrete. However, a trained petrographer can easily distinguish between the two reaction types based on differences in the crack pattern.

Freeze/thaw may cause further damage to concrete suffering from DEF (or ASR).

5 Conclusions

The following can be concluded about DEF:

- In order to avoid further confusion about DEF it is hereby suggested that the term DEF is only used when concrete has been subjected to improper heat curing at high temperatures causing later cracking and expansion during moist exposure.
- DEF is a type of internal sulphate attack.

- DEF is influenced by concrete composition, curing conditions and exposure conditions
- DEF is recognised by the appearance of gaps completely surrounding aggregates.
- Ettringite may be present in the gaps. However, the occurrence of ettringite in itself in the gaps or in any other types of cracks and voids is not a diagnostic feature of DEF.

6 Acknowledgements

Claus Pade is thanked for critically reviewing the manuscript.

7 References

1. Johansen, V., Thaulow, N., Jakobsen, U.H. and Palbøl, L. (1993) Heat Cured Induced Expansion, presented at the *3rd Beijing International Symposium on Cement & Concrete*
2. Johansen, V., Thaulow, N. and Skalny, J. (1993) Simultaneous Presence of Alkali-Silica Gel and Ettringite in Concrete, *Advances in Cement Research.* vol 5, No. 23. pp. 23-29
3. Johansen, V., Thaulow, N. and Idorn, G.M. (1994) Dehnungsreaktionen in Mörtel und Beton, *Zement-Kalk-Gips*, vol 3, pp 150 - 154.
4. Sylla, H.M. (1988) "The reactions in cements paste by heat curing (in German)" *Beton*, vol 38, No. 12. pp. 488-493.
5. Heinz, D. (1986) Schädigende Bildung ettringitähnlicher Phasen in wärmebehandelten Mörteln und Betonen, *Thesis,* RWTH Aachen.
6. Lawrence, C.D. (1995) Delayed Ettringite Formation: An Issue ? in *Material Science of Concrete IV*, (eds. J. Skalny & S. Mindess), American Ceramic Society, pp. 113-154.
7. Marusin, S (1995) Deterioration of Railroad Ties in the USA, *CANMET/ACI International Workshop on Alkali-Aggregate Reactions in Concrete*, Nova Scotia. pp 243-256.
8. Jones, T.N., Grabowski, E., Quinn, T., Gillott, J.E, Duggan, C.R. and Scott, J.F. (1990) Mechanism of Expansion in Duggan Test for Alkali Aggregate Reaction: Canadian Developments in Testing Concrete Aggregates for Alkali Aggregate Reactivity. *Engineering Materials Report 92*, Ministry of Transportation, Ontario, pp. 70-82
9. Lawrence, C.D. (1995) Mortar Expansion due to Delayed Ettringite Formation. Effects of Curing Period and Temperature, *Cement & Concrete Research*, vol 25, No. 4. pp. 903-914.
10. Tepponen, P. and Eriksson B.-E. (1987) Heat Treatment as the Cause of Concrete Failure, *Betonintuote*, vol 2.
11. Thaulow, N. (1990) Unpublished data
12. Kuzel, H.-J. (1990) Reaction of CO_2 with Calcium Aluminate Hydrates in Heat Treated Concrete, *Proceeding 12 International Conference on Cement Microscopy*, Vancouver, Canada, pp. 218-227

27 INFLUENCE OF ELEVATED AGGREGATE TEMPERATURE ON CONCRETE STRENGTH

M. MOURET, A. BASCOUL and G. ESCADEILLAS
Laboratory of Materials and Durability of Constructions,
Toulouse, France

Abstract
In summer, cement-workers and ready-mixed concrete manufacturers may be faced with drops in compressive strength observed on specimens for manufacturing control tested in laboratory at 28 days.

An experimental programme was carried out in order to evaluate the effect of the aggregate temperature on the performances of plain concrete.

This paper presents the results of tests for compressive and splitting tensile strengths conducted on normal strength concrete specimens. Cylinders (11cm x 22cm) were prepared with different aggregate temperatures ranging from 20°C to 70°C and cured under either controlled laboratory conditions (20°C) or simulated conditions of hot weather (35°C). In addition, the temperature evolution within specimens was monitored for a 24-hour period from the time of casting.

Results show that both the 28-day compressive strength and the splitting tensile strength of concrete are reduced with the increase in aggregate temperature by as much as 15% and 17% respectively. The aggregate temperature rise also implies an increase in the water demand which, in turn, cannot fully explain the drops in strength.
Keywords: aggregate temperature, curing and conservation modes, specimens temperature, compressive strength, increased water demand.

1 Introduction

Drops in compressive strength may occur for concrete produced in summer. They are observed at random on specimens for manufacturing control tested in laboratory at 28 days in order to verify the contractual compressive strength which is the most significant parameter for evaluating the general quality of concrete with regard to mechanical tests in general.

Mechanisms of Chemical Degradation of Cement-based Systems. Edited by K.L. Scrivener and J.F. Young.
Published in 1997 by E & FN Spon, 2–6 Boundary Row, London SE1 8HN. ISBN: 0419215700.

Here, the influence of elevated aggregate temperature is the basic assumption for drops in strength. Indeed, previous investigations [1] have shown a microstructure change in the paste-aggregate interface for both summer concrete specimens and aggregate-paste models. Particularly, for these latters, the higher the temperatures of cement or/and aggregate were, the greater the concentration of calcium hydroxide was. This observation leads us to assume that the transition zone might be weakened by chemical phenomenons due to the rise of the constituent temperature.

2 Experimental programme

Two series of normal strength concrete were performed.

- *Series A* (SA): 28-day nominal compressive strength of 35 MPa.
- *Series B* (SB): 28-day nominal compressive strength of 22 MPa.

For each series, three aggregate temperatures were chosen: 20°C, 35°C and 70°C. The cement temperature was kept constant at 70°C because cement is often delivered on sites at this temperature in summer.

The water content was adjusted in order to obtain the same workability:

- 80 mm slump for series A.
- 90 mm slump for series B.

2.1 Materials
Following materials were used for concrete:

- *cement*: OPC (CEM I 42.5) - Blended Portland cement with 20-25% of calcareous filler (CEM II/B 32.5 R). Both cements satisfied the requirements of revised NFP 15-301 (1994).
- *Fine aggregate*: siliceous sand 0/6 mm, specific gravity (SG) = 2.68, fineness modulus = 3.06, absorption (ABS) = 2.8%.
- *coarse aggregate*: siliceous fine gravel 6/10 mm, SG = 2.69, ABS = 0.70%. Siliceous gravel 10/20 mm, SG = 2.66, ABS = 0.70%.
  Aggregates were selected according to the specifications of NFP 18-541 (1994).
- *Admixture*: plasticiser based on naphtalene (series B only).

The denomination of all the mixes is presented in table 1. Mix proportions are shown in table 2.

Table 1. Mix denomination

Cement		Aggregate					
	Temperature (°C)	20°C		35°C		70°C	
		SA	SB	SA	SB	SA	SB
CEM I	70°C	A1	B1	A2	B2	A3	B3
CEM II/B	70°C	A4	B4	A5	B5	A6	B6

Table 2. Concrete mix data (quantities per m ^{3} of concrete)

constituent	CEM I 42.5		CEM II/B 32.5 R	
	SA	SB	SA	SB
0/6 mm	730	810	690	780
6/10 mm	300	340	300	340
10/20 mm	820	820	820	820
cement	350	240	400	270
water [*]	180	170	180	170
plasticizer (%) [**]	/	0.3	/	0.3

2.2 Curing procedures and conservation modes

As the case may be, following curing and conservation procedures were used:

- *Series A.* All the specimens were sealed and cured in air (20°C-60% RH) for a 24-hour period. Next, half of the specimens was immersed in a water tank under the temperature of 20°C ; the other half was placed in air (20°C-60% RH).
- *Series B.* Half of the sealed specimens was cured at 20°C, 60% RH for a 24-hour period (cure 1). The other half was kept in simulated conditions of hot weather: 35°C during 5 hours after casting, next progressive decrease in temperature down to 25°C till 24 hours after mixing (cure 2). Then, specimens cured at 20°C were kept immersed in water at 20°C. Specimens cured at 35°C for a 5-hour period were kept in air (20°C-60% RH).

3. Main results

3.1 Specimen temperature

Temperature-time curves derived from series B made with CEM I 42.5 are shown in figure 1. They are representative of the two curing procedures performed in this programme and whatever the cement was employed, the following discussion applies to series A as well as to series B.

3.1.1 Controlled laboratory conditions (fig. 1 a)
Within the 4 hours after casting (roughly the dormant period for hydration), the temperatures are dominated by the thermal exchanges with the ambience (20°C).
    Whatever the aggregate temperature, a change exists at about 4 hours after casting:

- *B1*: a more important growth rate of the temperature,
- *B2*: a subsequent increase in the temperature,
- *B3*: a stabilization of the temperature.

The heat generation due to hydration and setting of cement is followed by a cooling which appears at approximately the same time for B1 and B2 (15 hours and 16 hours after mixing respectively). But this cooling is more advanced for B3 (10 hours after mixing), thus traducing a more rapid hydration kinetics for B3 than B1 and B2.

---

[*] Theoretical amount of water corresponding to 80 mm slump (series A) or 90 mm slump (series B)
[**] Percent by weight of cement

During the 15-hour period after casting, the higher the aggregate temperature is, the higher the maturing temperature of concrete is: the gap is very pronounced between B3 and {B1, B2}. Beyond 15 hours after mixing, the temperatures tend towards the same values.

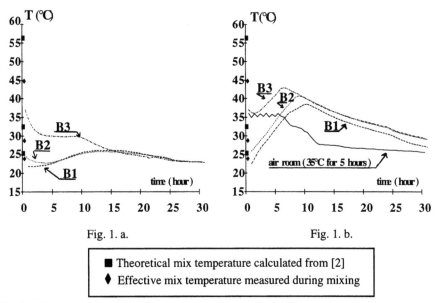

Fig. 1. a.   Fig. 1. b.

■ Theoretical mix temperature calculated from [2]
♦ Effective mix temperature measured during mixing

Fig. 1. Temperature evolution in concrete specimens: Series B.

3.1.2 Simulated conditions of hot weather (fig. 1 b)

A significant growth rate appears for B1 and B2 as soon as the temperature monitoring. In the case of B3, a cooling of concrete specimens exists and when the temperature gradient is vanished between these latters and the ambience, the temperature increases in specimens.

High aggregate temperature reduces the setting time which results in reaching peak earlier. This clearly underlines the distinction between the effect of the aggregate temperatures and the effect due to the immediate transfer of specimens at curing temperature of 35°C after placing.

In others words, the compensating effect of the greatest maturity at 35°C does not hide the influence of the aggregate temperature on time-temperature relationships of the specimens.

3.2 Mechanical test results

3.2.1 Compressive test results

In figures 2 and 3, the average 28-day compressive strengths and the corresponding standard deviation are plotted against aggregate temperature for series A and B respectively.

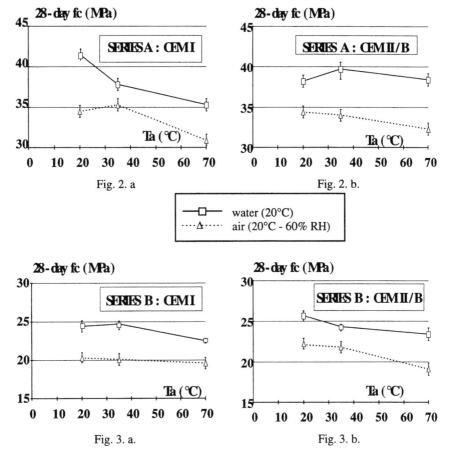

Fig. 2. a

Fig. 2. b.

Fig. 3. a.

Fig. 3. b.

Figs.3., figs.4. 28-day compressive strength - aggregate temperatures (Ta) relationships

3.2.2 Splitting tensile test results

Table 3. average 28-day splitting tensile strength (MPa)

Concrete	air (20°C-60%RH)	water (20°C)
A1	2.95	3.85
A2	3.35	3.72
A3	2.92	3.24
A4	2.96	3.40
A5	2.85	3.86
A6	3.08	3.36
B1	2.30	2.98
B2	2.27	2.47
B3	2.22	2.47
B4	2.46	2.93
B5	2.28	2.60
B6	2.08	2.57

3.2.3 Analysis

• *Influence of the conservation modes.* For each series and whatever the type of cement and the aggregate temperature, the 28-day values are greater for specimens conserved in water (20°C) than for the ones conserved in air (20°C-60% RH). In air, the desiccation is preponderant and involves the stopping of the hydration process as soon as the relative humidity in capillaries becomes lower than 75% [3].

• *Influence of the aggregate temperature.* In any case, the drops in compressive strength values occur when the aggregate temperature increases from 20°C to 70°C. Compressive strengths can be reduced by as much as 15% (relative variation of the mean value {1 - A3/A1} for water curing). On the other hand, they are similar between 20°C and 35°C in most cases.

As shown for series B specimens with CEM II/B (fig. 3. b.), the 70°C aggregate temperature might be an aggravating factor for the strength of a concrete cured at initial elevated temperatures. This is a somewhat unexpected observation because the initial curing temperature at 35°C inducing a more important maturity should have hidden the aggregate temperature contribution.

Concerning the splitting tensile strengths, the values globally follow the same trend as those of compressive strengths: they can be reduced by as much as 17% (relative variation of the mean value {1 - B3/B1} for water curing), but the dispersions are greater under tensile conditions than under compressive conditions. However, drops in 28-day splitting tensile strengths can be more important than the corresponding ones dealing with 28-day compressive strengths (series B). This is a phenomenon previously observed [4].

For normal strength concrete, fracture under tensile condition generally results from crack growth initiating at the transition zone. So, drops in 28-day splitting tensile strengths generally encountered with the rise of the aggregate temperature from 20°C to 35°C and 70°C may be attributed to a more pronounced weakness of the transition zone. This enhanced weakness in traction can explain in a some extent the drops in compressive strengths.

4. Discussion

4.1 Compressive strength - temperature relationships
During first ages, a higher maturing temperature in concrete specimens induced by heated aggregates (70°C) or high curing temperatures (35°C) involves drops in the later age compressive strength of these specimens. This fact is consistent with the occurrence of problems in regard to 28-day concrete strength if the mix temperature exceeds 32°C [2], as it is the case in this study when aggregates are brought to 70°C (see fig. 1, effective mix temperature measured during casting).

4.2 Increased water demand
Figure 4 shows that the amount of water required to produce a given consistency is increased with the rise of the aggregate temperature. Particularly, the water demand is systematically more important when the aggregate temperature reaches 70°C. At this temperature an evaporation of some amount of mixing water was observed during casting and placing.

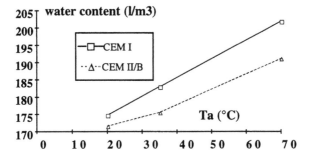

Fig. 4. Water content in concrete - aggregate temperature (Ta): series B

The extra water effect on the increase in the water-cement ratio per weight and the decrease in the 28-day compressive concrete strength is well-known [2]. Nevertheless, the application of the Feret's rule by fitting compressive strength references (A1, A4 ; B1, B4) to the water-cement ratio of concretes made up with heated aggregates cannot explain the drops in strength with regard to the increase in the water demand. This leads us to think that the extra-water due to higher aggregate temperatures is not the main reason of lower concrete strengths. Another explanation has to be investigated in the microstructure of concrete, i. e. the effect of aggregate temperature on the nature, the morphology, the distribution of hydration products in the internal space between cement grains and also the hydration state of concrete.

5. Concluding remarks and view

1. Drops in 28-day compressive and splitting tensile strengths were observed with the increase in the aggregate temperature, mainly from 20°C to 70°C.
2. Drops in 28-day strengths occured as well in controlled laboratory conditions as in initial simulated conditions of hot weather. The 28-day strength of concrete specimens conserved in water (20°C) was greater than the 28-day strength of concrete specimens placed in air (20°C-60%RH), whatever the initial curing conditions.
   Initial elevated curing temperatures did not hide the effect of elevated aggregate temperatures.
3. The application of the Feret's rule did not explain drops in the compressive strengths of concrete with regard to the increased water demand due to heated aggregates. Hence, a microstructural study on concretes made up at the laboratory has to be carried out.
   This investigation is also to undertake with another mineralogical nature of aggregates.

6. Acknowledgements

The authors gratefully acknowledge Association Technique de l'Industrie des Liants Hydrauliques (ATILH) for supporting this research. They are also indebted to Ciments Vicat for permission to use conclusions of an investigation cited as reference [1].

7.References

1. Escadeillas, G. (1990-1991) *Performances des bétons en période estivale*, private communication. Rapports Ciments Vicat - LMDC.
2. ACI committee 305 (1991) *Hot Weather Concreting*, revised, *ACI Materials Journal*, Vol. 88, No. 4 pp 417 - 36.
3. Abbasi, A.F., Al Tayyib, A.J. (1985) Effect of hot weather on modulus of rupture and splitting tensile strength of concrete. *Cement and Concrete Research*, Vol. 15, No. 2 pp 233 - 44.
4. Atlassi, E. (1991) Influence of cement type on the desorption isotherm of mortar. *Nordic Concrete Research*, Vol. 10 pp 25 - 36.

28 NUMERICAL SIMULATION OF THE EFFECT OF ELEVATED TEMPERATURE CURING ON POROSITY OF CEMENT-BASED SYSTEMS

K.VAN BREUGEL and E.A.B. KOENDERS
Delft University of Technology, Delft, The Netherlands

Abstract
Curing at elevated temperature of cement-based systems results in a lower ultimate strength than curing at low temperatures. On the microstructural level this strength reduction correlates with a higher capillary porosity. The higher capillary porosity promotes the ingress of aggressive substances and hence the degradation processes. In order to convince people of the importance the curing temperature on the porosity and permeability of the cement paste and concrete, numerical simulations of the temperature effects would be great value. Results of such numerical simulations are presented and discussed. Numerical simulations on the effect of water/cement ratio and curing temperature on microstructural development are presented. It is also shown how calculated porosity can be used in numerical predictions of autogenous shrinkage of pastes and concrete. Stress concentrations due of restraint autogenous shrinkage could indicate the probable areas of microcracking in concrete.
Keywords: Microstructural development, microcracking, modelling, numerical simulation, porosity, stress concentration, temperature.

1 Introduction

Both chemical and physical degradation processes in concrete structures are strongly dependent on the concrete's porosity and, more particularly, its permeability. Important parameters which determine these properties are the water/cement ratio and the degree of hydration. Curing at elevated temperatures has turned out to enhance the capillary porosity and most probably also the permeability of the concrete [1,2]. The presence of microcracks is another contributor to the porosity and permeability [3]. Speaking about porosity and permeability means speaking about the concrete's microstructure.

Mechanisms of Chemical Degradation of Cement-based Systems. Edited by K.L. Scrivener and J.F. Young. Published in 1997 by E & FN Spon, 2–6 Boundary Row, London SE1 8HN. ISBN: 0419215700.

Mathematical and numerical modelling of degradation of the microstructure presupposes a reliable description of the virgin microstructure. A possible way to characterize the microstructure is by performing porosity measurements. The interpretation of these measurement, however, is not free from difficulties [4]. Another line that can be followed is that of numerical simulation of the microstructure. This is not free from difficulties either, but has the advantage of flexibility is that these models can be refined and improved by inserting huge amounts of information available in different disciplines (physics, stereology, (colloid)chemistry). One of the first aims of these numerical simulation models is to check their consistency with available experimental data and to which extent they are able to allow for the effects of changes in curing parameters like the curing temperature. Once this consistency has been established, these models can be used for further parameter and sensitivity studies. Against the background of this philosophy the potentialities will be discussed of the simulation model HYMOSTRUC to simulate the effect of curing temperature on porosity. The calculated porosity will than be used as a basis for the prediction of autogenous shrinkage of cement paste. The calculated micro-level properties of the paste will than be inserted in a meso-level model for simulatiing shrinkage-induced deformations and stresses in a real concrete.

2 Hydration and microstructural development in cement-based materials

2.1 Basic features of the simulation model
In the computer-based simulation model HYMOSTRUC hydration curves are calculated as a function of the particle size distribution and the chemical composition of the cement, the water/cement ratio and the reaction temperature. Unlike most previously proposed models the effect of physical interactions between hydrating cement particles on the rate of hydration of individual cement particles is modelled explicitly.

2.2 Stereological aspects
In HYMOSTRUC cement particles are considered to be distributed in the paste. An arbitrary particle is considered to be located in the centre of a so called "cell" (Fig. 1)

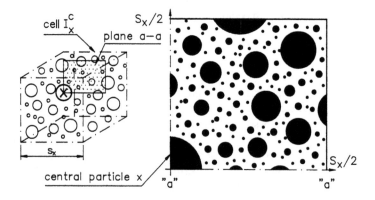

Fig. 1. Cell concept - schematic [5,6].

[5]. A cell is defined as a cubic space with a rib size S_x in which the central particle has a diameter of x mm. The rib size S_x further depends on the water/cement ratio of the paste and the particle size distribution of the cement. The amount of cement found in a fictitious shell with a certain thickness surrounding a central particle x will also depend on the water/cement ratio and the particle size distribution of the cement. This amount of cement plays an important role in the formation of interparticle contacts during hydration.

2.3 Particle expansion and particle interaction mechanism

The volume of the reaction products is about twice the volume of the reacting cement. When the larger portion of the reaction products are formed in the close vicinity of the anhydrous cement grains, the latter can be considered as gradually 'expanding' particles. Because of this 'expansion', interparticle contacts are made and a microstructure is formed. In order to obtain a workable algorithm for the determination of the interaction between hydrating cement particles the following assumptions were made [6]:

· Particles of the same size hydrate at the same rate.
· The ratio n between the volumes of the reaction products and the reactant decreases with increasing temperatures, i.e. n = n(T) (see section 2.4).
· Reaction products precipitate in the close vicinity of hydrating cement particles.

The expansion and interaction mechanisms are schematically shown in Fig. 2. On contact with water an arbitrary cement particle x starts to dissolve under formation of reaction products. These product are formed partly inside and partly outside the original surface of the particle. In this way cement particles exhibit an outward expansion, thereby making contacts with neighbouring particles. At a certain time the degree of hydration of this central particle x is a_x and the depth of penetration of the reaction front is d_{in}. Further hydration of this particle goes along with further expansion and embedding of neighbouring particles. The encapsulation of other particles causes and extra expansion and, as a consequence of this, the encapsulation of even more particles. For this expansion mechanism of continuous expansion and embedding of particles an algorithm has been developed with which the stereological aspect of structural development can be simulated.

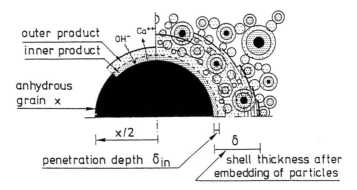

Fig. 2.   Interaction mechanism for expanding particles. Left part: free expansion, formation of inner and outer product. Right part: embedding of particles, several iteration steps.

2.4 Modelling of temperature effects on microstructural development
For quantification of the relationship between the curing temperature and the n-factor experimental test data of Bentur (1979) has been evaluated. In his tests on C₃S pastes Bentur has determined the capillary porosity of pastes as a function of the degree of hydration for isothermal curing at 4, 25 and 65°C (Fig. 3). The volume of the partly filled capillary pores V_{cp} [cm³/g cement] can be correlated to the degree of hydration a according to:

$$V_{cp} = \frac{\omega_0}{\rho_w} - \frac{\alpha_h}{\rho_s} * (\nu(T)-1) \tag{1}$$

with r_s and r_w the specific mass [g/cm³] of the unhydrated solid material and the water, respectively, and w_0 the w/c ratio. For the temperature dependency of the n-factor the following relationship was established [5]:

$$\nu(T) = \nu_0 * \exp(-28\ 10^{-6} * T^2) \tag{2}$$

in which n_0 is the n-factor that holds for isothermal curing at 20°C. In accordance with literature data a value of $n_0 = 2.2$ is adopted. Eq. (2) is assumed to cover the temperature range from 0 to about 80°C, as indicated in Fig. 4.

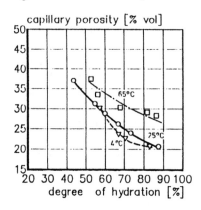

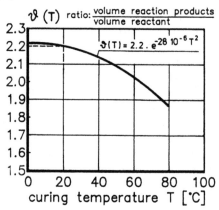

Fig. 3. Capillary porosity of C₃S pastes (Bentur (1979)).

Fig. 4. n-factor as a function of temperature (van Breugel (1991)).

3 Results of numerical simulations

For a random spatial distribution of the cement particles Fig. 5 shows the development of the simulated microstructure in pastes with w/c ratios 0.3 and 0.7, respectively. The pictures are taken in a cross section of a spherical paste volume with a diameter of 200 mm in which the cement grains are positioned randomly. In the low w/c-paste the sphere is filled with 250,000 particles, the biggest particle having a diameter of 37 mm. The rate of hydration of the randomly distributed particles is calculated with HYMOSTRUC. In the low w/c-paste a dense microstructure is developed, whereas the paste with w/c = 0.7 still exhibits a less dense pore system, even at a high degree of hydration of 79%.

The effect of the temperature dependence of the 'expansion factor' n on microstructural development is shown in Fig. 6. For these simulations a paste was considered used by Kjellsen et al. [2] in a research project on temperature effect on porosity and strength. Cement: Blaine 395 m^2/kg, w/c = 0.5. Fig. 6a and 6b show the simulated microstructure for curing temperatures of the paste of 20°C and 50°C, respectively. Fig. 6b. shows, as was to be expected from adopted temperature dependency of the expansion factor n(T), that in the warm-cured paste the interparticle contacts are less intensive than in the cold cured paste.

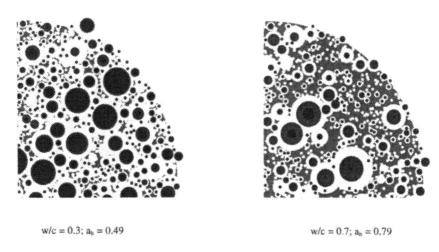

w/c = 0.3; a$_h$ = 0.49   w/c = 0.7; a$_h$ = 0.79

Fig. 5. Numerical simulation of microstructure in cement paste. w/c ratio: 0.3 and 0.7. Degree of hydration: 49% and 0.79%. White: outer product. Black: unhydrated cement. Grey: pore volume.

6a. T = 20°, a$_h$ = 0.70%   6b. T = 50°C, a$_h$ = 0.68%

Fig. 6. Simulated effect of curing temperature on microstructure. Grey: pore volume.

4 Towards numerical simulation of autogenous shrinkage and microcracking

4.1 Autogenous shrinkage

With progress of the hydration process the pore structure changes from a coarse to a more fine one. The gradual changes in the pore structure can be followed, in a way, by following the changes of the hydraulic radius, being the pore volume divided by the total pore wall area. Both the pore volume and the pore wall area can be determined with HYMOSTRUC as a function of the degree of hydration. From this, the distribution of the capillary pores can be determined. At the same time that the hydraulic radius decreases the amount of capillary water decreases. In the partly empty pore system the relative humidity matches with a certain pore pressure (Fig. 8, left). The relative humidity in the pore system can be calculated by the Kelvin equation [11]. In this equation, the pore pressure is related to the largest pore that is completly filled with water and with the surface tension. Considering a pore as schematized in Fig. 7, the changes in pore pressure can be modelled by using a thermodynamic approach. At a certain degree of hydration the pores are partly filled. Due to the (under-)pressure in the pore system, an adsorption layer will be formed at the internal surface of the pores. The thickness of this layer will change proportionally to the pressure in the pore system. The gradual emptying of the pores as a result of the hydration process will influence the pressure in the system and also the thickness of the adsorption layer. Thermodynamic equilibrium in the pore system can only be true for one unique set of parameters. In HYMOSTRUC this set of parameters is calculated by an iterative procedure where the internal work carried out by the pore pressure and by the volume changes are in equilibrium with the surface tension of the capillary water. Changes of the surface tension are linearly related to a certain volume change [7]. The volume changes of a paste can be calculated as a function of the degree of hydration. As the degree of hydration is known as a function of time, shrinkage can be presented as a function of time as well. Fig. 8 (right) shows an example of calculated and measured volume changes of a paste with a w/c ratio of 0.3. The good agreement between the simulated and measured shrinkage of pastes invites further work to implement this behaviour in meso-level models for the evaluation of the effect of autogenous shrinkage on the shrinkage-induced deformations of a real concrete.

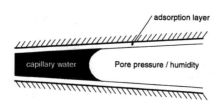

Fig. 7. Schematical representation of a pore.

4.2 Autogenous shrinkage of concrete

For numerical simulation of autogenous shrinkage of concrete the so called lattice model has been used [9]. In this model the paste is modeled by small bars, each bar having a length of approximately 1 mm. The bars built up a spatial formwork between the aggregate particles. The strength and stiffness properties of the paste, as well as its shrinkage properties, are

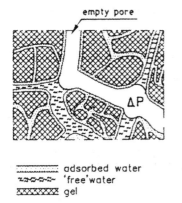

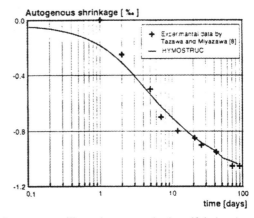

Fig. 8. Schematic presentation of partly filled pore system illustrating a stage in the self-desiccation process (left), and results of measured and simulated autogenous shrinkage of a paste with w/c = 0.3.

assigned to these bars. Fig. 9 shows the results of a simulation of the shrinkage-induced deformations of a small piece of concrete made with a paste (w/c = 0.3, cement content 475 kg/m³, max. aggregate size: 16 mm) with shrinkage properties shown in the diagram of Fig. 9. Hydration and shrinkage behaviour calculated with HYMOSTRUC. The calculated autogenous shrinkage of the concrete was about the 39% of the deformations of the plain paste. In areas where the paste is more or less free to deform the colour of the lattice structure changes from light to dark grew. Stress concentration are located in the light grey area, where the aggregate particles prevent the shrinkage deformations. In those areas microcracking is likely to occur.

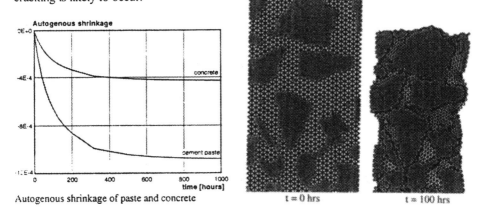

Autogenous shrinkage of paste and concrete

Fig. 9. Numerical simulation of autogenous shrinkage of high strength concrete sample with the help of the lattice model [9]. Sample size: 25 x 100 mm (2-dimensional).

5 Discussion

Development of microstructure/property relationships for concrete largely depends on a good characterization of the microstructure [10]. Similarly numerical simulations of degradation processes are based on an adequate characterization of the microstructure. In degradation processes the porosity and permeability of the concrete are of utmost importance. In this paper a survey has been presented of the potentialities of the simulation program HYMOSTRUC to simulate microstructural development in a paste as a function of the water/cement ratio and the curing temperature. In its present stage of development the 'resolution' that can be reached with the simulation program concerning microstructural characterization appears to describe autogenous shrinkage of non-blended cement pastes. By implementation of the predicted paste properties in a meso-level model, i.e. the lattice model, it was possible to describe the deformational behaviour of a concrete sample and to identify the areas to microcracking. Numerical simulations of microcracking caused by, self-desiccation drying or an imposed thermal loading, on the deformational properties of the concrete is the next area to be investigated. Combining micro- and meso-level models should enable us to define the resulting porosity and permeability of the concrete.

Acknowledgement
The authors like to acknowledge the assistance of Dr. E. Schlangen for performing the numerical analysis with the lattice model.

References

1. Bentur, A., Berger, R.L., Kung, J.H, Milestone, N.B. and Young, J.F. (1979) Structural Properties of Calcium Silicate Pastes. *J. American Ceramic Society*, Vol. 62, pp. 362-366.
2. Kjellsen, K.O., Detwiler, R.J. and Gjørv, O.E. (1991) Development of microstructures in plain cement pastes hydrated at different temperatures. *Cement and Concrete Research*, Vol. 21, pp. 179-189.
3. Häkkinen, T. (1993) The influence of slag content on the microstructure, permeability and mechanical properties of concrete. *Cement and Concrete Research*, Part 1: Vol. 23, No. 2, pp. 407-421, Part 2: Vol. 23, No. 3, pp. 518-530.
4. Beaudoin, J.J. and Brown, P.W. (1992) The stucture of hardened cement paste. 9th *Int. Conf. Chemistry of Cements*, Vol. I, pp. 485-525.
5. Van Breugel, K. (1991) *Simulation of hydration and formation of structure in hardening cement-based materials*. PhD, TU-Delft.
6. Van Breugel, K. (1995) Numerical simulation of hydration and microstructural development in hardening cement-based materials. *Cement and Concrete Research*, (I) Theory: Vol. 25, No. 2, pp. 319-331. (II) Applications: Vol. 25, No. 3, pp. 522-530
7. Setzer, M.J. (1972) *Oberflächenenergie und mechanische Eigenschaften des Zementsteins*. PhD, Munich.
8. Tazawa, E. and Miyazawa, S. (1992) Autogenous shrinkage caused by self-desiccation in cementitious material. *Proc. 9th Int. Conf. Chemistry of Cement*, New Delhi, Vol. IV, pp. 712-718.
9. Schlangen, E. (1993) *Experimental and numerical analysis of fracture processes in concrete*. PhD, Delft.
10. Scrivener, K.L. (1989) The Microstructure of Concrete, in *Materials science of concrete*, (Ed. J.P. Skalny), American Ceramic Society, pp. 127-161.
11. Defay, R. et. all. (1966) *Surface tension and adsorption*, Longmans, Green & Co ltd, London.

29 MICROCHEMICAL EFFECTS OF ELEVATED TEMPERATURE CURING AND DELAYED ETTINGITE FORMATION

M.C. LEWIS and K.L. SCRIVENER
Imperial College, London, UK

Abstract
To study the mechanism of expansion which occurs after curing at elevated temperatures, commonly called delayed ettringite formation (DEF), a series of mortars were cured at 90°C for 12 hours and then under water at 20°C. The mortars were made with cements prepared from the same clinker, with additions of calcium sulphate at 3, 4, and 5 wt.% SO_3 and also with additions of KOH. Analysis was carried out at times throughout the expansion process by scanning electron microscopy (SEM) and X-ray microanalysis. The results support the view that expansion is caused by relatively uniform expansion of the cement paste as opposed to by ettringite growth at the paste/aggregate interfaces. Sulphate and aluminate ions are incorporated within the C-S-H after heat curing, the concentration of these ions in the C-S-H immediately after heat curing appears to determine the potential for expansion.
Keywords: Delayed Ettringite Formation, microstructure, heat curing, temperature, X-ray microanalysis, expansion, threshold, sulphate incorporation.

1 Introduction

In recent years much research has focused on the phenomenon of expansion of concretes and mortars after curing at elevated temperatures. A number of workers [1,2,3] have observed that any ettringite originally present in the paste is destroyed above about 75°C and its ions, namely SO_4^{2-} and aluminate, become incorporated within the C-S-H gel. Wieker and Herr [4] also found that the concentration of SO_4^{2-} ions in the pore solution increases with increasing curing temperatures and with increasing alkali content of the cement.

Mechanisms of Chemical Degradation of Cement-based Systems. Edited by K.L. Scrivener and J.F. Young. Published in 1997 by E & FN Spon, 2–6 Boundary Row, London SE1 8HN. ISBN: 0419215700.

Microanalyses of the C-S-H at different periods of exposure after heat curing, previously reported [3], show that the incorporated SO_4^{2-} ions leave the C-S-H over time, regardless of whether or not expansion occurs.

Microstructural studies conducted on both field and laboratory specimens report the cement paste filled with deposits of ettringite in pores and cracks. In concretes and mortars which have expanded, relatively uniform rims, usually partly or wholly filled with ettringite, are observed around the aggregates particles, whose size is roughly proportional to the size of the aggregate [5,6]. These observations have led to the terminology of delayed ettringite formation. A number of researchers [7,8,9] have reported ASR in conjunction with DEF.

The mechanism through which expansion occurs is still unclear. Originally it was suggested that the crystallisation pressure of ettringite reforming at the paste/aggregate interface produces the gaps around the aggregate particles and causes the expansion [10]. Such a mechanism would require high levels of supersaturation within the pore solution far in excess of reported values [4]. Furthermore, it is unlikely that such a mechanism would lead to the formation of uniform rims, proportional to the aggregate size.

Subsequently it was proposed that expansion occurs relatively uniformly throughout the paste component, leaving gaps around the non-expanding aggregates [5,6,11]. This theory would also explain the observation that the thickness of the rims is proportional to the size of the aggregates.

The work reported here is a continuation of that presented earlier [3]. The behaviour of five cements made from the same clinker was studied at various times throughout the expansion process. The specimens were characterised by BSE images in the SEM and X-ray microanalyses.

2 Experimental

The cements used in this work were laboratory prepared grinds from a normal production clinker. The cements were ground to a specific surface area of 450 m^2/kg at sulphate levels of 3%, 4% and 5% (by weight) and to 350 m^2/kg at 4% sulphate. Four mortars were made, one from each of these cements. An additional mortar was produced from the cement containing 5% SO_3 to which KOH was added dissolved in the mixing water to increase the alkali level by the equivalent of 0.8 wt.% Na_2O.

Mortar specimens were cast at a w/c = 0.5 and sand/cement = 3. The samples were given a 4 hours precure and then cured at 90°C for 12 hours. The mortars were demoulded, stored in waters and removed periodically for testing. Further details of sample preparation and experimental techniques are given in [3].

Parallel mortar samples were cured at 20°C throughout.

3 Results

3.1 Characterisation Times

Microscopy and X-ray microanalysis were used to study the mortars throughout the period of expansion. The characterisation times were chosen to correspond to similar points on the expansion curves, namely:

(1): one day after heat treatment;
(2): just before expansion;
(3): midway through expansion;
(4): at the end of expansion.

3.2 Expansion

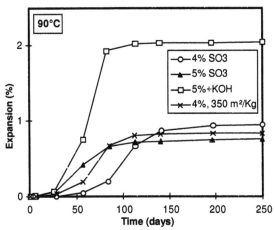

Figure 1:- Expansion of the different mortars.

None of the mortars cured at 20°C showed any significant expansion, even at the 5% SO_3 level. Figure 1 shows the expansion of the various mortars cured at 90°C. The mortar made with the cement containing only 3% SO_3 did not expand significantly and is not plotted. With an SO_3 content of 4% the ultimate expansion was just under 1%. Increasing the SO_3 content to 5% resulted in a lower expansion of about 0.8% - demonstrating a pessimum relation between sulphate content of cement and overall expansion reported elsewhere [12]. On the other hand the addition of KOH to the cement at 5% SO_3 almost tripled the ultimate expansion.

3.3 Microstructure

Figure 2 shows the microstructures of the mortars which were cured at 90°C and then in water at room temperature. In the mortar made from the cement containing 3% SO_3 (which did not expand - Fig. 2(a)) there were no ettringite rims around the aggregate particles. However, large deposits of ettringite were seen, infilling pores and voids (as indicated by the broken circles).

In the mortars which did expand, rims were seen around the aggregate particles (Fig.2(b)), usually, but not always filled with ettringite, (Fig. 2(c)). In some cases these rims also contained calcium carbonate. In addition, evidence was found of more localised expansion of the C-S-H gel. At the end of the expansion period, period 4, relic grains were seen in which the outer rim of C-S-H gel has expanded away from the now hydrated core (indicated by the broken circles in Fig. 2(b & c) and in Fig 2(d)).

In none of the specimens were there any signs of alkali silica reaction.

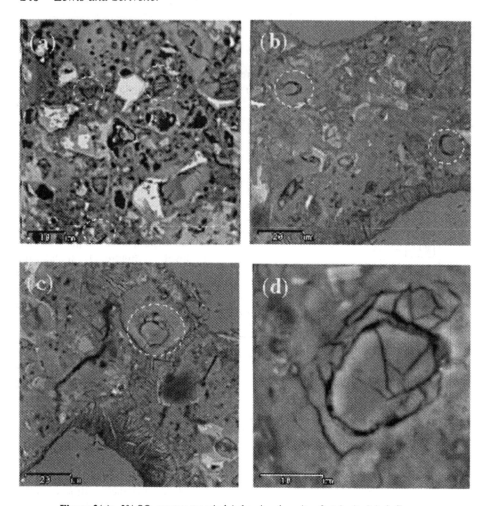

Figure 2(a):- 3% SO$_3$ mortar at period 4 showing deposits of ettringite [circled].
Figure 2(b):- Expansion gaps within the CSH [circled] at period 4.
Figure 2(c):- Ettringite needles at paste/aggregate interface and concentric multiple cracking in CSH grain [circled] at period 4.
Figure 2(d):- Multiple concentric cracking in fully hydrated anhydrous grain at period 4.

3.4 Microanalysis

Microanalyses were made selectively on the hydration rims of C-S-H formed during the heat curing. About 50 points were analysed at each period within the central region of the specimens. At the accelerating voltage used of 15kV, the microanalyses relate to a volume of about 1μm in each dimension. On this scale intimate mixtures of C-S-H with other hydrate phases cannot be differentiated. To interpret the likely composition of the sulfo-aluminate phases intermixed with the C-S-H, the analyses are plotted as atom ratios of S/Ca against Al/Ca, the lines indicate where mixtures of C-S-H with AFt or AFm respectively would lie, however, these lines do not imply that AFt or AFm are always present.

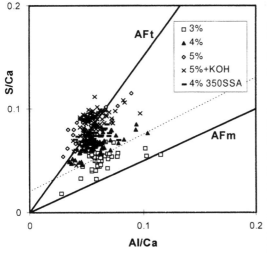

Figure 3:- Microanalysis of heat cured mortars at period 1.

The analysis points of the five cements at period 1 are presented in Figures 3. As the sulphate content of the cement increases, so does the amount of incorporated SO_4^{2-} ions within the C-S-H. Adding KOH to the 5% SO_3 system does not affect the incorporation of ions as the analysis points virtually overlap for these two cements. Although many of the points fall on the line which would represent a mixture of ettringite, AFt, within the C-S-H gel, X-ray evidence at this period indicates that no ettringite is present at this time.

There is a distinction between the composition of the C-S-H gel for those mortars which subsequently expanded and that (3% SO_3) which did not. Other mortars in the same study [3] also showed this same distinction at period 1, depending on whether they subsequently expanded or not. These results suggests that there is a threshold level of SO_4^{2-} and aluminate ion incorporation immediately after the heat curing, above which there exists a potential for expansion (indicated by the broken line drawn in Figure 3).

Figure 4 shows, for each cement, how the amounts of sulphate and aluminate ions intermixed with the C-S-H gel change during subsequent curing in water at 20°C for the four characterisations times. After extended curing (period 4), all the mortars show a reduction in the level of intermixed SO_4^{2-} ions , and to a lesser extent aluminate ions, until the analyses points lie on a line representing mixtures of calcium alumino monosulphate, AFm, with C-S-H gel.

The evolution of the analyses between periods 1 and 4, differs according to whether the mortar expands or not. With 3% SO_3 the level of SO_4^{2-} ions decreases continuously. However, in the mortars which expand the analysis points shift up toward the AFt line at the beginning of expansion (period 2). This suggests that ettringite reforms within the C-S-H gel before expansion occurs. This is particularly evident in the 4% SO_3 mortar, but less obvious in the cases of 5% SO_3 and 5% SO_3 with alkali, as the analyses at period 1 were already close to the C-S-H/AFt line.

There also appears to be some relation between the rate at which sulphate is lost from the C-S-H gel and the ultimate degree of expansion. Whereas at period 3 the analysis points of the 5% SO_3 mortar are already close to the C-S-H/AFm line, the points of the 5% SO_3 mortar with KOH, which expands considerably more, still lie in the region of the C-S-H/AFt line. A similar pattern was seen in other mortars examined in the study.

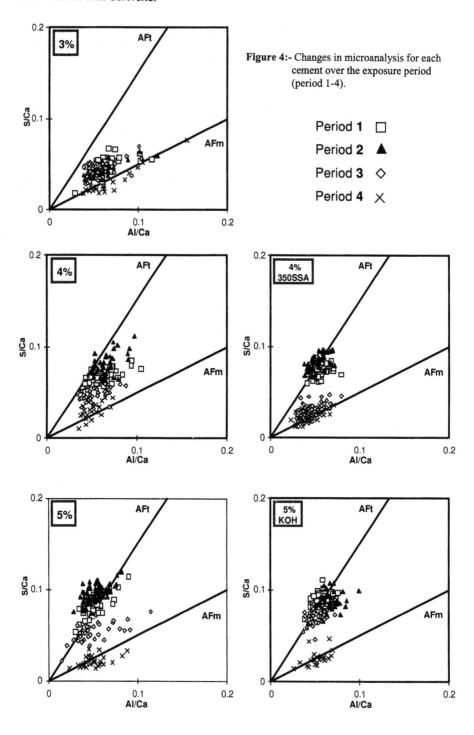

Figure 4:- Changes in microanalysis for each cement over the exposure period (period 1-4).

Period 1 □
Period 2 ▲
Period 3 ◇
Period 4 ✕

4 Discussion

The specimens in the current study, were relatively small and unrestrained, therefore any thermal stresses and cracking during heat curing are minimised and the subsequent ingress of water is comparatively easy. Under these conditions, relatively high expansions are observed and these take place relatively quickly. Consequently these specimens are well suited to the study of the effects of expansion on microstructure. The size of the gaps around the sand particles correspond well to the overall degree of expansion - i.e. approximately 10μm gaps are observed and the mortars expand by about 1%.

At these high degrees of expansion there is also evidence that the hydration rims of C-S-H gel formed during heat curing have themselves expanded away from the interior of the relic grains which hydrated at normal temperatures. In this case the degree of expansion appears to be considerably larger than the overall expansion of the bars, but this effect is probably accentuated by shrinkage of the C-S-H gel on drying of the sample for microstructural examination. This later observation and the confirmation of relatively even expansion of the paste, indicate that the C-S-H gel formed during the heat cure is the principal location of expansion. In all the mortars there is a considerable amount of pore space. In all cases, including mortars which do not expand, ettringite recrystallises in these pores, but there is no evidence that this recrystallisation causes expansion.

X-ray diffraction data [3,13] indicates that no ettringite is present in the mortars immediately after the heat cure (period 1). The microanalyses indicate that at this time the sulphate and aluminate ions are intimately mixed with the C-S-H gel, but the precise form of this incorporation is not clear. On subsequent curing under water, the microanalyses and XRD data [13] suggest that at Period 2 (just before expansion), ettringite is reformed within this C-S-H gel only in those mortars which subsequently expand. In the mortar which did not expand, the sulphate (and to a lesser extent aluminate) ions start to leave the C-S-H gel to recrystallise as ettringite in pores and voids.

The C-S-H gel contains all the necessary ions for the formation of ettringite, but the necessary water would have to be provided either by imbition of external water or by dehydration of the surrounding C-S-H. In either case, the uptake of water, directly by the ettringite, or in the later case by rehydration of the C-S-H, would provide the necessary pressure for expansion. During the expansion of the mortars the sulphate and aluminate ions leave the C-S-H gel as ettringite recrystallises in the pores of the cement paste and in the gaps at the paste aggregate interfaces, opened up by the expansion. As the gaps are a consequence, not a cause of expansion, other phases, such as calcium carbonate may also form within them. At the end of expansion the microanalyses of the C-S-H gel correspond to mixtures of C-S-H and AFm.

The analyses at Period 1 indicate a clear distinction between the C-S-H gel of those mortars which subsequently expand and those which did not. The broken line in Figure 3 indicates this distinction. In addition to the five mortars reported here, this threshold delineated all the mortars in the study in which the variables of curing temperature and slag additions were also investigated. However, the present results are limited to one clinker and the existent of a unique threshold would need to be confirmed by experiments on other cements.

5 Conclusions

- The microstructural development of the mortars was consistent with relatively uniform expansion of the paste component.
- The form of the grain relics in expanded mortars suggests that expansion is localised in the C-S-H gel formed during heat curing.
- Microanalysis indicates a distinction between the chemical composition of this constituent between mortars which subsequently expand and those which do not.
- In mortars which expand, microanalysis indicates the reformation of ettringite within the C-S-H gel formed during heat curing at the start of the expansion.
- In the long term the microanalyses of the C-S-H gel phase tend towards that of a mixture of C-S-H and AFm.

Acknowledgements
Matthew Lewis would like to acknowledge the financial support of Lafarge, the British Cement Association, and the Cementitious Slag Makers Federation. The authors would also like to thank Steve Kelham and Blue Circle for help in preparing the cements and mortars and giving access to previously recorded expansion data. We also gratefully acknowledge the advice of and discussion with Professor Hal Taylor, throughout this project.

6 References

1. Scrivener, K.L. and Taylor, H.F.W. (1993) Delayed Ettringite Formation: A Microstructural and Microanalytical Study. *Advances in Cement Research* Vol. 5:139-146.
2. Odler, I., Abdul-Maula, S., and Zhongya, L. (1987) Effect of Hydration Temperature on Cement Paste Structure. *Materials Research Society Symposium Proceedings* Vol. 85:139-144.
3. Lewis, M.C., Scrivener, K.L., and Kelham, S. (1995) Heat Curing and Delayed Ettringite Formation. *Materials Research Society Symposium Proceedings* Vol. 370:67-76.
4. Wieker.W, and Herr, R. (1989) Sulphate Ion Equilibria in Pore Solutions of Heat-Treated Mortars of Portland Cement with Respect of the Expansion Reaction by Secondary Ettringite Formation. *Proceedings of the Second International Symposium on Cement* 58-66.
5. Johansen V., Thaulow N., Jakobsen U. H. and Palbol L. (1993a) Heat Cured Induced Expansion, *present at The 3rd Beijing International Symposium on Cement and Concrete.*
6. Johansen N., Thaulow N. and Skalny J. (1993b) Simultaneous Presence of Alkali-Silica Gel and Ettringite in Concrete, *Advances in Cement Research* Vol. 5, 23-29.
7. Brown, P.W. and Bothe, J.V.(1993) The Stability of Ettringite. *Advances in Cement Research* Vol. 5:47-63.
8. Oberholster, R.E., Maree, H., and Brand, J.H.B.(1992) Cracked Prestressed concrete Railway Sleepers: Alkali-Silica Reaction or Delayed ettringite Formation. *The Proceeding of The 9th International Conference on Alkali-Aggregate Reaction in Concrete* Vol. 2:739-749.
9. Pettifer, K. and Nixon, P.J.(1980) Alkali Metal Sulphate - A Factor Common to Both AAR and Sulphate Attack on Concrete. *Cement and Concrete Research* Vol. 10:173-181.
10. Heinz D. and Ludwig U. (1987) Mechanism of Secondary Ettringite Formation in Mortars and Concretes Subjected to Heat Treatment, ACI SP 100, Vol. 2, 2059-2071.
11. Marusin, S.(1994) Concrete Failure Caused by Delayed Ettringite Formation Case Study, *3rd International Conference on Durability of Concrete*, Nice.
12. Kelham, S. (1996) The effect of cement composition and finess on expansion associated with DEF *Cement and Concrete Composites*, Vol. 18, no. 3. pp. 171 - 179.
13. Lewis, M.C. (1996) Delayed Ettringite Formation in Concretes, *PhD Thesis*, Imperial College, University of London.

PART FIVE
DURABILITY OF
NON-PORTLAND CEMENTS

30 DURABILITY OF CALCIUM ALUMINATE CEMENT CONCRETE: UNDERSTANDING THE EVIDENCE

C.M. GEORGE
Lafarge Aluminates, Paris, France

Abstract
Behaviour of calcium aluminate cement concretes (CACC) in sulphates and sea water is traditionally explained by absence of portlandite and presence of aluminum hydroxide in CAC, but there is increasing evidence that time related surface densification can play a critical role in enhancing the performance of CAC in practice.

The traditional view that CAC concrete is more vulnerable than PCC to reinforcement corrosion through carbonation is no longer supported either by accumulated field experience or theoretical analyses. The effect of the presence of alkalis remains controversial. However the fact remains that no evidence exists for alkali engendered, destructive carbonation of CAC when good practice in the use of this material is respected.

Under acidic conditions recent studies show that the usefulness of CAC can be extended to pH values well below the limit (0.4) traditionally related to the solubility of otherwise protective AH_3. The greater stability of C_3AH_6 over CAH_{10} can dominate over questions of porosity. Solubility of the reaction products of acid attack also plays a decisive role, and the acid neutralisation capacity of CAC, which greatly exceeds that of PC, provides a credible explanation of the greater service life of CAC observed in practice.

Resistance to sea water and sulphates

Calcium Aluminate Cement Concrete (CACC) has an excellent record of resistance to sea water and sulphated ground.

Lea [1] has well summarised a number of long-term exposure trials. All were initiated in the 1920's and 1930's and extended for up to 30 years in such diverse

Mechanisms of Chemical Degradation of Cement-based Systems. Edited by K.L. Scrivener and J.F. Young. Published in 1997 by E & FN Spon, 2–6 Boundary Row, London SE1 8HN. ISBN: 0419215700.

climatic conditions as Scandinavia, Western Europe, the Mediterranean and Florida in the USA.

With rare exceptions, CACC outperformed all other types of concrete, those based on Portland cements, slag cements, etc.

More recently Crammond [2] published results of a 15 year exposure test of CACC in heavily sulphated ground (0.26% SO_3 in the ground water) where the material showed very good resistance. Curiously, 100mm cubes of the same concrete tested in the laboratory performed less well.

The origins of this success are chemical in nature and can be traced back to pioneering work begun in the first half of the 19th century when marine structures built with the Portland cements of the time were deteriorating at an alarming rate. The earliest compositions studied were mixtures of lime and alumina (A) and silica (S) bearing clays, see Figure 1.

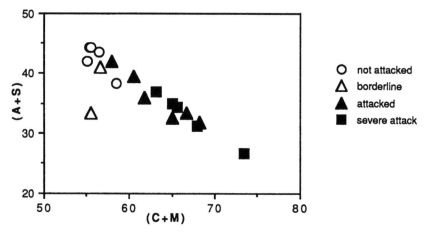

Fig. 1. Lime-calcined clay composition
exposed to $MgSO_4$ solution (Vicat, 1856)

As the proportion of clay was increased the amount of residual free lime after reaction diminished and the ability to resist sulphates and sea water improved. This led Vicat to the now well-known formula for a sulphate resistant cement:

$$\frac{[alumina] + [silica]}{[lime] + [magnesia]} > 1$$

It may be noted from the point of view of durability that the effectiveness or lack of it, of the compositions studied was totally unrelated to the levels of mechanical strength they developed.

It was Jules Bied [3] who later had the idea of precombining the lime, alumina and silica using limestone and red bauxite as raw materials. Unlike Portland cements, these could not be readily sintered because of the iron oxide in the bauxite used so a fusion process was developed and patented in France in 1908,

[4]. The French word for fusion being Fondu, the name Ciment Fondu came into use.

Bied extended Vicat's formula by requiring that the alumina content should greatly exceed the silica content because calcium aluminates are reactive whereas high silica content calcium silicates are not.

It is interesting to compare the range of compositions that Bied first suggested with Ciment Fondu as it is manufactured today.

Table 1. Original and modern compositions of aluminous cement

	Bied, 1922			1995
SiO_2%	5	10	15	4
Al_2O_3%	45	40	35	40
Fe_2O_3%	15	10	15	18
CaO%	35	40	35	38

The new cement proved virtually unattackable by all the laboratory tests habitually used to assess the resistance of cements to sulphates and sea water.

This success led in 1916 to a series of trials by a leading railway company (PLM) were sulphated waters draining into a tunnel had completely destroyed the structure. Blocks of concrete made with a variety of cements were immersed in the constantly flowing $CaSO_4$ solutions found in the tunnel. Only those made with the aluminous cement survived intact, even when anhydrite was used as the fine portion of the aggregate.

Equally successful trials in marine constructions were carried out and the reputation of Ciment Fondu was established.

The early success and widespread use of CACC was later diminished by a number of incidents which led to intensive investigations of the hydration products of this type of cement.

In effect standard laboratory tests (e.g. 24 hours, 20°C 100mm cubes) show deceptively high strengths and low porosities and this encouraged lean mixes and high water contents. In fact, CACC does not reach a mature stable state under these conditions and will convert in service to more modest levels of performance substantially equivalent to a Portland cement concrete of similar mix design.

Today this effect is well understood and CACC can be confidently designed for durability by keeping water contents low and using an adequate amount of cement, [5].

A 20 year programme of visual observation of CACC semi-immersed in sea water was reported by George [6]. A wide variety of curing conditions and mix designs were studied from which it was evident that at cement contents of at least 400kg/m^3 (to provide adequate compactability) and total water cement ratios not exceeding 0.4, specimens remained intact.

Thus the chemical composition of CAC is a necessary but not sufficient condition for long-term performance. There is also the physical requirement of impermeability.

With hindsight, all this seems logical enough but it does not explain some remarkable cases of longevity in structures built before the above rules of mix design were understood, much less respected. Two examples built in 1930, are noteworthy.

One of these, at Daggenham, in southern England [7], is a piled foundation where the piles were driven into sulphate contaminated soil in a tidal estuary. The other is a container terminal in the port of Halifax, Nova Scotia, [8]. Both contained steel reinforcement and both are still in good condition and in service today.

SEM examination of concrete samples taken from these structures show a dense surface region to depths of over 25 mm, gradually transforming at greater depths into the more porous structure typical of converted CACC. Samples of the submerged concrete from Halifax had a heavy deposit at the surface which analyses to $CaCO_3$ with a small amount of brucite below and a minor region rich in Mg, Al and Si immediately below that.

Beneath this in both structures the dense material is poorly crystalline aluminium hydroxide with some CAH_{10} and occasional areas of C_2AH_8. Friedel's salt is also detected in this region.

The porosity measured in the SEM is extremely low - about 3 to 4% - compared to about 17% deeper into the concrete. The O_2 permeability [9] of 50 mm discs cut from the surface of both the Halifax concrete and the Daggenham concrete measured between 10^{-16} and 10^{-17} and was at least an order of magnitude lower than samples taken from the interior.

We know that both mixes were cast at w/c ratios around 0.6 to 0.7 and that water cooling was applied during the initial cure. It would seem that over the years, moisture ingress has allowed residual anhydrous cement in the surface region to hydrate and further densify the matrix, gradually producing an ever thicker barrier to diffusion.

More work is in progress to establish the practical conditions under which a protective barrier develops at the surface of CACC but its presence in the concretes described here offers a very plausible explanation of their excellent durability.

Carbonation - labcrete and realcrete

The Building Research Establishment UK, is carrying out several studies on carbonation of CACC, [10], [11].

Carbonation rates of laboratory CACC under accelerated conditions appear to be approximately proportional to the square root of time, indicative of a diffusion controlled process. By contrast, samples carbonated naturally (by exposure to air) show a distinct levelling off of carbonation rates after 1 year. The natural carbonation of PCC obeys a $\sqrt{t}$ law. Yet after 2 years natural carbonation, CACC and PCC of comparable mix design show substantially the same overall depths of carbonation. This clearly implies that the initial rate of carbonation of CACC is more rapid than that of equivalent PCC but becomes considerably slower at later ages.

No microstructural examinations of the concretes accompanied these studies so it is a matter of speculation whether the differences in behaviour between the two types of concrete are associated with a heightened level of impermeability in the case of CACC due to the combined effects of carbonation accompanied by surface densification intrinsic to and specific to CAC. However, in support of such an hypothesis it may be noted that the depths of carbonation for a 0.4 W/C CACC after 2 years diminished according to storage conditions in the following order:

Internal 20°C, 65% RH	9.0 mm
Outdoor (sheltered)	5.5 mm
Outdoor (exposed)	1.5 mm

Additionally, the O_2 permeability after 8 weeks conditioning in N_2 (to avoid carbonation) is an order of magnitude lower in CACC than in the PCC's.

Field studies by the BRE of CACC buildings in the UK show that typically these concretes have cement contents of about 250 kg/m^3 and total water/cement ratios well above the recommended limit of 0.4. 90% of the samples show carbonation to the depth of the reinforcement but evidence of reinforcement corrosion is mostly minor and observed in only 10% of cases where the concrete was apparently dry and 11% of cases when the concrete was wet.

BRE Digest 405 (May 1995) contains the following paragraph in reference to CACC:

"The lower alkalinity of HAC concrete has led to a view that reinforcement in HAC concrete may be more susceptible than Portland cement concrete to corrosion but field evidence shows no quantifiable difference. As in Portland cement concretes, the risk and the rate of corrosion to reinforcement are determined mainly by a combination of depth of carbonation, chloride levels, relative humidity of the environment and the rate of transport of moisture and oxygen through the concrete."

Carbonation - thermodynamic considerations

During carbonation, the composition of the pore solution can be expected to correspond to the point of intersection of the solubility curves of the products of the carbonation reaction, viz. $Al(OH)_3$ and $CaCO_3$. As long as this point, (A) lies below the solubility curve of the calcium aluminate hydrate being carbonated, the carbonation reaction can proceed and with a continuous supply of CO_2 available will eventually go to completion.

The solubility curves change with the concentration CO_2 in solution. Figure 2 shows the respective curves at 3 levels of (CO_2).

At a (CO_2) level of 10^{-12}, point A is above the solubility curve of CAH_{10} so carbonation of this hydrate will not occur.

At a (CO_2) level of $10^{-11.6}$, all three curves intersect at point A. This thus represents the limit for carbonation, or the carbonation front. At or below this level, given an unlimited supply of CO_2, carbonation can continue to completion.

The calculation of the solubility curves is based on thermodynamic data for the equilibrium constants for the various chemical equilibria that exist in the pore solution [12]. In the example shown in figure 2, the value of the equilibrium constant for $Al(OH)_3$, K_2:

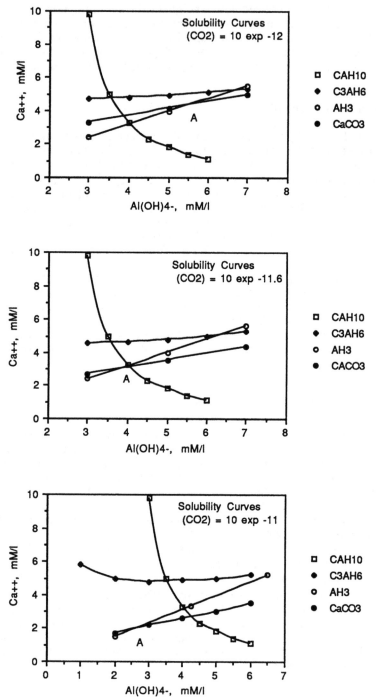

Fig. 2. Carbonation of CAH_{10} as a function of CO_2 in pore solution

$$\frac{[Al(OH)_4^-]\ [H_3O^+]}{[Al(OH)_3]} = Kw.K_2 \qquad (Kw = 10^{-14})$$

has been taken as 1.6 for alumina gel.

Equivalent calculations for the carbonation of C_3AH_6 should use a value for the AH_3 equilibrium constant corresponding to either microcrystalline bayerite or the fully matured hydroxide, gibbsite. These values are 0.3 and 0.06 respectively.

The concentration (CO_2) in the pore solution also affects the pH so it is possible to draw a diagram of (CO_2) against pH and show on it the carbonation front.

However, for a realistic representation of the behaviour of CAC it is necessary to also take into account the effect of alkali metal ions dissolved in the pore solution. These may be generated by ingress from external sources but even without this, the small amounts of releasable alkalis contained in the cement itself will have an effect. Goni and co-workers, [13] have reported concentrations after 24 hours hydration of: 0.019 moles/l Na^+, 0.030 moles/l K^+ from a CAC containing 0.046 Na_2O and 0.082 K_2O in weight percent, and a pH of 12.2.

The effect of the presence of alkali metal ions is to increase pH at the carbonation front. This is illustrated in figure 3 for C_3AH_6 (only Na^+ ion concentrations have been taken into account).

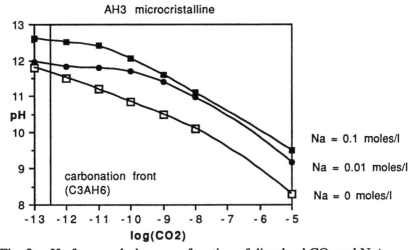

Fig. 3. pH of pore solutions as a function of dissolved CO_2 and Na^+

What emerges from this approach is a clear message that carbonation of CACC can proceed at levels of pH far above those traditionally associated with depassivation of steel. Note that the position of the carbonation front is independent of the alkali ion content.

The implication is that the carbonation front may need to move considerably beyond the depth of reinforcement cover before significant corrosion is likely to take place.

Coupling this conclusion with the known high electrical resistivity and low O_2 permeability of CACC relative to typical PCC may offer a more rational description of corrosion potential in CACC than the often cited fact that the initial pH of hydrating CAC is lower than that of PCC, (although still 2 orders of magnitude above the threshold value of 9.5 depicted in the Pourbaix diagram).

Resistance to Acidic Conditions

CAC is also widely used for acid resistant mortars and concretes, traditional applications have been for acidic conditions of pH down to about 3.5 to 4. The good performance obtained comparative to PC based systems is attributed to the protective action of AH_3 and the absence of $Ca(OH)_2$. There has been a reluctance in the past to recommend CAC for lower pH's where AH_3 becomes soluble.

It is also frequently argued that resistance will be impaired when conversion occurs because of the associated increase in porosity.

Recent work has created a different perspective.

J.-P. Letourneux and co-workers [14] have made an extensive study of neat CAC pastes and mortars of CAC cured at 20°C and at 50°C. 20 x 20 x 50 mm prism specimens were immersed in acid solutions of constant volume which were renewed twice daily. Weight loss of the specimens was then measured as a function of the number of renewals. The ratio of the weight losses of unconverted (20°C) to converted (50°C) specimens after 20 solution renewals is shown in table 4.

Table 4. Relative corrosion loss of unconverted and converted CAC pastes in acidic solutions

Acid	Weight loss ratio, (20°C/50°C)
Acetic	1.88
Citric	3.57
Lactic	2.10
Maleic	3.61
Oxalic	3.00 (wt. gain)
Succinic	0.78
EDTA Na	1.93
Hydrochloric	2.58
Nitric	3.31
Orthophosphoric	2.17
Sulphuric	0.85

With rare and minor exceptions the converted pastes are clearly more resistant despite a major difference in porosity (about 2% for the 20°C cure and about 15%

for the 50°C cure). The greater thermodynamic stability of the cubic hydrate obviously plays a dominant role over porosity.

Another parameter which has been highlighted by this study is the importance of the solubility of the reaction products of acid attack. The reaction of CAC with acids begins by the decomposition of the calcium aluminate hydrates forming a calcium salt and AH_3. If the calcium salt is soluble, attack occurs readily, and vice versa.

Calcium nitrate, for example, is highly soluble (1212 g/kg) and the weight loss after 20 solution renewals of 0.1 N nitric acid is 14.9% for the 20°C cure.

Calcium sulphate is much less soluble (2.4 g/.kg) and the associated weight loss under the same conditions is only 2.8%. With oxalic acid the calcium salt has a solubility of only 0.007 g/kg and samples actually gain in weight. Based on these findings it has been possible to recommend CAC bonded materials for pH values significantly below the traditional threshold of 4 in a number of cases.

Of particular interest here is the use of CACC in sewer systems, [15] where pH levels below 2 readily occur. The acid generated biogenically is sulphuric acid and the protective effect of the low solubility gypsum reaction product has allowed many such installations to perform well over many decades.

But this is not the complete explanation. Tests in a simulation chamber at 30°C in an H_2S atmosphere on mortar specimens inoculated with the bacteria that generate sulphuric acid in sewers [15] have demonstrated that CAC bonded systems are considerably less attacked than PC bonded systems and that concomitantly, in the case of CAC only, the bacterial cell count diminishes with time. This effect appears to be linked to the Al^{+++} ion and is still under investigation.

One further factor to be taken into consideration in the assessment of the suitability of materials for use in acidic environments is the neutralisation capacity of the material exposed to the acid. This is particularly important at low pH (< 4) were $Al(OH)_3$ protection is not present, as table 5 [16] shows.

Table 5. Comparative Neutralisation Capacity of CAC and PC systems

low iron CAC composition	mM/g	mM of H^+ required to neutralise 1mM	mM of H^+ consumed per g cement
Al_2O_3	5.10	6	31
Fe_2O_3	0.13	6	0.8
CaO	6.79	2	13
acid insoluble	-	0	0
			45 mM/g
Portland cement composition	mM/g	mM of H^+ required to neutralise 1mM	mM of H^+ consumed per g cement
Al_2O_3	0.29	6	1.7
Fe_2O_3	0.13	6	0.8
CaO	14.78	2	24
acid insoluble	-	0	0
			27 mM/g

It is clear that a given level of attack will require nearly 70% more acid in the case of CAC than for a PC based material, which in practice would translate into a much longer service life as is, in fact, observed.

Summary

Aluminous Cements (CAC) originated and were developed logically in response to needed improvements in resistance of concrete to sulphates and sea water. The vulnerability of $Ca(OH)_2$ in PC to both remains a key factor in explaining the better performance of CAC. However, the ability of CACC to develop a dense surface microstructure is emerging as an intrinsic advantage.

Carbonation of CACC has traditionally been regarded as potentially more dangerous for steel reinforcement corrosion than in PCC. Recent studies provide no support for this, either in the laboratory or in the field. Thermodynamic analyses indicate that carbonation progresses at alkalinities well above the threshold for steel depassivation.

Other studies have shown how the usefulness of CACC in acidic conditions can be extended to pH levels well below the traditional limit based simply on the pH related solubility of AH_3. It is quite clear that the greater thermodynamic stability of C_3AH_6 can dominate corrosion resistance. A factor more important than porosity is the solubility or insolubility of the reaction products of acid attack. Additionally, the neutralisation capacity of CAC bonded systems considerably exceeds that of similarly designed PC systems, which provides a simple and elegant explanation of the longer service life of CACC observed in practice.

Acknogledgements

I am extremely indebted to all those on whose published work I have drawn in preparing this presentation. I particularly appreciate my colleagues at Lafarge Research, J.-P. Letourneux, S. Marcdargent, K. Scrivener and M. Verschaeve discussing and making generously available their original work.

References

1.   Lea, F.M. (1970) *The Chemistry of Cement and Concrete*, Edward Arnold Publishers.
2.   Crammond, N.J. (1990) *Proceedings International Symposium on Calcium Aluminate Cements*, E. and F.N. Spon, p. 208.
3.   Bied, J. (1923) Bulletin Society Encourag. Pour Industrie Nationale, pp. 31-43.
4.   Br. Pat. (1909) 8193.
5.   George, C.M. (1990) *Proceedings International Symposium on Calcium Aluminate Cements*, E. and F.N. Spon, p. 181.
6.   George, C.M. (1980) ACI SP-65, pp. 331, 339, 341.

7.   Houghton, S.J. and Scrivener, K.L. (1994) *A microstructural study of a 60 year old Calcium Aluminate Cement concrete.* Proceedings of the 3rd CANMET/ACI International Conference on Durability of Concrete, Nice, France.
8.   Aitcin, P.C., Blais, F. and George, C.M. (1995) *Durability of Calcium Aluminate Cement Concrete: assessment of concrete from a 60 year old marine structure at Halifax,* NS, Canada, Symposium Advances Concrete Technology, Las Vegas, pp. 145-167.
9.   Dunster, A.M., Crammond, N.J., Pettifer, K. and Rayment, D.L. Presented at MRS Symposium R, Boston, Ma. Nov. 27-Dec. 1, 1995 and to be published by E. and F.N. Spon in the proceedings.
10.   Dunster, A.M., Bigland, D.J., Hollinshead, K. and Crammond, N.J. awaiting publication.
11.   Crammond, N.J. and Currie, R.J. (1993) Magazine Concrete Research, Vol. 45, No. 165, pp. 275-279.
12.   Verschaeve, M. Lafarge Aluminates, France. Previously unpublished work based on the approach developed by Barret, et al: Cement and Concrete Research (1983), Vol. 13, pp. 789-800.
13.   Gaztanaga, M.T., Goni, S. and Sagrera, J.L. (1992) Materiales de Construccion, Vol. 42, No. 228, pp. 65-77.
14.   Bayoux, J.P., Letourneux, J.P., Marcdargent, S. and Verschaeve, M. (1990) *Proceedings International Symposium on Calcium Aluminate Cements,* E. and F.N. Spon, p. 230.
15.   Sand, W., Dumas, T. and Marcdargent, S. (1994) ASTM STP 1232, pp. 234-239.
16.   Marcdargent, S. Lafarge Recherche, France. Previously unpublished work.

31 AN ASSESSMENT OF AGEING CALCIUM ALUMINATE CEMENT CONCRETE FROM MARINE STRUCTURES

A.M. DUNSTER, N.J. CRAMMOND, K. PETTIFER and D.L. RAYMENT
Building Research Establishment, Garston, Watford, UK

Abstract
The condition of calcium aluminate cement (CAC) concrete cores taken from two old marine structures has been assessed. Measurements of oxygen permeability and sulphate and chloride penetration were undertaken and the microstructure was examined.

Both concretes have exhibited good durability. Oxygen permeability coefficients for the surface were at least two orders of magnitude lower than the interior for both structures. Chloride levels at the depth of the steel were estimated to be < 0.5 % (by weight of cement) and the embedded steel was uncorroded.
Keywords: Calcium aluminate cement, chlorides, marine durability, microstructure.

1 Introduction

Calcium aluminate cements (CAC, also known as high alumina cements), were developed for use in locations where rapid hardening and/or sulphate resistance was required. The work described here focuses on two concrete structures built in the early 1930's using the calcium aluminate cement (CAC) Ciment Fondu. One, built in Essex, UK at Dagenham, is a large, estuarine piled foundation. The other, in the port of Halifax, Nova Scotia, Canada is a marine freight terminal. Both these structures are now more than sixty years old and provide the opportunity to study the condition of ageing CAC concrete following exposure in aggressive marine environments. The work forms a part of a co-operative study of CAC concrete durability involving several laboratories. Details are given in references [1] and [2].

Mechanisms of Chemical Degradation of Cement-based Systems. Edited by K.L. Scrivener and J.F. Young.
Published in 1997 by E & FN Spon, 2–6 Boundary Row, London SE1 8HN. ISBN: 0419215700.

2 Historical background

2.1 Halifax Pier B
Halifax Pier B consists of hollow, gravel filled concrete cribs (each with a wall thickness of approx. 0.53 m) which were sunk to the ocean bed following manufacture. After the cribs had been sunk, a protective layer of facing concrete (approx. 0.30 m thick) was cast over their outside faces in the tidal zone using either the same CAC as the cribs, or Portland cement. Records indicate that the CAC concrete was cast with a w/c ratio of 0.5 to 0.6 and a cement content of about 330 kg/m^3. As work progressed, the w/c ratio was increased to improve workability, to a maximum of 0.70. The climatic conditions at Halifax are extremely severe: it is estimated that the exposed concrete is subjected to about 100 freezing and thawing cycles per year.

2.2 Ford, Dagenham
Several thousand pre-cast reinforced hexagonal concrete piles were driven into the bank of the River Thames at Ford's Dagenham plant to support a power station. Records show that the mix for the piles had a cement content of approx 300 kg/m^3 and it is likely that the w/c ratio was above 0.5. As the site is an estuarine location, salt water covers much of the area at high tide. The diameter of the piles was 0.4 m and the length 25 m.

3 Experimental

3.1 Sampling and test procedures
All of the cores examined (approx. 150 mm and 100 mm diam), included the external surface of the structure. The cores from Pier B had been cut from the facing concrete in the intertidal zone as well as from the crib concrete under water. Details are given in ref [2].

A total of nine cores (five from Halifax Pier B; four from Ford Dagenham), were examined in the present study. Three of the Halifax cores (which were extracted by Lafarge in 1993), were taken from the inter-tidal zone and two from the submerged zone. All the Dagenham cores were taken by BRE in 1992 from a section of pile exposed in the inter-tidal zone. The concrete cores were cut in the laboratory using a diamond saw to produce suitable specimens for the following schedule of tests: oxygen permeability [3], chemical analysis, X-ray diffraction (XRD), Differential Scanning Calorimetry (DSC), optical microscopy and scanning electron microscopy (SEM).

Depth profiled samples of some of the cores from Pier B were obtained for chemical analysis by drilling. Powdered samples were collected to depths of up to 30 mm in increments of 5 mm. Cores from Dagenham were sampled in coarser increments to a greater depth by taking lump samples from sawn slices in increments of 25 mm closest to the outer surface and increments of 50 mm at depth. The powdered concrete samples were subjected to determination of cement and chloride content. The cement content of the powdered concrete samples was determined from the Al$_2$O$_3$ content in accordance with BRE Digest 392[4].

Petrographic thin sections were prepared from each structure perpendicular to the exposed surface. Two of these sections were subsequently carbon coated and examined under the SEM.

4   General condition of the structures and visual assessment of cores

4.1 Halifax Pier B
The authors were unable to inspect this structure. However, reports suggest that it was in good condition at the time of coring[1,2]. Plastic shrinkage and other macro cracks were identified in the cores taken from the cribs under water. Steel reinforcement was not encountered in the crib concrete cores. Where steel was encountered in the cores from the facing concrete (at three locations), the depth of cover was at least 90 mm. The diameter of the steel was 20 mm and there was no sign of reinforcement corrosion.

4.2 Ford Dagenham
The structure was apparently in good condition. No signs of cracking or spalling were noted during the site visit and the concrete was hard and dense. The pile examined was heavily reinforced longitudinally with bars 25 mm in diameter which were encircled by a series of steel hoops (10 mm diam.) perpendicular to the length of the main reinforcing bars. Three of the cores examined contained plain steel reinforcement. The minimum depth of cover to the main steel was 52 mm and the minimum depth of cover to the hoops was 32 mm. There was no sign of reinforcement corrosion in any of the steel examined.

5   Results of laboratory tests

5.1   Oxygen permeability
The coefficients of oxygen permeability k_0 for specimens taken from the surface (0-50 mm) and interior of both Halifax and Dagenham structures are summarised in Table 1. Each value represents the average of two Halifax Pier specimens and one Dagenham specimen. Permeability results from structures can be very variable and it is unwise to draw too many conclusions from the results of a few specimens. However, it is clear that the permeability of at least part of the surface (0-50 mm) is approximately two orders of magnitude less than for the interior. Values were of the order 0.6 to 1.7 x 10^{-17} m^2 for the surface, 187 to 385 x 10^{-17} m^2 for the interior. The permeability of the interior is in the range 200-400 x 10^{-17} m^2 which is rated as "high permeability" in Concrete Society Report No. 31[5].

5.2 Chemical analysis, mineralogy and chloride penetration
Samples from the surface and interior of the Halifax crib concrete and the Dagenham concrete were examined using XRD and DSC. The interior of both concretes was highly converted whereas the surface zone (0- 5 mm) had a low to medium degree of conversion (as defined by BRE Current Paper CP34/75[6]). The presence of chloroaluminate (C_3A $CaCl_2$. 10 H_2O) in both the surface and interior samples was indicated by XRD.

The variation of the chloride levels (by weight of cement) with depth in the powder samples is summarised in Figure 1 and the following sections.

5.2.1 Halifax Pier B

The chloride contents were greatest in the submerged environment and decreased with depth in both exposure environments (Figure 1). Although the chloride content in the surface was relatively high (> 3 % Cl by weight of cement) in both concretes, the levels declined rapidly with depth. The chloride levels near the surface of the submerged crib concrete were both between two and five times higher than the facing concrete in the tidal environment, (note the different y axis scales). At depth however, the levels were comparable in both environments.

5.2.2 Ford Dagenham

The samples for chemical analysis were obtained by lump sampling rather than drilling and therefore represent a much coarser profile than the Halifax samples but they extend to a much greater depth (Figure 1). The chloride levels at the depth of the steel were estimated to be approx. 0.5 % by weight of cement and there was no sign of corrosion in either the reinforcing bars or the hoops.

5.3   Petrography and Scanning Electron microscopy
The concrete samples examined were all in fairly good condition although the outer surfaces were slightly eroded. However, the cores from both structures contained two forms of chloro-aluminate which were products of chloride ingress. One exhibited low first order birefringence, which ranged from low first order to higher second order under the optical microscope (cross polars). The other exhibited high birefringence. Carbonation had penetrated 3 to 7 mm. No difference in the microstructure of the matrix between the surface and that at greater depth was discernable optically in any of the specimens examined.

Table 1: Oxygen permeability results

structure	coefficients of oxygen permeability ($x10^{-17}m^2$)					
	depth range (mm)					
	0-50	50-100	100-150	150-200	200-250	250-300
Halifax Pier B	6.3	257	-	-	-	-
Ford Dagenham	1.1	307.2	236.8	316.4	376.4	379.7

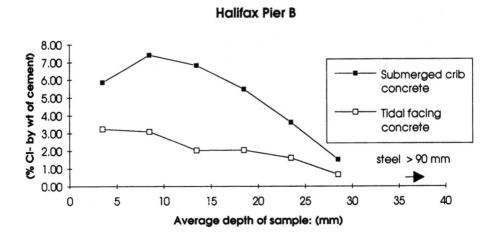

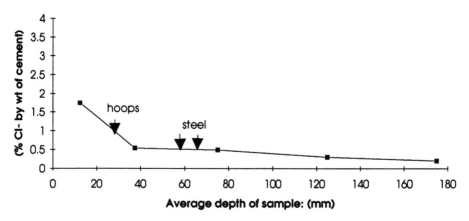

Fig. 1. chloride ion penetration data

5.3.1 Halifax Pier B- Crib wall (submerged)

The surface showed slight cracking infilled with calcite. More significant cracking had occurred at depth (40-50 mm). Many of the cracks contained the two forms of chloroaluminate. The amount of chloroaluminate decreased with depth from the surface and was present (albeit in small amounts) at 120 mm.

5.3.2 Halifax Pier B- Facing concrete (tidal zone)

There was slight cracking in the surface zone (1-2 mm depth) sub-parallel to the surface and infilled with the birefringent form of chloroaluminate. There was no evidence of large-scale macrocracking in the section and the amount of cracking (both micro and macro) was less compared with the fully immersed sample described above. Chloroaluminates were most abundant near the surface becoming less so (but still very evident) at depth. Generally chloroaluminates were less plentiful than in the fully immersed sample.

The same thin section was examined under the SEM. The facing concrete showed the presence of three distinct zones with differing hydrate compositions:

- Zone 1 hydrate The hydrate zone nearest to the surface was a magnesium aluminate hydrate with a maximum thickness of around 3 mm with a composition very similar to a magnesium aluminate hydrate $2\ MgO.Al_2O_3.9.9H_2O$ described by Lea[7]. The combination of high magnesium and low calcium suggests a process of ionic exchange between Mg in seawater and Ca in the CAC hydrates. Chloride levels were relatively low.
- Zone 2 (with an approximate thickness of 1 mm to 1.5 mm), was thinner than zone 1 and contained pockets of calcium aluminate hydrates and interstitial paste comprising magnesium aluminium hydrate together with small amounts of brucite. There was thus some evidence for a transition zone between the magnesium rich zone 1 and the bulk concrete.
- Zone 3 Comprised the bulk of sample from about 4 mm in from surface to 65 mm,(the full depth of the thin section). Both calcium and chloride levels were generally higher in zone 3. Optical microscopy had shown the presence of small crystals of chloroaluminate near the outer surface (in zones 1 and 2). This suggests that the chloride has tended to concentrate in the hydrate phases in zone 3 and in chloroaluminate rather than hydrate in zones 1 and 2 where greater chloride concentrations might be expected.

5.3.3 Ford Dagenham

Reaction products (the two forms of chloroaluminate and ettringite) were present mainly at aggregate-cement interfaces or in voids and cracks. Chloroaluminates were most common in the 35 mm nearest the original exposed surface although they were present throughout (to 200 mm depth).

6 Discussion

The CAC concrete in both structures is typified by a highly converted interior and an external skin with a lower degree of conversion. This is consistent with internal heating during casting causing conversion but leaving a surface layer of denser and apparently unconverted material some 5 to 15 mm thick. This "skin effect", which has been noted elsewhere[1,2], may have reduced sulphate and chloride ingress to the more permeable interior of these concretes. Further evidence of the skin effect is provided by the oxygen permeability results. This study has shown that there has been significant ingress of chloride into the concrete in both structures. Nevertheless , both structures had performed well for over 60 years and there was no corrosion of the reinforcement.

The concrete cores from Halifax Pier B examined were generally in good condition and relatively free of cracking. However, clear signs of cracking were observed in the interior of both cores from the submerged crib concrete. In contrast, the facing concrete (nominally to a similar mix design) was uncracked. These findings contrast with those reported by Blais et.al[2] who found significantly lower compressive strengths for the facing concrete compared with the crib concrete which were attributable to cracking caused by freeze thaw action. The origin of the cracking observed in the crib concrete during the present study is unclear although it may have a thermal origin associated with the greater wall thickness compared with the facing concrete. The cracking observed, and likely associated increased permeability, may be the cause of the higher chloride levels in outer 30 mm of the submerged crib concrete.

7 Conclusions

- The condition of calcium aluminate cement concrete cores from two 60 year old marine structures has been examined. These concretes were well made and have exhibited good durability.
- The hydration of the outer region of both structures produces a zone of low conversion which appears to have protected the interior from the penetration of deleterious ions.
- The permeability of at least part of the surface (probably the outer 5 mm to 15 mm) is approximately two orders of magnitude less than for the interior.
- One of the concretes (the crib concrete from Halifax Pier B), showed evidence of significant cracking in the interior. However, this has not seriously affected the long term durability.

8 Acknowledgements

The programme of work described is supported by the UK Department of Environment and Lafarge.

9 References

1. Capmas, A. and George, C.M. (1994) *Durability of calcium aluminate cement concretes.* Advances in Cement and Concrete Research. Proc. Eng. Foundation. Conf. (Ed. M.W. Grutzeck and S.L. Sarkar), Amer. Soc. Civ. Engnr. NY.
2. Blais, F. Aitcin, P. C. and George, C. M. (1995) *Durability of calcium aluminate cement concrete*: Assessment of concrete from a 60 year old marine structure at Halifax, NS, Canada. Proc. 2nd CANMET/ACI Int. Symp. Advances in Concrete Technology (Ed. V. M. Malhotra), Las Vegas, USA, ACI SP-154, pp 145-168.

3. Lawrence C. D. (1986), *Measurements of permeability*. Proc. 8th Int. Cong. Chemistry of cement, Rio de Janeiro,
4. Building Research Establishment (1994). *Assessment of existing high alumina cement concrete in the UK*, Digest 392. BRE, Garston.
5. *Permeability of concrete and its control* (1988) Concrete Society Technical report No. 31.
6. Building Research Establishment (1975). *HAC concrete in buildings*, BRE Current Paper CP34/75, Figure 8, p 13, (50pp), BRE, Garston.
7. Lea, F. M. (1983), *The chemistry of cement and concrete*, 3rd Edn. Edward Arnold, (London), p217.

32 BEHAVIOUR OF ALUMINOUS CEMENT IN HIGHLY ALKALINE MEDIA

S. GOÑI, C. ANDRADE, J.L. SAGRERA, M.S. HERMÁNDEZ and C. ALONSO
Institute "Eduardo Torroja" of Construciton Science (CSIC), Madrid, Spain

Abstract
In this work a hypothesis is presented to explain the alkaline degradation process of calcium aluminate cement concrete (CACC). It is based on x-ray diffraction (XRD) data of samples taken from real Spanish CACC structures. The identification from XRD data of a hydrated alkaline aluminate could serve as a guide to differentiate between processes: normal carbonation and alkaline hydrolysis.
Keywords: aluminous cement, carbonation, alkaline hydrolysis.

1 Introduction

Research on degradation processes of calcium aluminous cement, CAC, has been very limited because most countries forbade its use in structural concrete. That is why present knowledge on some of the intermediate steps and on the evolution of calcium aluminate phases in severe environments, is still very uncertain.

In Spain, failures in CAC structures were noticed from the 1960's when extensive use of this cement enabled quick erection of new buildings and apartments. Its use was therefore declining and although still allowed, its present consumption for structural applications is negligible.

However, a failure with human consequences of a floor under a terrace in Barcelona in November 1990, has led to extensive surveys in order to find out the degree of evolution or of damage, in buildings made with CAC. The number of apartments fabricated with CACC was evaluated at about 500.000 in the region of Barcelona alone.

Furthermore, the first chemical analysis of the concretes presenting dramatic loss

Mechanisms of Chemical Degradation of Cement-based Systems. Edited by K.L. Scrivener and J.F. Young. Published in 1997 by E & FN Spon, 2–6 Boundary Row, London SE1 8HN. ISBN: 0419215700.

of strength or reinforcement corrosion, disclosed that the evolution of CAC phases was not as had been reported in previous literature [1-5]. Multiple and different situations have being detected which do not fit well into previous classical interpretations. As a consequence, basic research is again developing in Spain to interpret the practical cases analyzed as part of the extensive surveys undertaken by local authorities and to assess the structural conditions of affected buildings.

1.1 Degradation processes of CAC

The most widely identified degradation process was that termed "**conversion**". It supposes the crystallographic evolution of calcium aluminate hydrate from hexagonal to cubic form:

$$3(CaO.Al_2O_3.10H_2O) \rightarrow 3CaO.Al_2O_3.6H_2O + 2(Al_2O_3.3H_2O) + 18H_2O \qquad (1)$$

This conversion was said to proceed with a loss of strength, to which were attributed the failures produced in U.K. during the sixties [5].

In addition to this kind of phase transformation, "**Alkaline Hydrolysis**", AH, is mentioned by Neville [1] although Lea [6] in 1940, identified the process. It occured following the reactions:

$$K_2CO_3 + CaO.Al_2O_3.aq. \rightarrow CaCO_3 + K_2O.Al_2O_3 \qquad (2)$$
$$CO_2 + K_2O.Al_2O_3 + aq. \rightarrow K_2CO_3 + Al_2O_3.3H_2O \qquad (3)$$
$$CO_2 + CaO.Al_2O_3.aq. \rightarrow CaCO_3 + Al_2O_3.3H_2O \qquad (4)$$

According to these reactions, the process leads to the complete destruction of the concrete due to the catalytic effect of carbonates.

Although the description of the process is frequently cited in the literature, only Neville reports two practical cases [1]. He describes structures made part with OPC and part with CAC which suffered from capillar absorption movements which led to the destruction of the contact zone.

Following observation of these cases it was forbidden to contact the two types of cement. However, in opposition to this finding, mixes of 50% of both types of cements are currently used as "instantaneous setting cement" or as repair materials, without reports of undesirable behaviour.

In addition to conversion and AH, the third process which significantly affects calcium aluminate phases is **carbonation.** This process has been extensively studied by Spanish researchers [2][3][9], following its detection in CAC structures. Reinforcement corrosion was also attributed to carbonation [3].

The carbonation may proceed following different intermediate steps [2] although the final products are always a mix of gibbsite and its polymorphs: nordstrandite and bayerite, and calcium carbonate in its polymorphic varieties: calcite, vaterite, aragonite:

$$CO_2 + CaO.Al_2O_3.aq. \rightarrow CaCO_3 + Al_2O_3.3H_2O \qquad (5)$$

Thus the final products of simple carbonation and of AH are the same. Therefore no chemical distinction on mechanisms can be made when the structure has reached its final state. The identification seems only feasible at intermediate steps.

However, the structural consequences reported in the literature are completely different depending on whether carbonation or AH is the evolutive process. While carbonation may significantly increase the mechanical strength to beyond 100 Mpa [9], the AH process may completely destroy the stony consistency.

In present paper an attempt to differentiate carbonation processes and alkaline hydrolysis is given. The differentiation is made by identifying by XRD the presence of an alkaline hydrated aluminate ($Na_2Al_2O_4 \cdot 6H_2O$)[10]. Although mentioned by Lea [6] as an intermediate product, no attention seems to have been paid to it, and so no steps towards its identification were taken in structures affected by CAC.

A classification of the different degradation processes will be proposed to avoid confusing descriptions and miss diagnosis.

2 Experimental

The identification of the alkaline aluminate hydrate was made in one of the beams of the failure aforementioned to happen in 1990 in Barcelona. In one beam a dark and wet zone, which could be disgregated with the fingers was identified, while in adjacent dry zones, the colour was lighter and the concrete kept a reasonable strength.

The aggregates were of granitic nature. Granite is known to be able to gradually liberate alkalies, and therefore, the situation was favourable to suspect that AH was the reason for failure.

Barcelona has a mild and humid climate due its location on the Mediterranean coast.

For comparative purposes, in addition to the wet and dry samples of the collapsed beam, a third sample was analyzed. It was taken from another structure located in Madrid (continental dry climate), also in an advanced state of degradation. The colour of this sample was brown (typical "chocolate") and its mechanical strength was acceptable.

XRD analyses were carried out with a PHILIPS PW 1730 diffractomer, using a graphite monochromator and Cu $K_{\alpha 1}$ radiation.

The aggregates were separated in order to analyze only the cement enriched part of the sample.

3 Results

3.1 X-ray diffraction analysis

The XRD patterns of the three samples are presented in Fig. 1. The following aggregate crystalline components were identified:

α-quartz: (α-SiO_2)   muscovite 2M: (KAl_2 (Si_3Al) O_{10} $(OH)_2$)
albite: ($NaAlSi_3O_8$)   microcline max: ($KAlSi_3O_8$)

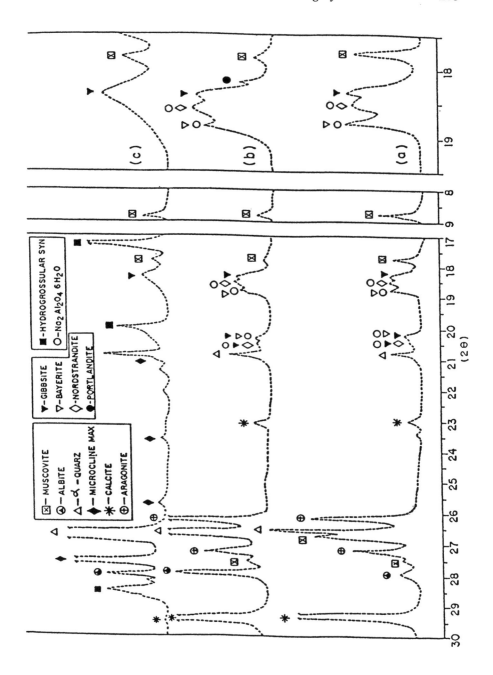

Fig. 1. XRD patterns of real Spanish concrete samples.

These results suggest that the aggregate could be granitic. In addition, in the two samples from Barcelona (exposed to a hot and moist environment) calcite and aragonite were detected. No calcium aluminate hydrates appear in these samples. The sample (c) from the village near Madrid (exposed to a drier environment) is not carbonated (see Fig. 1 (c)). In this sample only the cubic phase ($Ca_3Al_2O_6.6H_2O$) is detected, which seems to indicate that the only process which has happened has been conversion. Paying attention to the 18-19, 2θ angular zone, a wide ray peak develops, in which three maxima and shoulders are clearly manifested in samples (a) and (b). Nevertheless, for the sample (c) only an asymmetric peak with a maximum at 18.3, 2θ is observed. In this angular zone appears the most intense peak of three $Al(OH)_3$ polymorphic varieties as well as of $Na_2Al_2O_4 \cdot 6H_2O$.

Table 1. XRD-peaks appear in the 18-19, 2θ angular zone.

	2θ	Å	I
Gibbsite (γ-$Al(OH)_3$)	18.3	4.84	100
Nordstrandite ($Al(OH)_3$)	18.5	4.79	100
Bayerite (β-$Al(OH)_3$)	18.8	4.71	90
$Na_2Al_2O_4 \cdot 6H_2O$	18.6	4.71	100
	18.8	4.78	100

Therefore, all of these compounds could be present in samples (a) and (b). Assuming the possible presence of the alkaline aluminate hydrate, its formation could be due to an alkali-calcium-exchange from the albite of aggregates. This ionic exchange could be carried out via "dissolution", the liberated alkalis initiating an alkaline hydrolysis process: Calcium ions in the pore solution could react with the atmospheric CO_2 producing calcium carbonate, or could perhaps precipitate portlandite. So the shoulder appearing at 18.1, 2θ in the sample (b) could also be due to portlandite.

4 Discussion

4.1 Existence of an alkaline aluminate
In spite of its mention from a very early work, it has never been identified or considered as a chemical indication of the development of the alkaline hydrolysis process. However, this compound might explain the loss in the mechanical strength assuming that it has a lower cementitious nature.

The other alternative to justify the loss of mechanical strength linked to the presence of moisture, might be the swelling of the aluminium hydroxide. However, although this uptake of water is feasible, the dramatic loss of cementitious character seems to show at least the formation of a new compound, different from the calcium aluminates. On the other hand, the alkaline aluminate may also adsorb water, which also might justify the dramatic difference in mechanical strength between wet and dry parts of the same piece of the concrete beam.

4.2 Stability of aluminous species as a function of the pH value

Aluminium being an amphoteric ion, the pH value will influence the solubility of the different compounds and the type of cation present in the solution. The Pourbaix diagram shows that the $Al(OH)_4^-$ ionic species is stable at both acid or very alkaline pH values. Between pH values of approximately 5 and 10-11, $Al(OH)_3$ may precipitate. Also, calcium aluminate hydrate has largely proved to be stable in this intermediate pH range.

There are no data for the pore solutions of real CACC structures, but as is shown in Fig.2 (taken from previous results [7][8]), the aluminate content of the pore solution of synthetic CAC pastes, strongly depends on the pH value: Calcium aluminate becomes more soluble as the pH increases. This dissolution, leading to the increase of $Al(OH)_4^-$ ions, would continue until the solubility product of the alkaline aluminate is reached. The more feasible sequence of reactions is:

$$Ca(Al(OH)_4)_2 \, 6H_2O + 2Na,K(OH) \qquad\qquad (6)$$

$$\qquad\qquad \uparrow\downarrow \quad (6\text{-}I) \qquad\qquad\qquad\qquad\qquad (6\text{-}III)$$

$$2(Na^+, K^+) + Ca^{++} + 2OH^- + 6H_2O + 2Al(OH)_4^- \rightleftharpoons 2Al(OH)_3\downarrow + 2OH^-$$

$$\qquad \uparrow\downarrow \quad (6\text{-}II)$$

$$Ca(OH)_2\downarrow + (Na,K)_2 \, (Al(OH)_4)_2 \, 2H_2O\downarrow + 4H_2O$$

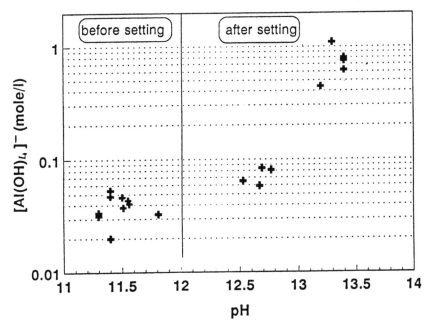

Fig. 2. Aluminate content versus pH values of pore solutions (taken from ref. [7] and [8]).

The $Al(OH)_4$ species may either initiate precipitation of the alkaline aluminate hydrate and $Ca(OH)_2$ [6-II] or, evolve towards the aluminium hydroxide [6-III], or both reactions may proceed simultaneously. While reaction 6-II produces a lowering of the pH value due to the precipitation of $Ca(OH)_2$ and an alkaline aluminate hydrate, a kind of buffering effect is given by the reaction 6-III which liberates OH^- ions.

This process, although chemically simple, has not been previously described in the literature concerning CACC and therefore the authors propose to name it **alkaline dissolution** (the same process involved in the activation of the aluminous phases of slags or fly ashes when they are in contact with high pH solutions eg from hydrated OPC), to distinguish it from **alkaline hydrolysis**, which will be considered next.

4.3 Alkaline Hydrolysis

What is then the action of atmospheric CO_2 on the CACC structures?. The process has to be split in two alternatives: 1) carbonation with high pH values in the pore-solution and 2) carbonation with relatively low pH values.

4.3.1 Carbonation with high pore-solution pH values

The presence of carbonate species will influence the reaction sequence of the alkaline dissolution by two effects: the low solubility of $CaCO_3$ and the buffer capacity of the $CO_3^=/HCO_3^-$ species. Thus, CO_2 will act on reactions [6] by precipitating $CaCO_3$ and leaving in solution soluble $CO_3^=$ ions as follows:

$$(Na,K)_2\,(Al(OH)_4)_2\,2H_2O\!\downarrow + Ca(OH)_2 + 2CO_2 + 4H_2O \qquad (7)$$
$$\uparrow\downarrow$$
$$CaCO_3\!\downarrow + 8H_2O + CO_3^=\,(Na^+,K^+)_2 + 2Al(OH)_3\!\downarrow \qquad (7\text{-}I)$$
$$\uparrow\downarrow$$
$$CO_3H^- + 2Na^+,K^+ + OH^- \qquad (7\text{-}II)$$

Therefore, inducing a decrease of the pH value tends to be buffered by the production of OH^- which in turn dissolve more calcium aluminates (6). Finally, the stable solid products will be $Al(OH)_3$ and $CaCO_3$. The liberation of $CO_3^=$ and OH^- will induce a cycling process, as was described by Lea [6].

Assuming the complete transformation of aluminates into these final species, the pH might decrease until reaching a value of 10.4, where the buffer $CO_3^=/HCO_3^-$ exists. Only higher partial pressures of CO_2 could aim at a pH decrease where reinforcement corrosion was feasible. This process is what has been named in the literature as **alkaline hydrolysis**, although it may be also named **alkaline carbonation**. The loss in strength could be attributed to both, an increase of the porosity and the alkaline aluminate hydrate formation, although in very late stages the mechanical strength might be recovered if this compound is finally converted into $CaCO_3$ and AH_3. The alkaline aluminate hydrate may also swell in wet conditions, inducing loss of strength.

4.3.2 Carbonation with low pore solution pH values

The same final compounds are formed from a carbonation process when the pH value of the pore-solution is insufficiently high for alkaline aluminate hydrate to form. In this case the relevant intermediate species are the carboaluminate reported in the literature:

$$Ca_3Al_2O_6 .6H_2O \quad + CO_2 \quad \rightarrow \quad Ca_3Al_2O_6 .4H_2O (CO_2) + 2H_2O \quad (8)$$
$$Ca_3Al_2O_6 .4H_2O (CO_2) + CO_2 \quad \rightarrow \quad Ca_3Al_2O_6 .2H_2O. 2(CO_2) +2H_2O \quad (9)$$
$$Ca_3Al_2O_6 .2H_2O. 2(CO_2) + CO_2 \quad \rightarrow \quad Ca_3Al_2O_6 . 3(CO_2) \qquad +2H_2O \quad (10)$$
$$\downarrow$$
$$Al_2O_3.xH_2O + CaCO_3 \text{ (aragonite)}$$

This process may be termed **normal carbonation**.

5 Conclusions

The evidence found in real structures that failed recently in Spain could not be fitted into the traditional description of the processes found in the literature relative to carbonation of CACC and to alkaline hydrolysis.

A new interpretation is presented here that is not yet complete, and needs extensive experimental confirmation, but explains much better the evidence available in Spain. The main proposals of this paper are:

1°) An alkaline hydrated aluminate could be identified by XRD linked to the cases where loss of strength is sometimes detected.

2°) Beyond the traditionally identified "conversion" the other processes which are likely to happen are:

- alkaline dissolution - dissolution of calcium aluminate hydrate in very highly alkaline media, with probable further precipitation of alkaline aluminate hydrate, and a mixture of $Ca(OH)_2$ and $Al(OH)_3$, which will depend on the pore solution composition.

- alkaline hydrolysis - occuring when CACC is carbonated in the presence of large amounts of alkaline ions, via the formation of sodium aluminate hydrate and $Ca(OH)_2$ as intermediate species. The presence of CO_2 means an acceleration of the dissolution process with formation of $CaCO_3$, $Al(OH)_3$ and alkaline carbonates as final products.

- normal carbonation - the process developed when the amount of alkalies is small, via the formation of carboaluminates as intermediate products.

3) The humidity seems to play a fundamental role in the degradation processes due to the particular character of aluminium gels and the hygroscopicity of the alkaline aluminate. The water absorption-desorption processes may contribute to reduce the mechanical strength by inducing a certain swelling.

Acknowledgments

The authors are grateful to the architects, V. Seguí and J. Aldamiz, who provided the samples analyzed. They are also grateful to the CICYT of Spain for the funding given to this research. Finally, they would like to recognize the useful discussions on the subject with M. Verschaeve, A. Capmas and M. George of Lafarge, as part of the meetings of the group "Durability of CACC" led by M. George.

5 References

1. Neville, A. (1975) High Alumina Cement Concrete. The Construction Press, Great Britain.
2. Vázquez, T., Triviño, F. and Ruiz de Gauna, A. (1976) *Estudio de las transformaciones del cemento aluminoso hidratado. Influencia del anhídrido carbónico, temperatura, humedad y adición de caliza en polvo.* Monography at the Eduardo Torroja Institute, Nº 334 ICCET, Spain.
3. Pérez, M., Triviño, F. and Andrade, C. (1981) Corrosion de armaduras galvanizadas y sin proteger embebidas en cemento aluminoso estabilizado. *Materiales de Construcción*, Vol. 182, pp. 49-68.
4. Capmas, A. and George, C.M. Durability of calcium aluminate cement concretes, Advances in Cement Concrete Research, Proc. of the Conference of Engineering Foundation, Wutzeck, M.W. and Sarkau, S.L. (Eds), Publishers, American Society of Civil Engineerings, New York, pp.377-405.
5. Midgley, H.G. and Midgley, A. (1975) The conversion of high alumina cement. *Magazine of Concrete Research*, Vol. 27, pp.59-77.
6. Lea,F.M.(1940) Effect of temperature on high-alumina cement, *Trans. Soc. Chem, Ind.*, Vol.59, pp.18-21.
7. Gaztañaga, M.T., Goñi, S. and Sagrera, J.L. (1993) Reactivity of high-alumina cement in water: pore-solution and solid phase characterization. *Solid State Ionics*, Vol. 63-65, pp.797-802.
8. Goñi, S., Gaztañaga, M.T., Sagrera, J.L. and Hernández, M.S. (1994) The influence of NaCl on the reactivity of high alumina cement in water: Pore-soluton and solid phase characterization, *Journal of Materials Research*, Vol. 9, pp.1533-1539.
9. Pérez, M., Vázquez, T. and Triviño, F. (1984) Study of stabilized phases in high alumina cement mortars. Part II. Effect of $CaCO_3$ added to high alumina cement mortar subjected to elevate temperaure curing and carbonation. *Cement and Concrete Research*, Vol.14, pp.1-10.
10. Grant Elliot, A. and Huggins Robert, R. (1975) Phases in the System $NaAlO_2$-Al_2O_3, *Journal of the American Ceramic Society*, Vol. 58, pp.497-500.

33 DURABILITY OF ALKALI-ACTIVATED SLAG CEMENTS

J. PERA and M. CHABANNET
Laboratoire Matériaux Minéraux, Institut National des Sciences
Appliquées, Lyon, France

Abstract
This paper presents some recent research work on the durability of alkali-activated slag cements placed in aggressive solutions (5 % Na_2SO_4 and 10 % $CaCl_2$). Two different slags were activated by waterglass and sodium hydroxide. Pastes of these cements were prepared and their microstructure was investigated by X-ray diffraction, differential thermal analysis, mercury intrusion porosimetry and scanning electron microscopy.

Mortar bars were cast and subjected to the aggressive solutions by means of wetting-drying cycles. The evolution of the compressive strength was measured as well as the length variation and loss of mass. The microstructure was also investigated.

The results obtained show a good durability of mortars cast using slags activated either by sodium hydroxide or waterglass. The durability also depends upon the nature of the slag used.

Keywords : Aggressive solution, alkali, durability, microstructure, slag, strength.

1 Introduction

Several papers have been published on the use of slag in cement and concrete based on its activation by Portland cement, sulphates and alkalies. The microstructure of such binders was investigated. In the cement-activated pastes, Regourd [1], Cook et al [2] have found that the main products of hydration are type I C-S-H gel, ettringite, and calcium hydroxide. With alkaline activators, C-S-H is formed and the aluminate-phase AFm varies in composition but include, C_4AH_{19} and C_2ASH_8 in connection with NaOH [1], [3]. A recent research work carried out by Wang et al [4] shows that during the hydration of alkali-activated slag (AAS) cements, the main hydration product is

Mechanisms of Chemical Degradation of Cement-based Systems. Edited by K.L. Scrivener and J.F. Young.
Published in 1997 by E & FN Spon, 2–6 Boundary Row, London SE1 8HN. ISBN: 0419215700.

calcium silicate hydrate with a low Ca/Si ratio. A hydrotalcite-type phase is also formed in the pastes activated with both NaOH and waterglass. A crystalline phase of AFm type is formed in slag paste activated with NaOH.

Since AAS cements do not contain any calcium hydroxide, they will perform better under severe conditions of sulphate or chloride attack [5]. Therefore it was interesting to verify the durability of AAS cements in aggressive solutions. This is the objective of the present paper.

Factors affecting the hydraulicity of slag are in short : glass content, chemical composition, fineness of grinding, type of activator, means of adding activator, curing conditions [6], [7]. Therefore two types slags were chosen and activated either by NaOH or waterglass. Their microstructure was investigated and they were used to cast standard mortars which were placed in aggressive solutions : 5 % Na_2SO_4 and 10 % $CaCl_2$. The durability of these mortars was assessed by both strength measurements and microstructural investigations.

2 Experimental

2.1 Raw materials

The characteristics of the granulated blast furnace slags used are given in Table 1. S1 is a ground pelletized slag, S2 is a ground granulated slag. X-ray diffraction testing (Fig. 1 and Fig. 2) indicated that these slags consist mainly of glassy phase associated with some crystalline merwinite and gehlenite in slag S1, merwinite and akermanite in slag S2. Slag S1 is less glassy and coarser than slag S2, even though they were ground in the same conditions. The Blaine fineness is 285 m^2/kg for S1 and 356 m^2/kg for S2. The mean particle sizes are respectively :15 μm for S1, and 5 μm for S2.

Table 1. Chemical composition of slags

Notation	Oxides						
	SiO_2	Al_2O_3	CaO	MgO	Fe_2O3	TiO_2	MnO
S1	35.5	10.8	43.7	7.2	0.4	0.5	0.3
S2	33.4	11.7	39.3	9.7	1.9	0.4	0.2

To predict the hydraulic activity of a blast furnace slag, various hydraulicity formulas have been proposed. Smolczyk [8], however, came to the conclusion that not one of those formulas was able to give generally valid information as to the effect that the chemical composition of a slag has on strength development. Two moduli were computed for S1 and S2 : a quality modulus $M1 = (CaO + MgO + Al_2O_3)/SiO_2$, and a basicity coefficient $M2 = (CaO + MgO)/(SiO_2 + Al_2O_3)$ which have to be higher than 1 for good performance [9]. For S1, the values obtained were $M1 = 1.09$ and $M2 = 1.77$; for S2, $M1 = 1.08$ and $M2 = 1.81$. Both types of slag are thus basic and high quality.

Alkaline solutions of sodium hydroxide (NaOH) and waterglass (wg) having a SiO_2/Na_2O ratio (i.e. modulus) of 2.3 were used as activators. Douglas et al [10] used a sodium silicate solution having a silicate modulus of 2.52 to activate different slags and obtained good results. The dosages of alkali components were respectively :

4 % NaOH and 6 % wg, by slag weight. These values were chosen for an economic point of view and to prevent efflorescence and brittleness.

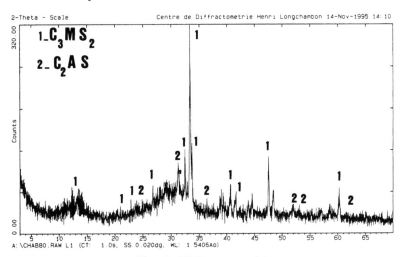

Fig. 1 XRD pattern of S1.

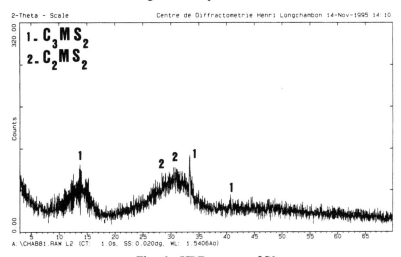

Fig. 2. XRD pattern of S2.

2.2 Sample preparation

Ground slag was mixed with the alkaline solutions with different liquid : solid ratios, in order to prepare pastes at normal consistency for microstructural investigations. These ratios were respectively : 0.27 for S1-NaOH, 0.25 for S2-NaOH, 0.20 for S1-wg, and 0.23 for S2-wg. The pastes were cast into cylinders (diameter = 20 mm ; height = 40 mm) for 24 hrs, then demoulded and cured in water at 20°C for different ages : 3, 7, 28, and 90 days. At these curing times, the compressive strength of pastes was determined on 6 specimens and the microstructure investigated.

For durability tests, mortar bars were prepared. Non-standard mortars were cast, they presented a sand : binder ratio of 3.86 instead of 3. This kind of mortar is more representative of the concrete behaviour than a standard one, as shown by Maximilien et al [11]. The water : binder ratio was adjusted in order that all the mortars presented the same workability. For S1 and S2 these ratios were 0.58 when NaOH was used as activator, and 0.50 for wg. The mortars were cast in prismatic moulds (40 x 40 x 160 mm), cured in the moulds for one day before demoulding and then immersed in de-ionized water at 20°C until 28 days. Then they were subjected to the aggressive solutions : 5 % Na_2SO_4 and 10 % $CaCl_2$. The control mortars were kept in de-ionized water.

They were subjected to wetting-drying cycles. The temperature of drying was 35°C. The samples were immersed for 16 hrs and dried for 8 hrs. Mechanical and microstructural data were carried out after 28, 60, and 90 cycles.

2.3 Microstructural investigations

For X-ray diffraction and differential thermal analysis, samples were ground to a fine powder less than 100 μm. A Siemens D500 automatic diffractometer, graphite monochromator, CuK_α radiation, was used ($2\theta = 1°$/min.). The DTA tests were carried out at a heating rate of 10°C/min. on a 600 mg sample.

For scanning electron microscopy (SEM), a JEOL 840 microscope was used, equipped with a Tracor energy dispersive (EDS) X-ray microanalysis system. Porosity and pore-size distribution were investigated by mercury intrusion porosimetry carried out with a Micromeritics Pore Sizer 9300.

2.4 Mechanical testing

Compressive strengths were measured both on pastes and mortars. The average value was calculated on 6 specimens. Length and mass variations were also recorded during the durability tests.

3 Results and discussion

3.1 Pastes

3.1.1 Compressive strength

Table 2 gives the strength development of pastes. When activated by NaOH, the two slags show the same behaviour. The use of waterglass solution is most favourable from the strength point of view for S2, specially for long-term performance. Such results are in agreement with previous results [7]. It can also be seen that the type of slag has a significant influence on strength but more so in the systems with a weak alkaline activator than in those with strong alkali such as NaOH. S1 and S2, having about the same chemical moduli show that these moduli are not sufficient to predict the strength development of AAS. In the present case, the fineness and the amount of glass phase are the most important parameters.

Table 2. Compressive strength of AAS pastes (MPa)

Notation	NaOH				wg			
	3 d.	7 d.	28 d.	90 d.	3 d.	7d.	28 d.	90 d.
S1	27	31	48	54	15	22	31	60
S2	29	33	51	63	29	32	55	90

3.1.2. Pore structure

Table 3 shows the pore structure characteristics of pastes cured for 28 and 90 days pastes. When activated by waterglass, S2 paste shows the finest pore distribution : this can explain the good mechanical performances obtained. The porosity decreases with hydration time for all pastes.

Table 3. Pore structure characteristics of AAS pastes

Notation	Age (days)	Porosity (%)	Pore volume (%)			
			< 8 μm	8-14 μm	14-33 μm	33-60 μm
S1-NaOH	28	21.2	10	20	28	10
	90	15.3	12	20	22	6
S1 wg	28	25.3	26	48	15	1
	90	25.3	27	55	8	1
S2-NaOH	28	23.0	3	8	34	13
	90	14.8	23	28	28	1
S2-wg	28	14.5	29	54	3	1
	90	10.6	38	53	1	1

3.1.3 Hydration products

From XRD, DTA and SEM data, it appears that the hydration products are mainly : C-S-H gel, C_2ASH_8, C_4AH_{13} in all pastes. In AAS2, some hydrogarnet (C_3AH_6) and hydrotalcite appear.

3.2 Mortars

3.2.1. Compressive strength

The strengths obtained for mortars cured in de-ionized water are reported in Table 4. The strength evolution is parallel to that found for pastes. Mortars cast with wg-activated slag show the highest strengths and the use of slag S2 leads to the best performances. Activation of slag by NaOH gives low strengths, irrespective of which slag was used.

Table 4.   Compressive strength obtained for mortars in different media (MPa)

Notation	De-ionized water			5 % Na$_2$SO$_4$			10 % CaCl$_2$		
	28 d.	60 d.	90 d.	28 d.	60 d.	90 d.	28 d.	60 d.	90 d.
S1-NaOH	15 ± 1	16 ± 1	18 ± 1	18 ± 1	20 ± 1	17 ± 0	18 ± 1	21 ± 1	18± 1
S1-wg	37 ± 2	38 ± 2	38 ± 2	34 ± 1	39 ± 1	42 ± 2	35 ± 1	37 ± 2	38 ± 2
S2-NaOH	16 ± 1	17 ± 1	18 ± 1	15 ± 1	18 ± 1	16 ± 1	16 ± 1	17 ± 1	18 ± 1
S2-wg	60 ± 3	61 ± 3	63 ± 3	57 ± 2	57 ± 2	60 ± 2	57 ± 2	59 ± 2	61± 2

When exposed to Na$_2$SO$_4$ solution, all mortars perform very well and there is no decrease in strength as seen in Table 6. These good results can be explained by the fact that Na$_2$SO$_4$ is also a slag activator [7], specially when finely ground basic slag is used.

Good performances are also obtained when mortars are subjected to the CaCl$_2$ solution, as shown in Table 4.

3.2.2   Length variation and mass loss.

No swelling was observed after 90 days of immersion in the aggressive solutions, whatever the mortar might be. When subjected to wetting-drying cycles, AAS mortars present a very low loss of mass : 0.25 % to 0.4 % after 90 cycles for the Na$_2$SO$_4$ solution ; 0.4 % for the CaCl$_2$ solution.

3.2.3.   Microstructure

As shown in Fig. 3, the platy AFm-type phase mixed with C-S-H is not affected by the CaCl$_2$ solution in S1-mortar activated with NaOH. No chloroettringite or chloroaluminate was detected by SEM-EDS analyses. The EDS analysis showed that Cl$^-$ ions are incorporated in plates of C$_2$ASH$_8$, as previously found by Jelidi et al[12].

Fig. 3. SEM images for S1-mortar activated with NaOH and exposed
to 10 % CaCl$_2$ solution.

The microstructure of S2-mortar activated with wg remains very dense when the mortar is exposed to 5 % Na_2SO_4 solution, as shown in Fig. 4. Slag grains are covered by C-S-H and no sulfate ions were detected by EDS.

Fig. 4. SEM images for S2-mortar activated with wg and exposed
to 5 % Na_2SO_4 solution.

4 Conclusion

Based on the above results, the following conclusions can be deduced : AAS mortars behave very well in aggressive solutions such as 5 % Na_2SO_4 or 10 % $CaCl_2$; their strength remains constant, and the length variations are very small. A glassy slag activated with 6 % waterglass performs especially well.

5 References

1.   Regourd, M.   (1986) Slags and slag cements, in *Cement replacement materials*, (ed.   R.N.   Swamy), Department of Mechanical Engineering, University of Sheffield, pp. 73-99.
2.   Cook, D.J.   (1988) Hydration and morphological characteristics of cement containing blast furnace slag, in *Concrete 88, International workshop on the use of fly ash, slag, silica fume and other siliceous materials in concrete* (ed. W.G. Ryan), Concrete Institute of Australia, Sydney, pp. 433-48.
3.   Regourd, M.   (1980) Structure et Comportement des Hydrates des Ciments au Laitier, in *Proceedings of the 7th International Congress on Chemistry of Cement*, Septima, Paris, Vol. I, pp. III.2.9.-III.2.26.
4.   Wang, S.D., and Scrivener, K.L.   (1995) Hydration products of alkali-activated slag cement, *Cement and Concrete Research*, Vol. 25, No. 3, pp. 561-71.

5. Mehta, PK (1992) Sulfate attack on Concrete - A Critical Review, in *Materials Science of Concrete III* (ed. J. Skalny), The American Ceramic Society, Westerville, pp. 105-130.

6. Sersale, R. (1992) Advances in Portland and blended cements, in *Proceedings of the 9th International Congress on Chemistry of Cement*, National Council for Cement and Building Materials, New Delhi, Vol. I, pp. 261-302

7. Wang, S.D., Scrivener, K.L., and Pratt, P.L. (1994) Factors affecting the strength of alkali-activated slag, *Cement and Concrete Research*, Vol. 24, No. 6, pp. 1033-43.

8. Smolczyk, M.G. (1979) Effect of the Chemistry of the Slag on the Strengths of Blast Furnace Slags, *Zement-kalk-Gips*, No. 6, pp. 294-96.

9. Mantel, D.G. (1994) Investigation into the Hydraulic Activity of Five Granulated Blast Furnace Slags with Eight Different Portland Cements, *ACI Materials Journal*, Vol. 91, No. 5, pp. 471-477.

10. Douglas, E , and Branstret, J., (1990) A preliminary study on the alkali activation of ground granulated blast-furnace slag, *Cement and Concrete Research*, Vol. 20, No. 5, pp. 746-56.

11. Maximilien, S, Ambroise, J, and Péra, J (1994) Influence of Acrylic Polymers on the Rheology of Mortars, in *Proceedings of the 4th CANMET/ACI International Conference on Superplasticizers and Other Chemical Admixtures in Concrete*, (ed. V.M. Malhotra), ACI SP-148, Detroit, pp. 89-104.

12. Jelidi, A., Chabannet, M., Ambroise, J, and Péra, J., (1992) Etude des interfaces dans des composites ciment-fibres de polyester, in *Proceedings of the RILEM International Conference on Interfaces in Cementitious Composites* (Ed. J.C. Maso), E & FN Spon, London, pp. 217-226.

34 ALKALI ACTIVATED CEMENTITIOUS MATERIALS IN CHEMICALLY AGGRESSIVE ENVIRONMENTS

W. JIANG, M.R. SILSBEE, E. BREVAL and D.M. ROY
Materials Research Laboratory, The Pennsylvania State University, University Park, PA, USA

Abstract
The objective of this paper is to present and discuss the nanostructural aspects and resistance to chemical attack of alkali activated cement (AAC), when subjected to extreme conditions. This exploratory work attempts to identify the nano-structure in alkali activated materials that distinguishes them from a control group of ordinary portland cements (OPC) using TEM. Tests for the purpose of making a general comparison of chemical resistance between the AAC and OPC were conducted.
Keywords: Alkali activated cements, aggressive environment, TEM, microstructure, hydrate, tobermorite.

1 INTRODUCTION

Recently, the value of alkali activated cementitious (AAC) materials which are "environmentally preferable products[1]", because of the utilization of solid wastes and by-products which may have potential or latent hydraulic properties, has been recognized[2]. In the past decades, the increasing interest in AAC has given rise to a number of research and application papers. However, probing deeply into the essence of the activation reactions is relatively recent. Richardson and Groves[3] used analytical transmission electron microscopy (TEM) and electron microprobe analysis (EMPA). Wang and Scrivener[4], Roy, et al.[5], and Roy and Silsbee[6] discussed different aspects of the reaction mechanisms. This paper focuses on AAC nanostructure and resistance to chemical attack.

2 EXPERIMENTAL

The starting materials are granulated slag, Class F fly ash, and alkali activator expressed as R_2O. Ordinary Portland cement (OPC) was used as control. The chemical compositions of the OPC, fly ash, and slag have been reported in an earlier paper[1]. Formulations selected in this work are shown in table 1. The specimens were cured under 98% relative humidity at 25°C. Hydration products from 90 day old hardened AAC and Ordinary Portland Cement (OPC) pastes were examined by means

Mechanisms of Chemical Degradation of Cement-based Systems. Edited by K.L. Scrivener and J.F. Young. Published in 1997 by E & FN Spon, 2–6 Boundary Row, London SE1 8HN. ISBN: 0419215700.

Table 1. Mixture proportions (by weight)

Mix	1	2	3	4	5	6	7	8	9	10
OPC type I	50	52	90	40	—	10	20	50	70	100
Slag	—	15	—	50	100	90	80	50	30	—
Fly ash	50	25	—	—	—	—	—	—	—	—
Silica fume	—	3	10	10	—	—	—	—	—	—
water/solid	0.22	0.25	0.22	0.22	0.30	0.22	0.22	0.22	0.22	0.22
KOH*	2	2	—	2	4	3.6	3.2	2	1.2	—

*P-14 was used NaOH instead of KOH, and 3 % Mighty 150 was use in all mixtures

of x-ray diffraction (XRD). The morphologies of hydrates were observed by TEM. The specimens were crushed in an agate mortar with isopropanol, which is often used as an inert liquid. A drop of the dispersion was then transferred to a carbon-coated TEM grid, and the liquid was allowed to evaporate. The TEM was a Philips EM 420 with acceleration voltage of 120 kV. Bright field images showed the morphology and contrast of the individualgrains. Selected area electron diffraction patterns (SAEDP) from regions of 100 to 200 nm showed if any crystalline or nano-crystalline phases were present. Crystalline phases would form either single crystal spot diffraction pattern or powder diffraction "rings", whereas completely amorphous material would only show a very broad and very weak ring. BY By using reflections from from the SAEDP (reference) it is possible to detect crystalline phases down to 1 nm, which would appear light on the dark field image.

3 RESULTS AND DISCUSSION

3.1 Microstructural aspect

According to Groves[8] and Richardson and Groves[9] during the hydration of tricalcium silicate, $Ca(OH)_2$ and two distinct silicate hydrate gels are formed. These C-S-H phases are designated as being either "inner" or "outer" products[7]. Richardson and Groves suggested that TEM reveals the presence of a poorly crystalline Fe, Al-rich phase, possibly a forerunner of hydrogarnet. Richardson and Groves[9] used an argon ion-beam milling method for sample preparation, but expressed their concern for thermal and irradiation damage of hydrated phases suffered during preparation and observation in the TEM[7]. When using ion beam milling the specimen is subjected to severe bombardment of argon ions. These ions may be harmless for strong materials such as silicon carbide, but hydrated cement paste and similar materials may be severely damaged. Damage of the original cement hydration structure during the ion beam thinning was recognized by Grudemo in his classical work[10], wherefore he recommended studing crushed particles of cement hydration products in the TEM. In this work, the sample crushing method was used in an agate mortar with isopropanol, then a drop of dispersion was transforred to TEM grid. The only damage to the samples is caused by the vacuum inside the electron microscope, so rapid viewing of the sample was used to protect it from damage.

Fig. 1 is a pure OPC paste (Mix 10). Fig. 2 and Fig. 3 show hydrated slag with KOH (Mix 5). The X-ray 'powder' diffraction pattern of AAC (mix 5, table1, activated by KOH) after 90 days of curing at 25°C is shown in Fig. 4, and the identifications of major peaks observed are also given. The main phases are C-S-H (I) (JCPDS file No. 9-210), C-S-H (JCPDS file No. 33-306), and calcium carbonate (JCPDS file No. 5-586). The intensity of those peaks increased with increasing cure time. These results suggest that the low-basic calcium silicate hydrate related to tobermorite is the major hydration products of AAC. A hydrotalcite-type phase is also found, and its formula is $Mg_6Al_2CO_3(OH)_{16}·4H_2O$ (JCPDS file No. 41-1428). Wang and Scivener[4] reported the same hydrotalcite-type phase, however in their case, the slag was activated by NaOH.

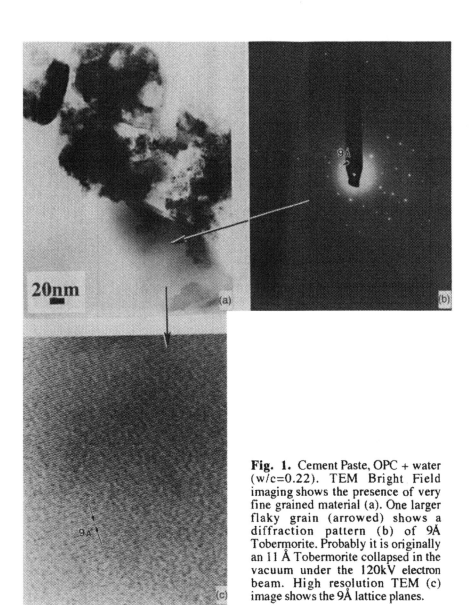

Fig. 1. Cement Paste, OPC + water (w/c=0.22). TEM Bright Field imaging shows the presence of very fine grained material (a). One larger flaky grain (arrowed) shows a diffraction pattern (b) of 9Å Tobermorite. Probably it is originally an 11 Å Tobermorite collapsed in the vacuum under the 120kV electron beam. High resolution TEM (c) image shows the 9Å lattice planes.

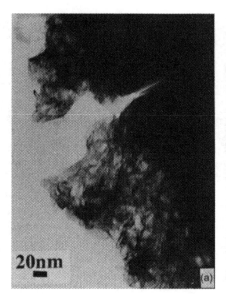

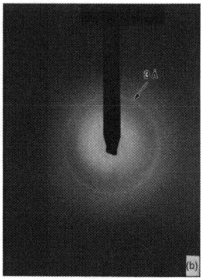

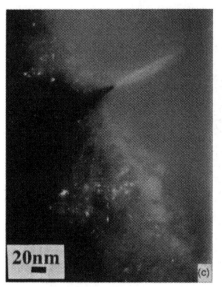

Fig. 2. Hydration product of slag with NaOH (w/c=0.3). Fragment of a larger amorphous glassy grain with a rim of reaction products. The TEM Bright Field image (a) shows a fine grained structure of the reaction products. The electron diffraction pattern (b) shows a strong diffuse reflexion at ~3.0Å, where most CSH gels usually give strong, diffuse X-ray diffraction. A few weaker rings are also seen. The Dark Field image (c) using the strong 3Å reflexion shows nm sized crystalline reaction products.

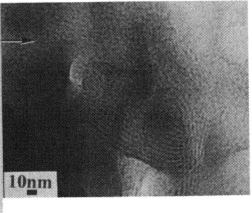

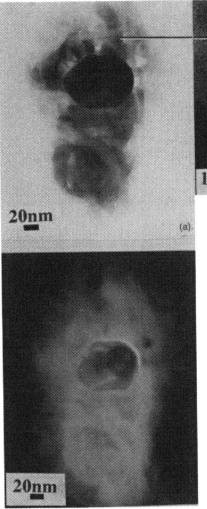

Fig. 3. Hydration product of slag with NaOH (w/c=0.3). A typical small amorphous glassy sphere with outer, nearly amorphous reaction products is shown. The TEM Bright Field image (a) shows the glass sphere (dark) surrounded by a cloud of reaction products. A close-up of these products reveals a structure which could be interpreted as precursors of crystalline reaction products. The Dark Field image (c) from a region in the reciprocal lattice of around 3Å shows the glass grain (dark) surrounded by reaction products, which appear light because they show diffuse diffraction in this region.

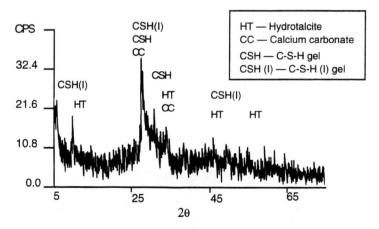

Fig. 4 X-ray pattern of AAC (mix 5, table1) after 90 days of curing at 25°C.

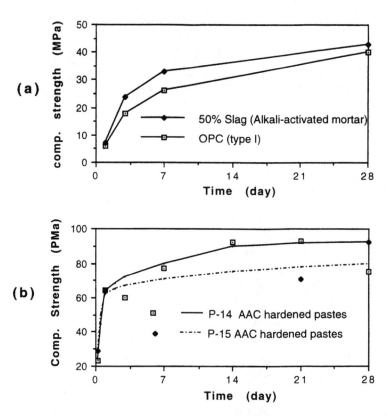

Fig. 5 (a) Compressive strength of AAC mortar which used on mix 8 AAC as the cementitious phase with cement to sand (1:2.75) at w/s =0.44, curing at 25°C, compared with OPC. (b) strengths of pastes based on mix 2 of P-15 activated by KOH and P-14 by NaOH, at w/s =0.22.

3.2 Mechanical Properties

Mortars were made using Mix 8 AAC as the cementitious phase with sand (1:2.75) at w/s =0.44 prepared according to ASTM C 109 procedure. OPC type I cements were used as control (Fig. 5a). The results show that the compressive strengths of AAC could compete with those of OPC. The pastes of batch P-15 and P-14 are based on Mix 2 with KOH or NaOH (Fig. 5b), at low w/s =0.22, the early paste strength development of AAC was rapid.

3.3 Damage Assessment after chemical attack

Table 2 summarizes an assessment of AAC mortar specimens conditions after chemical attack. The mortars were described in 3.2.

Table 2 Laboratory assessment of AAC specimens when subjected to chemical attack. The specimens were ranked as follows: Excellent (no change); Good (very slight change); Fair (moderate change); Poor (severe degradation).

The test specimens were immersed in solutions under the following conditions: (1) Solution temperature $20\pm2°C$; (2) specimens (50 x 50 x 200 mm) prisms were cast, cured at 20°C for 24 hours then demoulded and immersed in the specified solution; (3) The solutions were completely renewed every week; (4) a visual assessment of the condition of the sample was made at the end of 6 months.

Solution	Condition after 6 months	Comments
5 wt. % HCl	Fair	better than OPC
5 wt. % H_2SO_4	Poor	severe degradation
5 wt. % HNO_3	Fair	minor effect
5% citric acid	Good	negligible
10 wt. % $NaSO_4$	Excellent	sodium sulfate acts as an
10 wt. % $MgSO_4$	Poor	activator for AAC

To test the potential for efflorescene a sample was partially immersed in deionized water.

Solution	Condition after 6 months	Comments
Deionized water	white crust	Smith and Osborne[11] reported Na_2CO_3

Resistance to accelerated chemical attack was examined by exposing 6 month old specimens to boiling acid. (1) - a single 3 hours exposure. (2) - 6 times, 30 minute exposures with washing between.

Solution	Condition after 3 hours	Comments
Method 1		
5 wt. % H_2SO_4	Poor	potential as an accelerated
5 wt. % HCl	Fair	test method
Method 2		
deionized water	Good	Simulates industrial
10 wt% HCl	Poor	environment

To examine the resistance of AAC to attack by common deicer chemicals NaCl and $CaCl_2$ specimens were immersed in the designated solutions. At weekly intervals the specimens were removed washed with deionized water, then placed into fresh solution.

Solution	Condition after 6 months	Comments
20 wt. % NaCl	Fair	NaCl creates more damage
20 wt. % $CaCl_2$	Good	than $CaCl_2$

4. SUMMARY

• TEM studies of alkali-activated slag reveals the fragment of a larger amorphous glassy grain with a rim of reaction products. The electron diffraction pattern shows a strong diffuse reflection at 3.0 Å, characteristic of most CSH gels.

• The results suggest that the low-basic calcium silicate hydrate related to tobermorite-like C-S-H gel is the major hydration product of AAC, and a hydrotalcite phase is also formed.

• Preliminary results regarding accelerated methods of testing the for chemical resistance of concrete and cement paste in aggressive solutions are presented, and could be useful for making a general comparison.

5 REFERENCES

1. Roy, D.M. and Jiang, W. (1994) Microcharacteristics and properties of hardened alkali-activated cementitious materials, in *Cement Technology* (eds Gartner, E.M. Uchikawa, H.) Ceram. Trans. **40**, Am. Ceram. Soc. Westerville, OH, 257-267.
2. Jiang, W., Wu, X. and Roy, D.M. (1993) Alkali-activated fly ash-slag cement based nuclear waste forms, in *Scientific Basis for Nuclear Waste Management XVI*, (eds. Interrante, C.G. and Pabalan, R.T.) Mater. Res. Soc. Proc. **294**, Pittsburgh, PA., pp. 255-260
3. Richardson, I.G., Brough, A.R., Groves, G.W. and Dobson, C.M. (1994) The characterization of hardened alkali-activated blast-furnace slag paste and the nature of the calcium silicate hydrate (C-S-H). *Cem. Concr. Res.* **24**, 813-829.
4. Wang, S.D. and Scrivener, K.L. (1995) Hydration products of alkali activated Slag Cement. *Cem. Concr. Res.* **25**, 561-571.
5. Roy, A., Schilling, P.J., Eaton, H.C., Malone, P.G., Brabston, P.J. and Eaton, H.C. (1992) Activation of ground blast-furnace slag by alkali-metal and alkaline-earth hydroxides. J. Am. Ceram. Soc. **75**[12] 3233-40.
6. Roy, D.M. and Silsbee M. (1995) Overview of slag microstructure and alkali-activated slag in concrete, in *2nd CANMET/ACI Intl. Sym. on Advances in Concrete Technology*, Supplementary papers (ed. Malhotra, V.M.), pp. 699-716.
7. Richardson, I.G. and Groves, G.W. (1993) The microstructure and micro-analysis of hardened ordinary Portland cement pastes. *Mater. Sci.* **28**, 265-277.
8. Groves, G.W. (1987) The studies of cement hydration, in *Microstructure Development During Hydration of Cement* (eds.Struble, L.J. and Brown, P.W.) Mater. Res. Soc. Proc. **85**, Pittsburgh, PA., pp. 3-12
9. Richardson, I.G. and Groves, G.W. (1992) Microstructure and microanalysis of hardened cement pastes involving ground granulated blast-furnace slag. *Mater. Sci.* **27**, 6204-6212.
10. Grudemo, Å. (1965) *The Microstructure of Cement Gel Phases*. Swedish Cement and Concrete Research Institute. Transactions of the Royal Institute of Technology, No. 242, Civil Eng. 13, Stockholm, Elander, Göteborg, Sweden.
11. Smith, M.A. and Osborne, G.J. (1977) Slag/fly ash cements. World Cement Technology. Novermber/December, pp. 223-233.

ACKNOWLEDGEMENTS—This research was partly supported by the National Science Foundation grant MSS-9123239 and EPA grant R 819482.

35 CHARACTERISTICS OF DELAYED ETTRINGITE DEPOSITES IN ASR-AFFECTED STEAM CURED CONCRETES

D. BONEN
Northwestern University, Evanston, IL USA
S. DIAMOND
Purdue University, West Lafayette, IN, USA

Abstract
Examinations were carried out on a large number of specimens of steam cured concrete that had experienced both delayed ettringite formation (DEF) and mild to severe alkali silica reaction. A consistent pattern was found relating the chemical composition of the ettringite to its morphology and to the type of site in which it was deposited. Ettringite deposited along aggregate boundaries and in filled cracks and air voids was found to be of almost ideal ettringite composition. In contrast, sulfoaluminate particles typically appearing as nondescript or acicular particles embedded within the paste itself had much lower SO_3 contents. Slightly distressed concrete was largely characterized by such low SO_3 content particles; highly distressed concrete, in which extensive indications of DEF are present, contained mostly particles of almost ideal ettringite composition.
Keywords: Alkali silica reaction, chemical composition, DEF, EDS analyses, ettringite, scanning electron microscopy, steam cured concrete, stoichiometry.

1 Introduction

Ettringite is a natural mineral, usually found as prismatic crystals of hexagonal cross section. The ideal composition is $\{Ca_6[Al(OH)_6]_2 \cdot 24H_2O\}[(SO_4)_3 \cdot 2H_2O]$, although solid solutions with iron, and chloride and carbonate-bearing analogs are known. The principal ionic substitutions include Fe^{+3} for Al^{+3}, and CO_3^{2-}, OH^-, and Cl^- for SO_4^{2-}. Ettringite is produced in ordinary concrete and is the first phase that precipitates on mixing, through an almost instantaneous reaction of dissolved sulfate and tricalcium aluminate according to:

$$Ca_3Al_2O_6 + 3CaSO_4 + 32H_2O \longrightarrow Ca_6Al_2S_3O_{18} \cdot 32H_2O \tag{1}$$

Mechanisms of Chemical Degradation of Cement-based Systems. Edited by K.L. Scrivener and J.F. Young. Published in 1997 by E & FN Spon, 2-6 Boundary Row, London SE1 8HN. ISBN: 0419215700.

The overall mole ratio of SO_3 to C_3A in most cements is far below 3; typical values range from about 0.7 to 1.2. Thus, according to the classical view of portland cement hydration, the primary ettringite that is formed in the early stage of cement hydration tends to be converted to monosulfate according to the reaction:

$$C_6A\bar{S}_3H_{32} + C_4AH_{13}+1/2H_2O \rightarrow 2C_4A\bar{S}H_{12} + 2Ca^{2+} + SO_4^{2-} + 2(OH^-) + 20H_2O \quad (2)$$

Kuzel and Meyer[1], however, have shown that in the presence of small amounts of CO_2 the conversion to monosulfate is to a large extent suppressed. In our experience, the ettringite content in many concretes is substantially larger than the monosulfate present.

The primary ettringite that crystallizes out early in the hydration process has no known adverse effect on concrete properties. However, ettringite generated within a rigid, hardened concrete, due for example to sulfate attack or to extensive delayed ettringite formation (DEF) may produce expansion and cracking.

Recent failures of various structures, especially steam cured concrete railroad ties, have increased interest in the processes of DEF. On a number of occasions[2,3,4] DEF and alkali silica reactions (ASR) have been found together in the same concrete. Although both DEF and ASR can be produced at ambient temperature, Ong[5] has shown experimentally that heat-curing cycles similar to industrial steam curing processes facilitate the occurrence of both DEF and of ASR when reactive aggregates are present. While maximum temperatures reached in steam curing cycles vary, they are often high enough that the primary ettringite produced prior to heating is entirely destroyed. New ettringite is gradually deposited on subsequent exposure to moisture at ordinary temperatures, usually over a period of months or years.

Diamond et al.[2] recently pointed out the large compositional variations accompanying the several different modes of occurrence of secondary ettringite produced in steam cured laboratory mortars, and the similarity of these variations to those found in field-exposed steam cured concretes. The purpose of the present paper is to further elaborate on this subject and to relate the compositional variations of the ettringite found in steam cured concretes to sites of deposition and to overall intensity of the DEF process in a given concrete.

2 Experimental

Examinations were carried out on specimens obtained from a number of steam cured concrete ties that had been exposed to field conditions for varying periods up to 7 or 8 years. It was found that essentially all of these concretes exhibited at least mild indications of ASR. Some were extensively cracked and had significant deposits of both ASR gel and of secondary ettringite.

Specimens for SEM examination and EDS analyses were prepared by impregnating slices of the concrete with ultra low viscosity epoxy mix, and polishing with successively finer grades of diamond grit. The fiat surfaces so produced were sputter coated with a 5 nm-thick layer of palladium, and examined in an Akashi Beam Technology 55A SEM equipped with a Tracor Northern EDS system. Fully ZAF-corrected analyses were carried out at 15 kV for the following elements: Ca, Si, Al, Fe, Mg, Ti, Na, K, and S. The EDS procedure was validated by checking results of duplicate analysis against the known compositions of a number of standards, including NISTcertified standard portland cement (No. 1114), basalt glass, jadeite 35, orthoclase, wollastonite, diopside glass, pyrope glass, pyrrhotite, and anhydrite. The estimated analytical error for the major components Ca, Si, Al, S, and Fe is within 4% of the amount found; for the others, within 7%. The analysis results were automatically normalized to a 100% water-free oxide basis.

3 Results and Discussion

Figure 1 provides an indication of the typical appearance of many of the specimens examined. Evidences of both DEF and ASR can be found in the field presented. In particular, a string of ettringite are found in a band approximately 20 µm wide immediately beneath the complex coarse aggregate grain in the upper left portion of the figure. A small air void entirely filled with a deposit of ettringite occurs near the center of the micrograph, and subsidiary cracks occur in various locations. Evidences of ASR can be observed around the fractured and partly dissolved aggregate grain in the right hand margin.

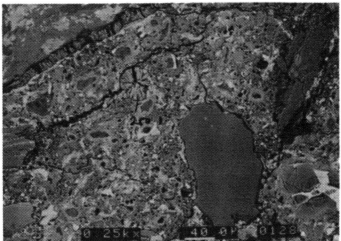

Fig. 1. Micrograph illustrating typical appearance of many DEF-affected specimens.

An analysis carried out for the ettringite in the filled air void yielded the following composition: SO_3 - 37.8%, CaO - 51.8%, Al_2O_3 - 9.6%, SiO_2 - 0.16%, and Fe_2O_3 - 0.0%. Except for the somewhat low content of Al_2O_3, this analysis is very close to that expected for the ideal ettringite composition , as indicated in Table 1.

A total of 192 individual spot analyses of calcium aluminate sulfate hydrate deposits found in these concretes were carried out. All of the particles analyzed were individually uniform in gray level and appearance in the SEM; particles showing locally varying gray levels were excluded from analysis. Each analysis reported represents in general the average of two or more individual spot analyses made at different spots within the particle or local deposit concerned. The particles analyzed were large enough to permit analysis at spots sufficiently far from the surroundings to avoid obvious analytical contamination, and were separated from the surroundings by clear boundaries. All of the analyses were taken far away from any visible ASR products, and it is assumed that none of the results reflect contamination from such products. However, contamination of analyses by underlying material can never be ruled out entirely, since the material below the plane of observation cannot be seen. The SiO_2 content of some particles, typically those embedded in the groundmass of the hydrated cement paste, can be interpreted as indication of some contamination of the analysis from underlying C-S-H.

The 192 analyses produced in this work sorted into six categories based on their SO_3 contents, variation in SO_3 content being the major departure from ideal ettringite composition. The numbers of particles and the categories chosen were as follows: 40 particles with SO_3 contents > 33% that were close to the ideal ettringite composition; 25 particles with SO_3 contents between 30% and 33%, that were reasonably close to the ideal composition; 25 particles with SO_3 contents between 25% and 30% that were somewhat sulfate deficient when compared to ideal ettringite; 15 particles with SO_3 contents between 20% and 25%, 45 particles with SO_3 contents between 15% and 20%, and finally, 42 particles with SO_3 contents between 10% and 15%.

Table 1 provides the average composition for each of these classes of particles on a water free basis, and for comparison, the ideal compositions of pure ettringite and of monosulfate.

Category	I	II	III	IV	V	VI	Stoich. ettringite	Stoich. mono-sulfate
SO_3 Content	> 33%	30-33%	25-30%	20-25%	15-20%	10-15%		
CaO	49.97	52.67	49.12	52.35	57.06	59.28	49.58	55.20
SiO_2	1.60	3.50	8.03	9.36	7.26	10.97		
Al_2O_3	11.37	10.08	10.95	11.96	13.86	11.95	15.02	25.09
Fe_2O_3	0.39	0.81	1.59	1.03	1.68	1.88		
MgO	0.24	0.39	0.73	1.05	0.60	1.01		
TiO_2	0.16	0.21	0.45	0.31	0.34	0.47		
Na_2O	0.69	0.33	0.39	0.23	0.35	0.28		
K_2O	0.44	0.51	0.82	1.33	1.10	0.71		
SO_3	35.12	31.49	27.91	22.40	17.74	13.44	35.39	19.70
Fe_2O_3/Al_2O_3	0.03	0.08	0.14	0.09	0.12	0.16		
No.	40	25	25	15	45	42		

In view of the high degree of substitution of other ions for sulfate in sulfoaluminate phases found in ordinary cement pastes[6], the sulfate content by itself is not necessarily an adequate measure for distinguishing between ettringite and monosulfate. The Al_2O_3 content would appear to be a more reliable parameter; it should be near 25% for monosulfate and 15% for ettringite. Iron substitution seems to be a minor factor.

Examination of Table I indicates that the average Al_2O_3 contents of each of the categories of particles is between 10% and 14%. The analyses also show small contents of Fe_2O_3; all are less than 2%. The Na_2O and K_2O contents observed are also small, being below 0.7% for Na_2O and between 0.5% and 1.3% for K_2O. Alkali contents at such low levels suggest casual incorporation of alkalis into these sulfoaluminates, and are not consistent with appreciable contamination of the these delayed ettringite products with ASR gel. It should be noted that ASR gel was detected in some areas that were not selected for analysis.

It was thought instructive to plot the individual analyses of each category separately on a ternary field consisting of Ca, S, and Al+Fe atomic percents. It should be emphasized that the quantities plotted here are atomic percents, not oxide percents, and that the entries are necessarily normalized to 100% - i.e. other elements not on the field are ignored. Thus the appreciable contents of Si in some of the analyses do not appear in the plots. Individual ternary plots for all of the analyses in each of the six sulfate content categories are provided in Figures 2a through 2f.

To assist in interpretation, in each figure we plot a dashed triangle, the apices of which are the projections on the field of the stoichiometric compositions for ettringite, for monosulfate, and for the amorphic form of groundmass C-S-H as taken from [6]. The triangle is introduced to illustrate one possible interpretation of the analyses. Points lying on a tie line between C-S-H and either ettringite or monosulfate could represent an

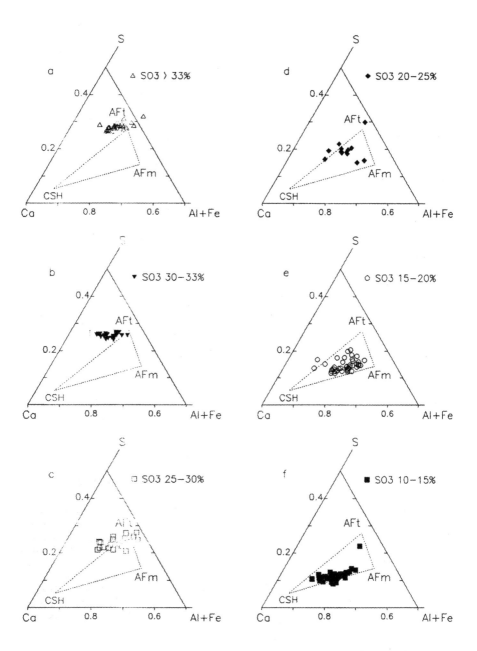

Figure 2. Plots of atom compositions of analyzed particles on a Ca · S · Al+Fe field by groups based on SO$_3$ contents.

intimate mixture between the end members of that tie line. Similarly, points plotted within the triangle could represent a mixture of all three phases on submicroscopic scale.

Alternatively, points plotting at lower SO_3 contents than that of normal ettringite might simply represent ettringite of reduced SO_3 content, reflecting perhaps substitution of OH^- ions for sulfate ions. Also, points that fall within the triangle, i.e. that are of higher Ca proportion than expected for ettringite, might reflect contamination by underlying C-S-H that was not visible in the SEM field.

It appears that the analyses of the particles of the first three categories are consistent with ettringite, with some fluctuation of the points around the theoretical composition. There is little Si present in these particles (Table 1).

The particles in the fourth and fifth category plot essentially within the dashed triangles. The points in the sixth category, those for the lowest SO_3 content particles, plot along the C-S-H - monosulfate tie line. Interpretations of these analyses will be discussed later. However it should be noted here that the particles of the last two categories were primarily found within the groundmass of the hydrated cement, and were thus most susceptible to analytical contamination from unseen C-S-H.

The relationships between sulfate content and particle morphology, and between sulfate content and specific deposition site within the concrete were alluded to previously, but deserves further elucidation. Many of the ettringite particles that have high sulfate contents and nearly ideal ettringite compositions appear as massive tubular deposits comprised of parallel laths of low aspect ratio. These form thick strings of ettringite along aggregate boundaries such as that seen in Figure 1, and many continue to run in cracks extending into and through the surrounding cement paste. Strings of such ettringite particles are sometimes found wholly within the paste, without any association with a visible crack. Near-stoichiometric ettringite is also found filling air voids. Usually the apparent c-axis of these ettringite particles maintain a consistent orientation normal to the aggregate surface or to the vein walls; often in air voids they are normal to the local boundary of the void. In contrast, the low sulfate content particles commonly appear in these backscatter SEM views as nondescript or acicular, or sometimes short rod shaped particles. They are commonly found as small nodules within the solid cement paste itself, or as fillings in capillary pores or sometimes as linings of air voids.

In this study a record was kept of the site of deposition of each of the particles analyzed. Figure 3 show the observed tally of deposition sites as a function of sulfate content category. In this figure, particles indicated as being "within the paste" include nodules within the solid paste, fillings in capillary pores, and linings, (but not complete fillings), of air voids. Particles forming part of the filling material of air voids that were completely filled are separately categorized in the figure.

It is evident that there is a relationship between sulfate content and deposition site in these concretes. Almost 70% of the particles making up the lowest sulfate content category were found within the paste, and another 22% were found in filled air voids. The proportions changed somewhat for the next two sulfate content categories, and then stabilized for the three categories of highest sulfate contents. For the latter, i.e. for particles of sulfate contents greater than 25%, a significantly different distribution of deposition sites was tallied. Of these third-group particles, more than half were found in cracks (including "bond cracks" around aggregate particles), and less than 10% were found within the paste.

There was also a general relationship between the distribution of deposition sites and the overall degree of distress observed in the concrete. Highly distressed concrete was characterized by closely-spaced microcracking and an abundance of ettringite deposited in strings around aggregate grains, as veins within the paste, and as entirely filled air voids. Concrete exhibiting little overall distress featured only narrow and sporadic cracks, most of which were empty, and air voids, lined, but not filled, with the sulfoaluminate particles.

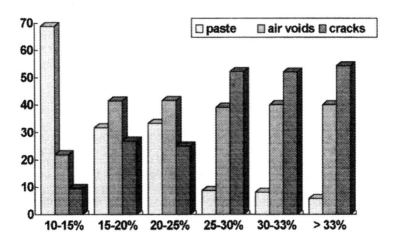

Figure 3. Locations of particles analyzed as a function of SO_3 content class.

4. Discussion

Interpretation of low sulfate content analyses
Interpretation of the nature of the relatively sulfate-rich particles assigned to categories 1 - 3 is straightforward. Consideration of the plots presented in Figures 2a through 2c and of their overall chemical compositions indicated in Table 1 suggests that they are all ettringite, although of varying degrees of substitution. Similarly, the particles of the fourth category appear to be ettringite, although some contamination of the analyses with signal from adjoining or underlying C-S-H seems indicated.

Interpretation of the particles of the two lowest sulfate content categories, 5 and 6, is somewhat uncertain. Judging from Figures 2e and 2f, the former might represent mixtures of ettringite, monosulfate, and C-S-H, and the latter mixtures of monosulfate and C-S-H. Alternatively, the apparent C-S-H component may represent analytical contamination from adjacent or underlying C-S-H rather than the presence of C-S-H in a physical mixture of phases.

Regardless of the details, the findings indicate quite clearly that the DEF processes taking place in these concretes involves a major redistribution from sulfate-poor aluminosulfates to ettringite of near-stoichiometric composition.

Co-existence with ASR depositions
Comment should be made concerning the relation, or lack of it, between the ettringite deposited in these concretes and the evidences of ASR in the same concretes. As indicated previously, the low alkali contents found in these analyses appear to preclude any systematic local relation between the gel produced by the ASR process and the specific sulfoaluminate particles analyzed. However, instances have been observed of apparent co-deposition of both products in the same general area (especially in air void linings). In addition to co-deposition, occasionally separate but parallel streams of the two kinds of products have been observed.

These observations do not preclude the existence of a functional relationship between the results of ASR processes that may be stimulated by steam curing, and the subsequent deposition of ettringite. Diamond et al.[2] reported that in controlled laboratory investigations steam cured mortars containing ASR reactive aggregates cracked extensively and produced copious deposits of ASR gel during the steam curing process itself. On subsequent exposure to ordinary temperatures delayed ettringite formation was significantly greater in these mortars than in similarly treated mortars lacking reactive aggregates. The latter did not crack during the steam curing process, expanded only marginally on subsequent exposure, and on exposure generated much smaller contents of ettringite.

4 Concluding Remarks

It is evident that sulfoaluminate of widely varying composition exists in concrete undergoing DEF, despite the fact that in highly affected concrete most of the prominent ettringite displayed in the characteristic DEF features is nearly of ideal ettringite composition. Sulfoaluminates found within the body of the paste have very much lower SO_3 contents and the analyses reflect some SiO_2.

It is evident that DEF is a gradual process that involves ion transport through the hydrated cement paste and deposition of the product in porous and moisture-rich zones such as cracks, paste-aggregate boundaries, and water-filled or partly water-filled air voids. Preferred deposition sites may also involve local sources of calcium hydroxide, common along aggregate or air-void interfaces. Because of the extensive ion transport, it is not possible to trace the specific source of the sulfate and other components of the ettringite being deposited in any given locality.

A feature not often dwelled on in considerations of DEF is the fact that it seemingly involves extensive movement of aluminum ions from other areas to the sites of ettringite deposition. Aluminum concentrations of pore solutions are extremely small, and one might consider that aluminum mobility in hydrating cements is ordinarily quite restricted. How aluminum is transported in sufficient amounts to completely fill previously empty sites such as air voids with ettringite would seem to require some explanation.

5 References

1.   Kuzel, H-J. and Meyer, H. (1993) Mechanisms of Ettringite and Monosulfate Formation in Cement and Concrete in the Presence of CO_2 *Proceedings*, 15th International Conference on Cement Microscopy, ICMA, Duncanville, TX. pp. 191-203.
2.   Diamond, S., Ong S., and Bonen, D. (1994) Characteristics of Secondary Ettringite Deposited in Steam Cured Concrete Undergoing ASR *Proceedings*, 16th International Conference on Cement Microscopy, ICMA, Duncanville, TX., pp. 294-305.
3.   Shayan , L. and Quick, G.W. (1992) Microscopic Features of Cracked and Uncracked Concrete Railway Sleepers *ACI Materials Journal*, Vol. 89, pp. 348-362.
4.   Oberholster, R.E., Maree, H., and Brand, J.H.B. (1992) Cracked Prestressed Concrete Railway Sleepers: Alkali Silica Reaction or Delayed Ettringite Formation? *Proceedings*, 9th International Conference on Alkali Aggregate Reaction, London, pp. 739-749.
5.   Ong, S. (1993) Effects of Steam Curing and Alkali Hydroxide Additions on Pore Solution Chemistry, Microstructure, and Alkali Silica Reactions, Ph.D. Thesis, Purdue University, 497 pp.
6.   Bonen, D. and Diamond, S. (1994) Interpretation of Compositional Patterns Found by Quantitative Energy-Dispersive X-ray Analysis for Cement Paste Constituents, *Journal, American Ceramic Society*, Vol. 77, pp. 1875-1882.
7.   Bonen , D. and Sarkar, S.L (1993) Replacement of Portlandite by Gypsum in the Interfacial Zone and Cracking Related to Crystallization Pressure, *Cement-Based Materials: Present, Future and Environment Implications*, Ceramic Transactions, Vol. 40, American Ceramic Society, Westerville, OH, pp. 49-59.

PART SIX

CEMENTITIOUS WASTE FORMS AND PERFORMANCE OF CONCRETE BARRIERS FOR NUCLEAR WASTE MANAGEMENT

36 CHEMICAL EVOLUTION OF CEMENTITIOUS MATERIALS WITH HIGH PROPORTION OF FLY ASH AND SLAG

T. BAKHAREV, A.R. BROUGH, R.J. KIRKPATRICK, L.J. STRUBLE and J.F. YOUNG
University of Illinois, Urbana, Il, USA

Abstract
Cement mixtures containing high proportions of slag and fly ash were tested to assess their suitability to immobilize simulated off–gas waste solutions after vitrification of low–level radioactive tank wastes stored at Hanford. Materials were mixed with carbonated or alkaline solutions and cured initially adiabatically, then at 70°C. Chemical changes were monitored for 7 months using X–ray diffraction, selective dissolution and SEM; NMR was utilized to follow the polymerization of silicate species. The process of hydration during the first months of curing was characterized by formation of quite crystalline Al–substituted C–S–H structurally related to 1.1 nm tobermorite and traces of zeolites in some materials. A low content of calcium hydroxide was found in all materials after 1 month of curing. The SEM examination demonstrated rapidly decreasing porosity, making the mixtures favorable for long–term durability.
Keywords: hydration, chemistry, cement, fly ash, slag, wasteforms.

1 Introduction

To predict the performance of cementitious waste forms in different environments, it is necessary to consider the development of phase composition and microstructure due to hydration reactions. The main product of hydration, calcium silicate hydrate (C–S–H), is the matrix–forming phase in cementitious systems. Its properties determine to a large extent the properties of the matrix itself (permeability, diffusivity, and thermodynamic stability). The research presented here is part of a study of the chemistry, microstructure and durability of cement–stabilized low level radioactive wastes. A companion paper on durability is included in this symposium [1].

Several factors make these wasteforms quite different from ordinary portland cement paste hydrated at room temperature. The liquid phase contains appreciable alkali and hydroxide ions. These provide early activation of slag and fly ash. Such activation

Mechanisms of Chemical Degradation of Cement-based Systems. Edited by K.L. Scrivener and J.F. Young.
Published in 1997 by E & FN Spon, 2–6 Boundary Row, London SE1 8HN. ISBN: 0419215700.

Table 1. Composition of solutions

Table 2. Oxide composition (% weight) of dry materials by XRF

	A, g/l	C, g/l
$Al(NO_3)_3 \cdot 9\,H_2O$	7.03	7.03
$Na_3(PO_4) \cdot 12H_2O$	24.53	24.53
$NaNO_2$	12.17	12.17
$NaOH$	53.76	—
$NaNO_3$	32.73	32.73
Na_2CO_3	—	39.97

Oxide	Cement[1]	Fly ash[2]	Slag[3]	Clay[4]
SiO_2	22.17	46.13	37.0	59.71
Al_2O_3	3.24	25.02	8.0	9.05
Fe_2O_3	4.24	7.25	0.2	3.17
CaO	64.48	8.02	39.0	3.13
MgO	1.13	1.81	11.0	11.2
K_2O	0.52	0.63	0.4	0.83
Na_2O	0.15	4.74	0.3	0.07
TiO_2	0.23	4.70	0.4	0.42
P_2O_5	0.12	0.42	0.01	1.47
MnO	0.06	0.03	0.5	0.05
SO_3	2.14	0.12	2.8	<0.1
Loss on ignition	1.20	0.53	0.39	10.64

1 Type I/II, Ash Grove, Durkee, OR
2 Type F, Centralia, Ross Sand and Gravel Co., Portland, OR
3 Koch Minerals Co., Chicago, IL
4 Attapulgite, Engelhard Co., Iselin, NJ

plus adiabatic curing conditions produce quite high reaction temperatures. These factors cause considerable changes in hydration chemistry.

2 Experimental

The solid materials (cement, fly ash, slag and clay) were mixed at room temperature with either alkaline (A) or carbonated (C) solution. Solids were mixed with liquid at w/s ratio of 1 (1 liter of liquid mixed with 1 kg of solid). The compositions of solutions are shown in Table 1. The mixtures tested are designated by their proportion (by weight) of cement, fly ash , slag and clay. For example, the material 3:3:3:1 contains 30% cement, 30% fly ash, 30% slag and 10% clay. The oxide compositions of the cement, fly ash, slag and attapulgite clay are shown in Table 2.

The materials were sealed in tubes and cured adiabatically for the first 4 days, after which they were cured isothermally at 70°C for 7 months. Adiabatic calorimetry results are shown in Figure 1. At the desired age hydration was stopped by crushing the pastes and immersing in methanol for 7 days. The samples were then dried in vacuo and kept under vacuum until analyzed.

The weight percent of slag and fly ash reacted at different ages was determined by selective dissolution of materials in salicylic acid and methanol. In this method C–S–H and cement are largely dissolved, while ~97% of fly ash and ~85% slag are pre-

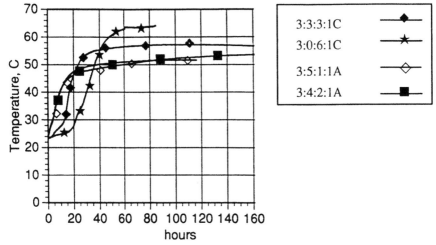

Figure 1. Adiabatic calorimetry.

served. The method is not suitable for quantitative determination of unreacted cement phases. The calculations were based on ignited weight. The procedure utilized can be found in [2, 3].

Specimens for ^{29}Si MAS NMR spectroscopy were ground to a powder and packed into zirconia rotors. The single –pulse ^{29}Si MAS NMR spectra were acquired using a GE WD–300 spectrometer equipped with an Oxford Instruments superconducting magnet (magnetic field 7 T and operating frequency of 59.62 MHz for ^{29}Si). The ^{29}Si chemical shifts are given relative to tetramethylsilane (TMS) at 0 ppm.

3 Results

3.1 XRD
Powder XRD spectra for hydrated materials are shown in Figures 2–3. The main hydration products at 1 month were C–S–H(I), hydrotalcite, AFm (which appears to be hemicarboaluminate), and calcite; trace amounts of calcium hydroxide, a sodalite and Na–P1 were present in some of the mixtures. The rate of pozzolanic reaction was observed to be different for the two solutions. The maximum intensity for the calcium hydroxide peaks for the carbonated solution was observed at 3 days, while for the alkaline solution it was observed at 7 days. At 28 days only trace amounts of calcium hydroxide were present in all materials. Traces of a sodalite were observed in 3:4:2:1A and 3:5:1:1A at early ages (3–5 days). Later traces of Na–X and Na–P1 zeolite (gismondine) were also observed (Figures 2–3). Crystallization of zeolites in the matrix of materials 3:4:2:1A and 3:5:1:1A did not cause loss of strength [1]. In most samples, the C–S–H appeared to be largely amorphous. More crystalline C–S–H, structurally related to 1.1 nm tobermorite, was observed in 3:4:2:1A and in 3:3:3:1C after 2 months (Figure 3).

3.2 SEM
The examination of fracture surfaces of 3:3:3:1C and 3:0:6:1C at 1 month showed a considerable amount of the AFm and hydrotalcite–type phases. The back scattered electron

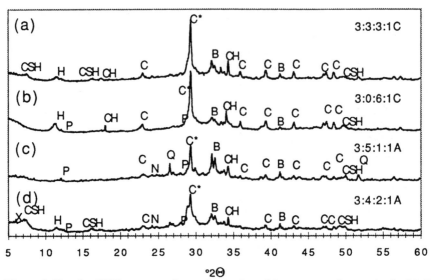

Figure 2. Powder XRD spectra of samples hydrated for one month, acquired with Cu Kα radiation; B=belite, C–S–H=calcium silicate hydrate, C=calcium carbonate (calcite), C*=overlapping C–S–H and calcite, H=hydrotalcite ($[Mg_6Al_2(OH)_{16}] \cdot CO_3 \cdot 4H_2O$) and AFm–type phase, hemicarbonate ($C_4A\overline{C}_{1/2}H_{12}$), N=a sodalite type zeolite, P=Na–P1, zeolite (gismondine), Q=quartz, X=Na–X zeolite; (a) 3:3:3:1C, (b) 3:0:6:1C, (c) 3:5:1:1A, (d) 3:4:2:1A.

image (BEI) of these materials showed a reaction rim surrounding slag and fly ash particles, and in the space between them plate–like AFm and hydrotalcite were crystallized (Figure 4). There was much unreacted material and unfilled space between the particles. Examination at later ages showed that the course of reaction differed between the two solutions. In the alkaline solution (A), slag and fly ash were activated earlier and dissolved from the surface more rapidly than in the carbonated solution (C) (for example, see Figure 5 at 2 months). In the latter we observed precipitated C–S–H on the surface of particles (Figure 6). Dissolution of fly ash in the carbonated case occured only at 4 months, when the glassy phase of some of fly ash particles has dissolved leaving distorted shells of outer product C–S–H, which had presumably formed earlier (see Figure 7). A lot of reaction seams to take place in the hardened matrix through dissolution and precipitation. As a result of the continuing hydration a finer porosity and more uniform matrix is observed at later ages in all the materials (for example, see Figure 8 at 7 months).

3.3 Selective Dissolution
Table 3 shows the weight percent of slag and fly ash reacted at different ages determined by selective dissolution. The results are consistent with other invetigations of blended cements [2, 3, 7]. The total amount of fly ash and slag was ~60% in all mixtures. The results show that hydration continued throughout the time period.

3.4 NMR
Figure 9 shows the Si MAS NMR spectra for 3:3:3:1C and 3:0:6:1C at the ages of 1 month and 7 months. We observed considerable intensity of $Q^2(1Al)$ at ~–82 ppm and

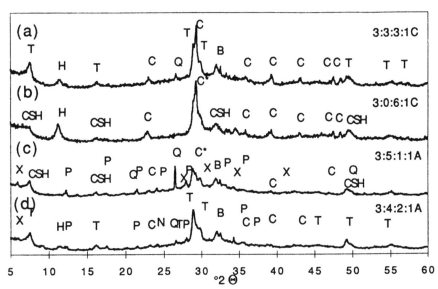

Figure 3. Powder XRD spectra of samples hydrated for seven months, acquired with Cu Kα radiation; B=belite, C–S–H=calcium silicate hydrate, C=calcium carbonate (calcite), C*=overlapping C–S–H and calcite, H=hydrotalcite ($[Mg_6Al_2(OH)_{16}] \cdot CO_3 \cdot 4H_2O$) and AFm–type phase, hemicarbonate ($C_4A\overline{C}_{1/2}H_{12}$), N= a sodalite type zeolite, P= Na–P1, zeolite (gismondine), Q=quartz, T= more crystalline C–S–H, X=Na–X zeolite; (a) 3:3:3:1C, (b) 3:0:6:1C, (c) 3:5:1:1A, (d) 3:4:2:1A.

$Q^2(0Al)$ at ~–85.5 ppm for all the samples. The results are in agreement with observations of other authors for ordinary Portland cement [4], alite [5] and slag hydration [6].

4 Discussion

As expected, the course of reaction was different in the carbonate and alkaline solutions. Larger amounts of portlandite were observed in the alkaline solution and were consumed later. The amounts of AFm and hydrotalcite were found to relate to the amount of slag, because slag is activated more rapidly than fly ash and releases Al and Mg. In 3:4:2:1A and 3:5:1:1A some Al crystallized in sodalite, Na–P1 and Na–X zeolites instead of AFm due to

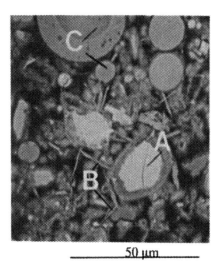

Figure 4. BEI of 3:3:3:1C at 1 month; A–slag, B–hydrotalcite and AFm phases, C–fly ash particle.

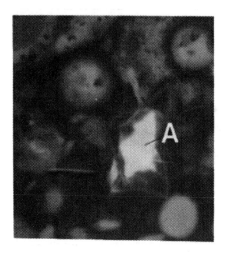

20 µm

Figure 5. BEI 3:5:1:1A at 2 months;
A–activated slag particle.

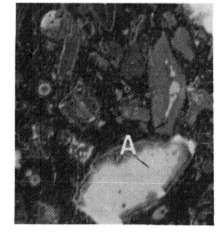

25 µm

Figure 6. BEI 3:3:3:1C at 4 months;
A–slag particle.

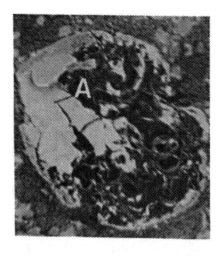

50 µm

Figure 7. BEI 3:3:3:1C at 4 months;
A–activated fly ash sphere.

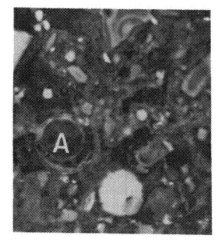

50 µm

Figure 8. BEI of 3:3:3:1C at 7 months;
A–fully reacted fly ash sphere.

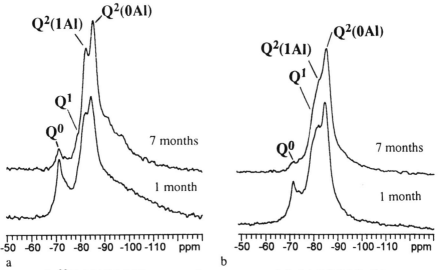

Figure 9. ^{29}Si MAS NMR spectra for waste materials (a) 3:3:3:1C, (b) 3:0:6:1C. The single pulse spectra were acquired (~4000 scans) using a pulse recycle delay of 10 s and a flip angle of 90°.

the low C/S ratio of solids and the presence of considerable alkali. At early ages we observed traces of a sodalite, and later a combination of sodalite, Na–P1 and Na–X zeolites, although Na–P1 was prevailing.

By NMR high intensity of Q^2(1Al) at ~–82 ppm was observed for all materials from the early stage of hydration, showing that the precipitated C–S–H had a considerable substitution of Al. For all the samples at 7 months we observed a high intensity of Q^2(1Al) and Q^2(0Al), and very little intensity for Q^1 at ~–79.9 ppm (chain terminating groups), which indicates the formation of C–S–H with quite long chains. The 3:0:6:1C spectrum had the highest intensity of Q^1(~–79.9 ppm), so the average chain length in 3:0:6:1C was the lowest.

Table 3. Weight percent of slag and fly ash reacted determined by selective dissolution

Age	3:3:3:1C	3:0:6:1C	3:5:1:1A	3:4:2:1A
10 days	29	49	24	42
1 month	42	62	36	55
5 month	nd*	nd	42	65
7 month	54	75	nd	nd

* not determined

5 Conclusions

The reactions observed in alkali activated cementitious materials cured at elevated temperatures resulted in the formation of Al–substituted C–S–H structurally related to 1.1 nm tobermorite, and traces of sodalite, Na–P1, and Na–X zeolites. The Na–P1 zeolite was found to replace a sodalite at later ages. A considerable substitution of Al in C–S–H was observed from a very early age. In some samples considerable crystallinity was found in C–S–H at 7 months. These reactions can be explained by the low C/S ratio of the mix and the large amount of Al that was activated by alkaline solution. The rates of fly ash and slag activation were found to depend on the initial pH and composition of the solution. The system is moving to a more stable phase assemblage that is characterized by crystallization of Al substituted C–S–H that has quite long chain lengths and is more crystalline and structurally related to 1.1 nm tobermorite. From SEM and dissolution work it is evident that the degree of reaction is increasing with time, creating more fine microporosity and filling the larger pores. As the microstructure becomes more uniform, its permeability is decreasing, improving the resistance of these materials to deterioration.

6 Acknowledgement

The research was funded by grant MTS–SVV–097600 from the Westinghouse Hanford Company. We thank the Ross Sand and Gravel Co., Koch Mineral Co., Engelhard Co., and the Ash Grove Cement Co. for materials.

7 References

1. Bakharev, T., Brough, A.R., Kirkpatrick, R. J., Struble,L. J. and Young, J. F. (1996) Durability of cement stabilized low level wastes, in *Mechanisms of Chemical Degradation of Cement – Based Systems*, (ed. K.L. Scrivener and J.F. Young), E & FN Spon, London, current volume.
2. Ohsawa, S., Asaga, K., Goto, S., and Daimon, M. (1985) Quantitative determination of fly ash in the hydrated fly ash $-CaSO_4 \cdot 2H_2O - Ca(OH)_2$ system. *Cem. Concr. Res.*, Vol. 15, pp. 357–66.
3. Luke, K. and Glasser, F.P. (1987) Selective dissolution of hydrated blast furnace slag cements. *Cem. Concr. Res.*, Vol. 17, pp. 273–82.
4. Barnes, J.R., Clague, A.D.H., Clayden, N.J., Dobson, C.M., Hayes, C.J., Groves, G.W. and Rodger, S.A. (1985) Hydration of Portland cement followed by ^{29}Si solid–state NMR spectroscopy. *Journal of Material Science Letter*, Vol. 4, p. 1293–95.
5. Young, J.F. (1988) Investigation of calcium silicate hydrate structure using silicon–29 nuclear magnetic resonance spectroscopy. *J. Am. Ceram. Soc.*, Vol. 71, No. 3. pp. C–118–C–120.
6. Richardson, I.G., Brough, A.R., Groves G.W. and Dobson, C.M. (1994) The characterization of hardened alkali–activated blast–furnace slag pastes and the nature of the calcium silicate hydrate phase. *Cem. Concr. Res.*, Vol. 24, pp. 813–29.
7. Luke, K. and Glasser, F.P. (1988) Internal chemical evolution of the constitution of blended cements. *Cem. Concr. Res.*, Vol. 18, pp. 495–502.

37 ROLE OF CARBONATION IN LONG-TERM PERFORMANCE OF CEMENTITIOUS WASTEFORMS

J.C. WALTON
Department of Civil Engineering, University of Texas at El Paso, Texas, USA
R.W. SMITH
Biotechnologies Department, Idaho National Engineering Laboratory, USA
A. TARQUIN, P. SHEELEY, R. KALYANA, J. GWYNNE, N. GUTIERREZ, M.S. BIN-SHAFIQUE, M. RODRIGUEZ and R. ANDRADE
Department of Civil Engineering, University of Texas at El Paso, USA

Abstract
Cement-based wasteforms are one of the most widely applied site remediation and waste disposal options, yet definitive understanding and quantification of processes controlling long-term performance is lacking. In the U.S., wastes are required by law to be placed in the unsaturated zone but wasteform performance is evaluated using laboratory tests (TCLP, ANSI/ANS-16.1) that specify fully water saturated conditions. These tests ignore the potentially important role of reaction with soil gases in influencing wasteform performance. The presented research represents a combined experimental and modeling program approach to evaluate the role of the carbon dioxide reactive component of soil gas on the long-term performance of cementitious wasteforms. A series of wasteforms were exposed to an accelerated environment for carbonation and then subjected to leaching tests. Results are analyzed by comparison of experimental data with theoretical models of the leaching process. The results indicate that carbonation increases the apparent diffusion coefficient for unreactive species while resulting in stronger chemical binding of metals through solid solution in calcite. The net effect can be either positive or negative and depends upon the species being considered.
Keywords: Cementitious wasteform, concrete carbonation, long-term performance

1 Introduction

Solidification and stabilization with cementitious materials is perhaps the most widely accepted and economically attractive waste management option for disposal of low-level radioactive wastes. Advantages of cementitious wasteforms include (a) availability of materials locally on a worldwide basis, (b) low cost, (c) ability to tailor the mixture for different wastes, and (d) high physical strength of the resulting wasteform. The solidified wasteform can be developed in place (grouting of wastes), or can be poured

Mechanisms of Chemical Degradation of Cement-based Systems. Edited by K.L. Scrivener and J.F. Young.
Published in 1997 by E & FN Spon, 2–6 Boundary Row, London SE1 8HN. ISBN: 0419215700.

into steel drums or concrete canisters for later disposal, or formed directly into large monoliths. Although cementitious material is used primarily for its structural and physical properties, the effect of cementitious materials on the geochemical environment in the immediate vicinity of the concrete is pronounced.

The geochemistry of the pore fluid in contact with hydrated cementitious materials is characterized by persistent alkaline pH values buffered by the presence of hydrated calcium silicates [C-S-H] and portlandite [$Ca(OH)_2$]. The high pH and large internal surface area for sorption make cementitious wasteforms ideal for sequestering radionuclides. Wasteforms to which blast furnace slag [BFS] has been added are additionally characterized by the presence of high concentrations of reduced sulfur species in the pore fluid and by strongly reducing redox potential (Eh values), both controlled by reduced sulfur in the BFS. These highly reducing conditions are effective in sequestering multivalent radionuclides such as Tc-99 in insoluble reduced forms. Because the geochemical environment within the wasteform pores is profoundly different from the ambient environment within the vadose zone, the wasteform reacts irreversibly with and is altered by reactive agents within the vadose zone. Chief among the reactive agents are oxygen and carbon dioxide, both components of soil gases.

Although soil gas composition can be highly variable (Table 1), increased levels (relative to the atmosphere) of CO_2 and reduced levels of O_2 are expected because of respiration.

1.1 Limitations of current test methods

Although significant research has been expended upon design of cementitious wasteforms, relatively little effort has been put into evaluation of how the wasteforms will perform within the context of the overall disposal system. Standardized leach tests provide only a limited amount of information and are insufficient for estimation of long-term performance. The standard test for radioactive wastes is the ANSI/ANS-16.1 leach test. ANSI/ANS-16.1 generates a single parameter called the leachability index. The leachability index is derived by fitting a diffusion-based equation to short-term leaching results. The diffusion equation is generally successful in fitting the results of short-term tests because several processes, in addition to simple diffusional release, lead to a square root of time dependence for total contaminant release. For example, adsorption-controlled release [1], solubility-controlled release, and oxidation-controlled release [2] all lead to a square root of time dependence for release in short-term tests. Although a variety of phenomena lead to similar short-term leaching behavior, and can thus be described by a leachability index, over longer time periods, performance diverges depending upon the processes controlling release rate. For this reason, simple leach tests, combined with "black box" analyses, are inadequate to predict absolute or relative long-term performance of wasteforms.

The Toxicity Characteristic Leaching Procedure (TCLP), promoted by the U.S. Environmental Protection Agency (EPA), is based on overnight extraction using a pH 3 acetic acid solution and a crushed wasteform. Quantitative interpretation of these tests are hampered because of the lack of correspondence between test conditions and conditions encountered in the disposal environment. Three broad areas are identified [3] where further research is needed in cementitious wasteforms: (a) correlation of physical properties to performance, (b) fundamental chemistry and microstructure, and (c) performance tests.

Table 1. Soil gas partial pressures

Gas	Log P_i (atm)	
	Air	Soil
N_2	−0.1	−0.1
O_2	−0.7	−1.7 to −1.5
CO_2	−3.5	−2.5 to −1.5
H_2O	−1.9	−1.8

One of the limitations of current tests for wasteform leaching is that water-saturated conditions are specified. The specification of water-saturated conditions for leach tests contrasts with current regulatory requirements and engineering practice of placing stabilized wastes in the unsaturated zone. In terms of short-term, diffusion-controlled leach rates, an argument can be made that leach tests under saturated conditions represent a worst case and therefore provide a conservative estimate of *in situ* performance. However, more careful consideration of the geochemical processes involved in *in situ* leaching suggests that water-saturated leach tests do not necessarily provide conservative estimates of long-term wasteform performance. The most stable and optimal environment for long-term survival of cementitious materials is in a pool of stagnant water. The vadose zone, in general, provides a much more aggressive long-term environment for cementitious materials. The aggressive environment, as discussed below, can be related to increased contact with carbon dioxide and oxygen in the soil air, and variable contact with water. More generally, enhanced design of remediation systems requires improved fundamental understanding of wasteform leaching in likely service environments, and this information cannot be produced from standardized leach tests. To this end, we have evaluated the role of soil gas CO_2 on the performance characteristics of cementitious wasteforms.

1.2 Reaction With Carbon Dioxide

Carbon dioxide interacts with cementitious wasteforms in a process known as carbonation. Carbonation occurs when portlandite and other calcium-bearing phases in the wasteforms react with CO_2 to form calcite [$CaCO_3$]. Major mineralogical changes associated with carbonation include conversion of portlandite [4]:

$$Ca(OH)_2(s) + CO_2(g) \Rightarrow CaCO_3(s) + H_2O(l) \tag{1}$$

and calcium silicate hydrate gel:

$$C - S - H(s) + CO_2(g) \Rightarrow CaCO_3(s) + SiO_2 \bullet nH_2O(s) + H_2O(l) \tag{2}$$

Portlandite, as shown in Equation (1), reacts most rapidly with CO_2; however, all the cement phases are subject to carbonation, leading ultimately to an assemblage of $CaCO_3$, SiO_2 gel, and metallic oxides.

The conceptual model for the reaction of soil gas with a wasteform is given in Figure 1. Reactive components present in soil gas diffuse into the wasteform, resulting in the alteration. An inner zone of intact concrete is surrounded by a growing rind of altered concrete. Because the geochemical and physical environment in the altered rind is significantly different from the environment of the intact core, the behavior of radionuclides and toxic metals is expected to be different. Quantifying the effects of these differences on radionuclide retention requires that the rate of rind formation be known and the salient chemical and physical properties of the altered concrete be available.

The rate of penetration of the carbonated reaction front can be approximated with a shrinking core model [5, 6]. Experimental investigation has demonstrated that the migration of the front is most rapid at a relative humidity of 50% and declines at greater and lower relative humidity. Carbonation has several influences on the wasteform, both physical and chemical. Porosity tends to drop as previously open, large pores fill with calcium carbonate [4]. With progressive carbonation, the pH of the system gradually drops [7], leading to altered solubility (sometimes higher, sometimes lower) of waste metals previously immobilized by the high pH of the wasteform. Another change in wasteform behavior caused by carbonation is the potential for solid solution of some ions in the calcium carbonate present in the carbonated zone. Initial calculations of the importance of calcite solid solution on contaminant release, from cementitious wasteforms, suggest that the process is potentially important in limiting the release of some ions [6]. The carbonated shell is likely to be critically important in long-term performance, but is not addressed in current leach tests.

2 Experimental method

Small wasteforms (diameter 3.1 cm, height 6.7 cm) were cast with synthetic wastewater containing dissolved metal ions. Synthetic wastewater was prepared by dissolving nitrate, chloride, or oxide salts of cadmium, cobalt, lead, strontium, zinc, and antimony in deionized water to yield concentrations of 3,000 mg/L for Cd, Co, Pb, and Zn, 1,000 mg/L for Sr, <670 mg/L for Sb (solid Sb_2O_3 added to synthetic wastewater did not fully dissolve), and 3,200 for NO_3^-.

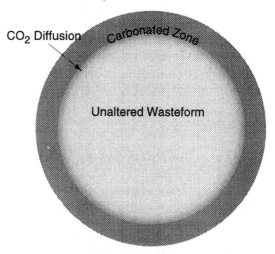

Fig. 1. Conceptual model for alteration of a cementitious wasteform by reaction with CO_2 gas.

The wasteforms were cured for 1 week at 100% relative humidity followed by placement into a controlled environment chamber intended to accelerate the rate of carbonation. The experimental wasteforms were exposed to a 50% CO_2 atmosphere at 75% - 50% relative humidity and a temperature of 50° ±1° C to enhance rate of carbonation. The controls were placed in identical environments, except CO_2 free air was passed through the system.

After 26 days exposure, the wasteforms were leached in 150 mL of deionized water at room temperature. The leachate was changed on each of the samples at intervals of 2 hours, 7 hours, 1 day, 2 days, 4 days, 7 days, 14 days, 24 days, and 39 days. The leachates were analyzed by atomic absorption spectrophotometry using standard methods. Five replicates of each of the experimental and control wasteforms were leached. All metals except Sr were found to be at or below the detection limits of the standard methods employed in both the carbonated and control samples.

3 Results and discussion

The wasteforms exposed to a 50% CO_2 atmosphere were found to be fully carbonated with an average bulk calcium carbonate concentration of 9×10^{-4} mol/cm^3. Calcium carbonate amount was measured by digesting the wasteforms in acid solution and measuring the CO_2 given off in the gas phase. The final porosity was approximately 48% for the controls and 35% for the carbonated wasteforms. The reduction of porosity results from the larger molar volume of calcite relative to the initial portlandite in the cement.

The leaching results for NO_3^- are shown in Figure 2. The plotted data points represent the average of five replicates. Leaching of NO_3^- is rapid, giving apparent diffusion coefficients of 8.5×10^{-8} cm^2/s for the controls and 8.5×10^{-7} cm^2/s for the carbonated wasteforms based upon the fitted straight lines. Note that the straight line semi-infinite

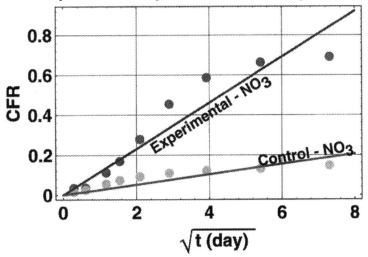

Fig. 2. Cumulative release of NO_3^- from leaching of carbonated (experimental) and uncarbonated (control) wasteforms.

domain assumption is not appropriate for NO_3^- leaching from the carbonated waste-forms because approximately 70% of the initial NO_3^- is leached. Fitting the line to the linear portion of the data will give an even higher apparent diffusion coefficient.

Nitrate was placed in the wasteforms to assess the diffusion rate of chemical species for which chemical interactions would not occur. Thus the NO_3^- results are indicative of physical changes in wasteform performance as a result of carbonation. Theoretically, the lower porosity of the carbonated samples should lead to lower diffusion coefficients – just the opposite of what is observed in Figure 2. The discrepancy between expected and observed releases may result from increased microcracking in the carbonated samples as a result of expansive forces generated during carbonation or perhaps a modification of the pore distribution toward larger average pore size and/or removal of occluded pores.

The leaching results for Sr^{2+} are shown in Figure 3. Leaching of Sr^{2+} is much slower, giving apparent diffusion coefficients of 9.0×10^{-9} cm^2/s for the controls and 1.1×10^{-9} cm^2/s for the carbonated wasteforms based upon the fitted straight lines. Strontium leaching is slower from the carbonated samples, suggesting that the formation of solid solutions with calcite controls release rate.

Strontium and nitrate data are plotted in Figure 4. Comparison of the apparent diffusion coefficients allows estimation of physical versus chemical changes in leaching as a result of carbonation. Walton [1] describes the relationship between the apparent diffusion coefficient (i.e., the parameter derived from leach tests) and the properties of the system such as distribution coefficient, porosity, tortuosity, and retardation factor:

$$D_a = \frac{\tau D}{R_d} \tag{3}$$

where

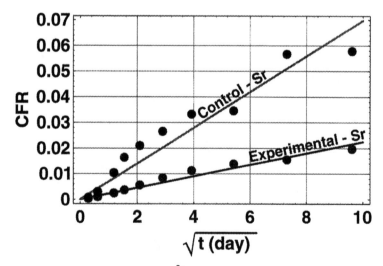

Fig. 3. Cumulative release of Sr^{2+} from leaching of carbonated (experimental) and uncarbonated (control) wasteforms.

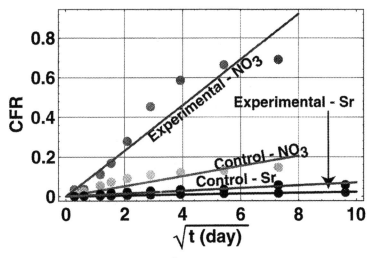

Fig. 4. Cumulative release of Sr^{2+} and NO_3^- from leaching of carbonated (experimental) and uncarbonated (control) wasteforms.

τ = tortuosity/geometry constant that is independent of diffusing species
D_a = apparent diffusion coefficient (cm^2/s)
D = diffusion coefficient in water (1.9×10^{-5} cm^2/s for NO_3^- and 7.9×10^{-6} cm^2/s for Sr^{2+})
R_d = retardation factor for an ion = $1 + \rho_b K_d / \theta$
ρ_b = bulk density (g/cm^3)
K_d = distribution coefficient = concentration in solid phase/concentration in liquid phase (mL/g)
θ = volumetric water content.

Because the K_d for NO_3^- is 0, its leachability index can be used to estimate a tortuosity factor of 4.5×10^{-2} for the carbonated and 4.5×10^{-3} for the control wasteforms, respectively. The derived value of τ can be used to estimate a retardation factor for Sr^{2+} of 3.9 in the controls and 330 in the carbonated wasteforms. The corresponding distribution coefficients are 0.63 mL/g for the controls and 70 mL/g for the carbonated wasteforms. The observed results of partitioning are consistent with the theoretical predictions of Smith and Walton [6] based upon a model for solid solution of Sr^{2+} in calcium carbonate.

4 Conclusion

The goal of the work was to better elucidate the importance of carbonation on long-term performance of cementitious wasteforms. The experimental data indicate that carbonation leads to decreased physical retention of contaminants despite the fact that carbonation causes a significant decrease in porosity. The decrease in physical retention may reflect microcracking and/or the rearrangement of pores to remove or decrease occluded areas. The physical changes are evidenced by more rapid release of NO_3^- from the carbonated wasteforms in relation to the controls. However, some metals, including Sr^{2+}, are more tightly bound in the carbonated wasteforms through solid solution in calcium carbonate. The net result of carbonation may be to either increase or decrease release rate and depends upon the specific radionuclide of concern. The importance of carbonation in determining long-term performance is ignored in current standardized test methodologies and severely limits the usefulness of those tests in predicting long-term performance.

5 Acknowledgment

This work was supported by the University of Texas at El Paso's Center for Environmental Resource Management (CERM) through funding from the Office of Exploratory Research of the U.S. Environmental Protection Agency (cooperative agreement CR-819849-01-4).

6 References

1. Walton, J.C. (1992) *Performance of Intact and Partially Degraded Concrete Barriers in Limiting Mass Transport*, NUREG/CR-5445, U.S. Nuclear Regulatory Commission, Washington, D.C., 1992.
2. Smith, R.W. and Walton, J.C. (1993) The role of oxygen diffusion in the release of technetium from reducing cementitious waste forms. *Materials Research Society Symposia Proceedings,* Vol. 294, pp. 731-736.
3. McDaniel, E.W., Spence, R.D., and Tallent, O.K. (1991) Research Needs in Cement-Based Waste Forms. *Materials Research Society Symposia Proceedings,* Vol. 212, pp. 449-456.
4. Dayal, R. and Reardon, E.J. (1992) Cement-based engineered barriers for carbon-14 isolation. *Waste Management,* Vol. 12, No. 2/3. pp. 189-200.
5. Papadakis, V.G., Vayenas, C.G., and Fardis, M.N. (1989) A reaction engineering approach to the problem of concrete carbonation. *American Institute of Chemical Engineers Journal,* Vol. 35, No. 10. pp. 1639-1650.
6. Smith, R.W. and Walton, J.C. (1991) The effect of calcite solid solution formation on the transient release of radionuclides from concrete barriers. *Materials Research Society Symposia Proceedings,* Vol. 212, pp. 403-409.
7. Berner, U.R. (1992) Evolution of pore water chemistry during degradation of cement in a radioactive waste repository environment. *Waste Management,* Vol. 12, No. 2/3. pp. 201-219.

38 CHEMICAL ATTACK ON CEMENT IN NUCLEAR REPOSITORIES

F.P. GLASSER
Department of Chemistry, University of Aberdeen, Old Aberdeen, Scotland

Abstract
The performance lifetime of cements intended for chemical conditioning of nuclear wastes is limited by internal and external degradation. Internal degradation occurs by interaction with waste components; external degradation, mainly by reaction with ground water constituents. Much knowledge has been accumulated about attack mechanisms but less is known about their kinetics. A combination of practical experience, testing and theoretical calculations, using computer-based solubility models, are suggested as the best way to obtain validated lifetime-performance predictions.

1. Introduction

The chemical deterioration of cements differs in many important respects from that of other materials. Comparing cements with metals, for example, the time to physical perforation of metal barriers is a crucial parameter whereas cements depend to an important extent on chemical conditioning. This conditioning action has been the subject of numerous papers at MRS Symposia [1] and elsewhere [2]. There are, however, a number of other important differences between barrier materials.

- Cementitious barriers are typically more permeable than metal; thus not only is the external geometric surface available for reaction but internal surface, controlled by porosity-permeability relationships, may be available for reaction.
- The corrosion products of cement, like those of metal, may range in solubility but where they are soluble, they typically have a much greater conditioning influence on the pH of attackant solutions than metal corrosion products. Cements condition high pH, often ≥12.

Mechanisms of Chemical Degradation of Cement-based Systems. Edited by K.L. Scrivener and J.F. Young. Published in 1997 by E & FN Spon, 2–6 Boundary Row, London SE1 8HN. ISBN: 0419215700.

- Whereas attack on cements involves ion exchange, dissolution, precipitation, etc., it does not normally involve chemical oxidation-reduction couples characteristic of a metal undergoing corrosion.

Since the long-term performance of cement systems relies on chemical conditioning, it is appropriate to define chemical deterioration as follows. *Chemical deterioration is a process, or series of processes, induced by chemical substances which, upon contact with cement cause dissolution with mass wastage, neutralize or reduce system pH and Eh buffering capacities, adversely affect the sorption characteristics or affect the physical performance of the system-as by altering permeability or by causing significant dimensional changes with cracking or disruption.* This wide definition embraces chemical as well as physical characteristics.

The characterisation of cement deterioration has had three stages. First, the purely phenomenological   stage, at which cements are observed not to perform well in certain environments. It should be noted that the definition offered does not exclude water as a chemically aggressive agent and indeed, performance in aggressive environments is usually judged relative to performance in water. With more than 140 years of experience of the use of Portland cements in a wide range of environments, a vast store of empirical knowledge on performance has been assembled [3].

The second stage identifies the specific mechanisms responsible for deterioration. Examples include numerous studies of the rates of carbonation of cements, of dissolution in "aggressive" carbonic acid solutions and of chloride and sulfate penetration. The third stage, which is just being commenced, is to acquire data leading to formulation of physicochemical models which would enable cement systems to be modelled and future performance to be predicted.

2. Analysis of the Problems

The time scale for performance in nuclear waste conditioning much exceeds normal expectation of performance in conventional civil engineering applications. This has meant an urgent need to move to stage 3, as defined herein. Coincidentally, as civil engineers push for longer design lifetimes, they have also begun to respond to these pressures. For nuclear wastes I assume that cement barriers will be required to perform underground in flooded environments. Depending upon the integrity of the geological barriers, cements may have as their primary role a physical barrier action, or a chemical conditioning action, or both. Table 1 prioritises some of the factors which must be quantified in assessing performance. It is of course essential to have complete characterisation parameters for the cementitious formulations, many of which are unconventional by normal civil engineering standards.

In shallow repositories as well as in storage and transportation, the physical barrier action is important. During the closure phase, numerous openings into a repository require to be sealed; dimensional stability of seals is clearly important to prevent 'short circuit' pathways.

Performance in the repository is complex; internal reactions arising from reaction between wastes and degraded waste components may occur. For example, decay processes of organic wastes leading to methane formation will not significantly affect cement performance, although over pressure may need to be avoided, whereas CO_2,

Table 1. Aspects of Chemical Deterioration of Cement Barriers

Intended Purpose/Application	Factors Leading to In-Service Degradation
Physical barrier	Chemical dissolution, increase in permeability arising from corrosion, spalling and cracking.
Seal (shafts, tunnels)	As above: also annular shrinkage
Chemical conditioning of wastes in the repository	Loss of high pH, more oxidizing E_h. Crystallization with possible loss of high sorption. Reaction with complexants: enhanced dissolution. Potential for colloid generation

generated under more oxidative conditions, reacts with cement forming $CaCO_3$. This in turn leads to reduction of pH buffering capacity. Clearly, degradation of wastes, especially of organics, and their impact is a major and as yet incomplete area of research. Ashing much reduces problems associated with disposal of organics. However, many immobilization strategies seek to incorporate organics directly, without molecular transformation. The large range of solubilities and molecular structures of organics result in a range of reactions only a few of which have been characterized. Table 2 shows selected examples which have thus far been studied.

Perhaps the best explored features of future performance are those associated with cement-ground water interactions. Table 3 shows some phenomenological features of attack. Of course special circumstances may greatly extend this list; only the

Table 2. Interaction Between Cements and Selected Organics

Organic Substance	Characteristic Reaction
Low m.w. alcohols and acids, R-COOH	Form Ca salts, e.g. Ca methoxide (1) Ca oxalate (2); alternatively decompose with formation of $CaCO_3$ e.g. citric acid (2).
Complexing agents .e.g. EDTA	Not affected. Remain in pore fluid and potentially available for complexation (2)
Higher m.w., water soluble polymers, e.g. superplasticizers	Initially sorbed. Subsequently hydrolized. Loss of solubilizing functional groups results in precipitation (3)
Ion exchange polymers	Imbibe water with swelling. Long term fate unknown.

most destructive. Fig. 1 adapted from [7], shows the mechanism of attack on a relatively good quality matrix, i.e., with low permeability. In the near surface layers, extensive decalcification occurs while gypsum and brucite develop: the latter is a product of magnesium replacing calcium. Replacement need not be isovolumetric; hence this zone is marked by cracking, resulting from physical expansion. In severe cases, spalling and disruption of the near surface layers occur constantly exposing fresh surface to attack. In the absence of severe spalling, a secondary zone of ettringite and gypsum develops. These mineralogical trends are anticipated by calculated data: for example, the thermodynamics of Mg for Ca replacement are calculated to be favorable at all realistic Mg concentrations in natural waters [8]. Moreover, ettringite is known to coexist stably with gypsum [9] although it is not itself stable below pH ~10. Thus the transition between the zones with and without ettringite also marks the position at which the permeating pore fluid has pH ~10, the minimum necesssary to stabilize ettringite.

In many realistic situations carbonate will also be present in ground water. Under a wide range of conditions, a skin composed of calcium carbonate and brucite, with perhaps a little gypsum, forms spontaneously. This skin typically has very low permeability which, once developed, greatly slows the rate of attack and hence the rate of permeation. Precipitation of $CaCO_3$, even at low a_{CO2} and at skin pH's, likely to be in the range 8-9, is also anticipated by modelling studies. What cannot be anticipated from models is the physical properties of the mineralogical zonation: for example, net dissolution from one zone but physical expansion in those where net accumulation of solids occurs.

Zonation is least likely to develop in very permeable backfills which will respond more homogeneously to chemical attack. Transport may still occur within the pore network, but is difficult to model, given the present state of knowledge.

Many of the experiments thus far described have been conducted in relatively concentrated solutions, e.g. $MgSO_4$ concentrations, in the range $\geq 0.25M$. This may be realistic of certain repository environments but salt-rich attackants tend to be used in laboratory simulations to accelerate reaction. Modelling, coupled with experimental results, enables extrapolation of the mechanisms to other, usually more dilute, attackants and to determine thresholds below which attack will no longer occur or where the attack mechanism undergoes change.

Magnesium sulfate attack is characterized by essentially the same mechanisms occurring over large concentrations of aggressive ions. Within this envelope the intensity of attack correlates well with sulfate concentration; moreover reduction of pH buffering reserves are proportional to the mass of $MgSO_4$ which reacts. Kinetics are however complex; even low carbonate concentrations enable self-protecting skins to be established. But even without appeal to purely kinetic factors, the chemistry of other mechanisms of attack is more complex. For example, CO_2 dissolved in water may, under some circumstances, be aggressive: a regime of skin formation gives way to one of active dissolution. The complex chemistry has been analysed, although has not yet been extended to include more complex natural waters [10].

4.GENERAL APPROACH TO MODELLING CHEMICAL ATTACK

Fig. 2 shows conceptually a model for cement deterioration which expresses the

Table 3. Chemical Constituents in Ground Water which
Impact on Cement Performance.

Constituent	Impact
CO_2	Accelerated dissolution: loss of high pH
SO_4	Cracking, spalling; loss of pH
Cl	Accelerated corrosion of embedded metals
Acids	Accelerated matrix corrosion

commonest constituents are listed. The conventional approach has been to explore each constituent individually, although it must be accepted that in the majority of natural environments, several may occur together: for example, $MgSO_4$-containing waters. The acids referred to are miscellaneous: humic acid in ground water, acetic acid from oxidative decompositon of organic wastes, etc. Examples will be explored in more detail, using sulfate attack.

3. SULFATE ATTACK

Decades of phenomenological experience show that sulfate attacks Portland cement. The mechanism also depends on the associated cation: magnesium sulfate being the

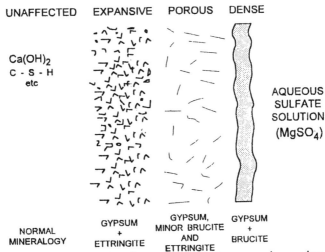

Fig. 1. Normal Portland cements develop a zonal structure upon immersion in $MgSO_4$. The Figure shows some of the mineralogical and physical characteristics.

328   *Glasser*

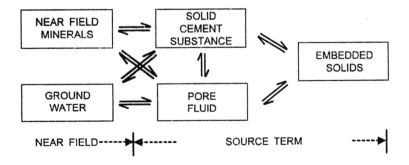

Fig. 2. Conceptual model of cement interactions with included wastes and the nearfield.   Arrows indicate transport reactions.

possible interactions.   Note that, besides reaction with impinging or permeating solutions, the composition of which is buffered by the near-field geochemistry, the embedded waste materials may themselves react with cement. These reactions include the radioactive species as well as inactive components.   Amongst the active components thus far studied in any detail only Cs and Sr do not react strongly with cement to form compounds. Thus Cr [11] M[12] and U[13] react and in all cases, compound formation lowers the solution activity, by several orders of magnitude, relative to expectations based on solubility calculations which employ the relevant hydrous oxides/hydroxides at pH 12 to13.   There are relatively few data to show what happens to these complex, solubility limiting salts during degradation, but it is shown in this Symposium [14] that carbonation of the solubility-limiting phases which condition Cr(III) may continue to immobilise chromium.   Owing to the very variable complexing power of HCO_3^- or CO_3^{2-} ions for radwaste species it would be unwise to assume that carbonated, weakly alkaline cements will generally perform as well as fresh, high pH matrices.

5. DISCUSSION

It is asserted that a combination of empirical testing and theoretical solubility-reaction modelling are likely to provide the best approach to predicting the future immobilisation properties of cement systems.   The time scales for performance are governed by the half-lives of the radioactive species: these being fixed, it is impossible to compress the time scale, often 10^5 - 10^6 years, into empirical tests.   Cements do not respond well to accelerated tests: their mineralogical nature, and hence their reaction with other components in the near field, is irreversibly altered by elevated temperature. Moreover, the mechanism of attack may undergo change, with transition to a new regime, as concentrations of the attackant species are increased.   There is still need for accelerated testing, but with greater sensitivity paid to the underlying mechanisms.

Modelling does have its drawbacks. The number of possible interactions in complex systems tends to increase factorially with the number of components. Hence, given a cement system of five or six components which interacts with ground water containing perhaps another four or five additional components, it is not possible to assess all possible interactions even by computation. It is not a case of lacking computational capability, but of assimilating the significance of the calculations. Thus insight and judgement, the latter often based on experience, must be used to highlight and define aspects of key processes for subsequent calculation. Data shortage is also a major limitation to the scope and accuracy of the calculations. Unfortunately, acquisition of new data and critical evaluation of existing data are not generally afforded a high priority in the allocation of funds.

Results of both experiment and calculation confirm that cements will alter in the natural environment. This is not unexpected; all man-made materials are likely to degrade as indeed will many naturally-occurring materials. Cements respond readily to changes in their local environment. In some respects this response is rapid; cements contain metastable and somewhat soluble matrix-forming materials many of which are characterised by having high specific surfaces and exchangeable ionic constituents. Perhaps the most surprising feature is that cements often manage to isolate themselves form their local environment, usually by forming low permeability skins. This has both advantages and disadvantages to the construction of performance-related scenarios. On the one hand, the rates of cement deterioration in potentially aggressive environments may be reduced but on the other, their conditioning action could be marginalized by failure to react with and raise the pH of percolating water, or inability to provide surfaces for sorption or both. Probably a combination of computation and targeted experimentation provides the way forward to predict the effective lifetimes of cements used as chemical barriers to retard the transport of radionuclides.

6. ACKNOWLEDGEMENT

Financial support from Commission of the European Communities and the UK Department of the Environment have enabled the writer to conduct research and gain an appreciation of the problems. Their support is acknowledged. However the contents do not necessarily reflect UK Government policy, although they may be used in the formulation of policy.

7. REFERENCES

1. Bavkatt, A. and Van Konynenburg, R.A. (1994) *Scientific Basis for Nuclear Waste Management XVII*. Mater. Res.Soc., Pittsburg PA. NB there are also 16 previous volumes in this series.
2. European Materials Research Society E-MRS) *Chemistry of Cements for Nuclear Applications*. (ed. P. Barret and F.P. Glasser), North Holland - Pergamon (1992).
3. Lea, F.M.(1970) *The Chemistry of Cement and Concrete*. 3rd Ed., Ed. Arnold, London.
4. Beaudoin, J.J., cited in Taylor, H.F.W. *Cement Chemistry* 2nd printing (1990) Academic Press, London and New York p124-125.

5. Smillie, S. MSc Thesis, University of Aberdecen (1996).
6. Yilmaz, V.T., Kindness, A. and Glasser, F.P. *A Determination of Sulphonated Naphthalene Formaldehyde Superplasticizer in Cement: A New Spectrofluorimetric Method and Assessment of the UV Method.* Cement and Concr. Res.., 27, No. 4, 653-660 (1992).
7. Taylor, H.F.W. Sulfate Reactions in Concrete - Microstructural and Chemical Aspects, in *Cement Technology.* (ed. E.M. Gartner and H. Uchikawa) Ceramic Transactions, Vol. 40, Amer. Ceram. Soc., Westerville OH (1994) pp 61-78.
8. Stronach, S. PhD Thesis, University of Aberdeen (in preparation).
9. Damidot, D. and Glasser, F.P. Sulfate Attack on Concrete: AF_t Stability From Phase Equilibria. In *Proc. 9th Intl. Congress on the Chemistry of Cement* (New Delhi, 1992) (Nat. Council for Building Materials, New Delhi) Vol. 5, pp 316-321.
10. Cowie, J. and Glasser, F.P. *The Reaction Between Cement and Natural Waters Containing Dissolved Carbon Dioxide.* Advances in Cement Res., 4, No. 15, 119-134 (1993).
11. Kindness, A., Macias, A. and Glasser, F.P. *Immobilisation of Chromium in Cement Matrices.* Waste Management, 4, No. 1, 3-11 (1994).
12. Kindness, A., Lachowski, E.E., Minocha, A.K. and Glasser, F.P., *Immobilisation and Fixation of Molybdenum(VI) by Portland Cement.* Waste Management 14, No. 2, 1-6 (1994).
13. Moroni, L.P. and Glasser, F.P. *Reactions Between Cement Components and U(VI) Oxide.* Waste Management (in press) 1995.
14. Macias, A., Kindness, A. and Glasser, F.P. *Impact of Carbon Dioxide on the Immoblization Potential of Cemented Wastes: Chromium.* MRS Proceedings (submitted).

39 MODELLING THE LONG-TERM DURABILITY OF CONCRETE FOR RADIOACTIVE WASTE DISPOSALS

B. GERARD[1-3], O. DIDRY[1], J. MARCHAND[2], D. BREYSSE[4] and H. HORNAIN[5]

1 Electricité de France, Research Division, France. 2 Centre de Recherche Interuniversitaire sur le Béton, Laval University, Canada
3 Laboratoire de Mécanique et Technologie de Cachan, France
4 Centre de Développement des Géosciences Appliquées, Universite Bordeaux I. 5 Laboratoire d'Etudes et Recherches sur les Matériaux, France

Abstract
In the past decades, cement-based materials have been increasingly used for the construction of radioactive-waste containment barriers. The design of durable structures for this specific application requires a precise knowledge of the evolution of the engineering properties of the materials over a 1000-year period. The features of a new modelling approach are presented. The model considers the various chemical and mechanical loads and their eventual couplings. The physical bases behind the development of the model are detailed. A section of the paper is specifically devoted to the experimental techniques designed to validate the model : accelerated leaching test, assessment of the permeability and mechanical behavior of concretes under tensile loading, characterization of the concrete microstructure and microcracking by image-analysis. Results of selected numerical simulations are presented in the last section of the paper.
Keywords : cement, concrete, coupling, damage, durability, electric field, leaching, model, nuclear waste.

Notations

a	geometric parameter (m^{-1})	g	gravity (m/s^2)
C	solid calcium concentration ($mmol/dm^3$)	I	unit tensor
Ca^{2+}	calcium concentration in pore solution ($mmol/dm^3$)	K	permeation factor (m/s)
		K_0	permeation factor of sound material (m/s)
d	scalar damage variables	V	ageing function
D	diffusion coefficient of the uncracked medium (dm^2/s)	w_i	crack width (m)
		$[\varepsilon]$	strain tensor
D_0	diffusion coefficient of sound material (dm^2/s)	λ, μ	Lame coefficient (N/m^2)
D_{fw}	diffusion coefficient in free water (dm^2/s)	$[\sigma]$	stress tensor (N/m^2)
E	elastic modulus (N/m^2)	ς	crack surface factor
E_0	elastic modulus of the sound material (N/m^2)	θ	porosity
f	crack closure function	τ	crack collapse function

1 Introduction

Concrete has been widely used in the construction of low level nuclear waste containment structures (concrete packages, surface vaults). Since the durability of these concrete barriers has to be assessed over long periods of time, typically 300 years for

Mechanisms of Chemical Degradation of Cement-based Systems. Edited by K.L. Scrivener and J.F. Young. Published in 1997 by E & FN Spon, 2–6 Boundary Row, London SE1 8HN. ISBN: 0419215700.

surface repositories. In that context, the reliable prediction of their long-term behavior requires a sound knowledge of the various deterioration mechanisms that will affect the structures over their life time, and the extensive use of simulation tools. This paper presents the main features of a new modelling approach that has been developed to predict the long-term behavior of low-level nuclear waste containers. The main originality of the approach resides in the fact that the model considers various chemical and mechanical loads and their eventual couplings. The first stage of this work was thus to thoroughly analyze the environmental conditions of a typical surface repository, and to subsequently define a "reference scenario" of degradation. After a description of the "reference scenario", and a brief literature survey in which the main aspects of the problem are outlined, the principal features of the model are presented. A section of the paper is specifically devoted to the experimental techniques designed to validate the model.

2 Identification of the reference scenario for concrete degradation

Short-lived wastes in France typically contain radionuclides with half-lifes of less than 30 years which fixes the repository lifetime at 300 years. Irradiation effects and residual heat effects are negligible. Since low-level nuclear waste containers are generally kept in surface repositories, it is also reasonable to believe that these barriers are not subjected to any significant hydraulic gradient over their lifetime.

First of all, a constant flow of pure water along the external surfaces of the concrete barriers has been identified as the most critical aggression.This initiates a slow leaching of calcium from the cement paste which is followed by microstructural changes like an increase of porosity. These microstructural alterations are likely to increase both the diffusion and the permeation, and detrimentally affect the material properties.

Even if, in the worst degradation scenario, nuclear waste containers placed in surface repositories are not expected to be subjected to severe mechanical loading, several thermo-mechanical studies have clearly demonstrated that the cement hydration can initiate cracks in these structures [1-4]. Obviously, the development of these cracks should be influenced by the pore structure alterations induced by the calcium leaching process. Simultaneously, the presence of these cracks should affect the mass transfer properties of the material, and therefore have a certain influence on the evolution of the calcium leaching process.

As can be seen in Figure 1, six different physical interactions (Fig. 1) are included in this reference scenario.

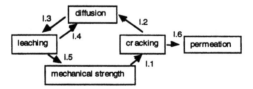

Fig. 1 - Physical interactions from the reference degradation scenario

3 Literature review

The development of the predictive model required to build up knowledge on each of the physical interactions identified in Fig. 1. The following paragraphs present a brief summary of the literature review by focusing on research needs.

• The damage theory is a successful application of a continuous analysis to the description of the mechanical behavior of concrete [5].

- For instance, simple considerations show that cracking significantly affects the permeation, as flow rate increases with the crack width in accordance with Poiseuille's law [6]. As regard diffusion, the available references all indicate that cracking has a less significant influence. The processes involved in the cracking-diffusion interactions seemed not fully understood, we did not find models on the subject [6].
- Since the calcium is the dominant element of the hydrated phases, the calcium concentration has often been chosen as a state variable while considering the pore solution [7,8]. The equilibrium between the solid calcium phases and the calcium concentration of the pore solution is shown in Fig. 2. As can be seen, the equilibrium is characterized by a two-step curve, one step being associated with the dissolution of Portlandite, the second with the dissolution of C-S-H.
- The leaching of concrete increases its porosity, and decreases its mechanical properties [9]. Unfortunately, the interaction between the chemical degradation of concrete and its mechanical behavior has not been modelled so far.

The survey shows a lack of existing data to identify the model parameters involved in the interactions between cracking and mass transfer properties (diffusion and permeation), as well as that between leaching porosity and mechanical properties (elastic modulus). The development of original procedures in accordance with modelling objectives was then necessary.

4. Model description and fundamental bases

For each of the six interactions identified in section 2, representative variables (or state variables) were identified and connected together through a set of constitutive laws.

4.1 Interactions 1 and 5 - mechanical behavior of concretes
The damage theory can be applied using scalar state variables or a damage tensor for anisotropic behavior. In the model, the scalar state version is used [5]. The stress tensor σ is written as a function of the damage d and the strain tensor ε. In order to account for the influence of the chemical degradation on the Young modulus of the material, an ageing function ($V[Ca^{++}]$) has been introduced in the model :

$$[\sigma] = (1-d)\{2\mu[\varepsilon] + \lambda \, tr[\varepsilon]I\} \qquad E = E_o(1-V)$$

It should be emphasized that the function V, which varies from 0 (no degradation) to 1 (full degradation). The evolution of V is a function of the remaining solid described from the calcium content and it can be determined from experimental data (see section 6). The Poisson's ratio is set constant.

4.2 Interaction 2, 4 and 6 - water tightness and diffusion evolutions
The evolution of the material water permeability is based on the application of Poiseuille's law to describe crack effect and an emprical law to describe the effect of the increase of porosity [10].
The pessimistic approach assumes that the damage zone has a diffusion coefficient equal to that of free water. The proposed evolution for the diffusion coefficient D_i is :

$$D_i = D_o\left(\frac{D_{max}}{D_o}\right)^{d^n} + D \quad \text{and} \quad D = D_o\left(\frac{D_{fw}}{D_o}\right)^{\frac{\theta - \theta_o}{\theta_{cr} - \theta_o}}$$

θ_{cr} is a critical porosity. Do is the sounnd diffusion coefficient.

4.3 Interaction 3 - Degradation

The capacity of the pore solution to dissolve a certain amount of hydrates depends on its saturation rate, which in turn depends on the diffusion rate. In the model, the kinetics of dissolution are considered to be much quicker than the diffusion kinetics. Thus, the evolution of the material degradation is solely conditioned by the diffusion of calcium ions, and the solid matrix evolution is derived from the equilibrium diagram (Fig. 2) [7,8].

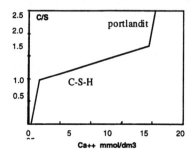
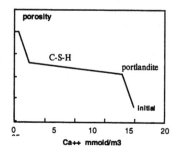

Fig. 2 : Diagram of solid-pore solution equilibrium for modeling

Fig. 3 : Diagram of the evolution of the porosity as a function of calcium

The model accounts for leaching through a single variable : the concentration of Ca^{2+} in the liquid phase [11]. The kinetics is derived from mass conservation equation :

$$\frac{\partial\left(\theta\, C_{Ca^{++}}\right)}{\partial t} + \frac{\partial C}{\partial t} = div\left\{[D]\overline{grad}\left(C_{Ca^{++}}\right)\right\}$$

The phase change induces high non-linearities in this equation. The first term is essentially dominated by the solid calcium C variation within 1 to 3 orders of magnitude, much higher than the porosity θ rise. D follows the above variation as a function of porosity, which in turn is determined by integrating to the initial porosity the different contributions of each leached hydrate volume determined in each point from the C/S value. Fig. 3 shows the evolution of the porosity as a function of the calcium in the pore solution.

5. Experimental needs

5.1   Development of procedures

The elaboration of a numerical model often requires the determination of empirical laws through laboratory experiments, and ultimately a full scale validation. Since the coupling of chemical and mechanical phenomena has been rarely done in the concrete field, the development of our model required the design of four specific experimental procedures and experimental devices. Table 1 briefly presents the developed procedures. The **BIPEDE** procedure has been developed to contribute to the understanding of the damage process and the permeation evolution under a tensile strain. The **LIFT** procedure have been designed to accelerate the chemical degradation of concrete by calcium leaching. The aim of this paper is not to discuss the several electrochemical processes occuring in the sample which has been studied during the design of the procedure. **Image analysis** developments have been made to investigate the evolution of the concrete microstructure and cracking. A correlation between **micro-hardness** and degradation depth (from LIFT samples) has been evidenced. Besides, a Young's modulus evolution with respect to hardness can be derived using the empirical equation proposed by Beaudoin and Feldman [15]. Thus the ageing function V can be identified.

Table 1 : Experimental procedures and devices developed to assess the modelling data.

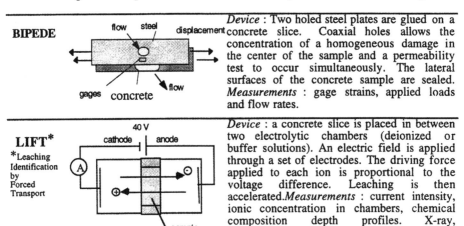

BIPEDE		*Device* : Two holed steel plates are glued on a concrete slice. Coaxial holes allows the concentration of a homogeneous damage in the center of the sample and a permeability test to occur simultaneously. The lateral surfaces of the concrete sample are sealed. *Measurements* : gage strains, applied loads and flow rates.
LIFT* *Leaching Identification by Forced Transport		*Device* : a concrete slice is placed in between two electrolytic chambers (deionized or buffer solutions). An electric field is applied through a set of electrodes. The driving force applied to each ion is proportional to the voltage difference. Leaching is then accelerated.*Measurements* : current intensity, ionic concentration in chambers, chemical composition depth profiles. X-ray, thermogravimetry, NMR, SEM.

IMAGE ANALYSIS

The developments have been realized with VISILOG software. This procedure requires adjustment of sample preparation techniques (cutting, grinding operations with emery, polishing using a 6 mm diamond paste). A red dye impregnation technique has been used for ordinary cement matrices [12]. Alternatively, a fluorescent liquid replacement technique has been used for darker matrices [13, 14]. The revealed defects are classified by applying appropriate shape criteria (cracks, porous zones, spherical air voids, etc.).

5.2 Results and discussions

Four concrete mixtures have been tested in uniaxial conditions following the BIPEDE procedure : C30, C50, C65, C90 (compression strength at 28 days : 32, 50, 65, 90 MPa). Fig. 4 show the tensile constitutive laws (for each mixture, curves indicate the average on 4-6 samples).

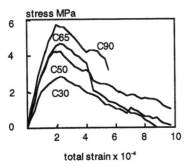

Fig. 4: Tensile constitutive law for several concretes: C30, C50, C65, C90.

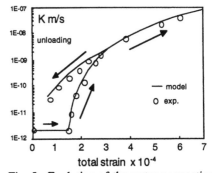

Fig. 5 : Evolution of the water permeation factor as a function of the applied strain. C65 concrete.

The procedure has proved to perform good reproducibility between samples. The modelling evolution parameters can be identified with respect to the mechanical model used. Fig. 5 introduces the evolution of the water permeation coefficient (from Darcy's law) as a function of the strain applied for C65 concretes. Up to $1.5 \ 10^{-4}$ (before the stress-peak) the permeation factor remains constant. The post-peak behavior is characterized by a sudden increase in the apparent permeation factor. Cyclic loads show an irreversible process between strains of 2.10^{-4} and 4.10^{-4} which is correlated with the development of one or two localized cracks. Beyond 5.10^{-4}, the permeation factor follows a quasi-reversible process upon unloading. Loading opens or closes major cracks. In this area, the crack collapse function is equal to one, which means that the flow is totally governed by the localized cracks (one or two). This tends to support our approach to model the permeation factor with a fictitious crack aperture w.

Fig. 6 shows the solid calcium content in a sample w/c = 0.45 after 3 weeks. The calcium content in the reference sample is 32 % by weight. The zero depth corresponds to the cathode surface (negative pole). At 3 mm depth, the reference level is reached. This concentration distribution is very similar with the calcium profile which could be found in a deteriorated sample subjected to natural conditions. It also validates the choice of using Fig. 2 to model the chemical equilibrium. At the anode, a similar degradation is also found, although significantly lower. The equivalent time under natural conditions is estimated at 6 years.

Following Vicker's procedure, Fig. 7 shows the obtained profile of micro-hardness as a function of the depth. The relationship between the solid-calcium content and the micro-hardness both at the cathode and the anode has been clearly shown. Assuming that the relationship between elastic modulus (E) and hardness (H) is $E/Eo=(H/Ho)^{0.7}$ [15], Fig. 8 shows how the ageing function V can be identified through these results. The elastic modulus is affected by the loss of Portlandite. Between C/S = 1.65 and C/S = 1.2 a quasi constant modulus is found, followed by a decrease of the stiffness below 1.2. The resin disturbs the results near the surface.

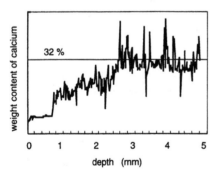

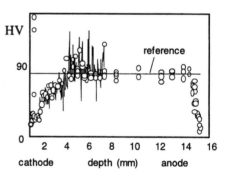

Fig. 6 : calcium content profile in the solid body of an OPC cement paste after 3 weeks and 20-25 V between the sample ends. 32 % is the reference content in sound sample. w/c = 0.45.

Fig. 7 : Evolution of microhardness in a LIFT sample. The calcium content of Fig. 10 is superimposed.

Fig. 9 shows the crack maps of two concrete BIPEDE samples. Two main cracks have been generated in the samples. Apart from these localized defaults, microcracks have also developed. The higher the concrete strength, the lower the microcrack density. It seems that microcracks could be responsible for the damage in the lowest strains. However, their connectivity appears to be low which is also coherent with the slow evolution of the permeation factor in this range of strain $(0-2.10^{-4})$. The localization appears after the stress-peak. However, their connectivity appears to be low according to the slow evolution of the permeation factor in this range of strain $(0 - 2.10^{-4})$. From Image-Analysis, the increase of porosity can also be quantified.

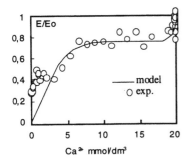

Fig. 8 : Elastic modulus as a function of the calcium concentration

Fig. 9 : Crack map obtained from Image-analysis of BIPEDE C90 for an applied strain of 8.10^{-4}

6 Associated numerical developments

The numerical implementation associated with the model requires some specific developments. Some basic concepts have been numerically performed to assess the influence of cracking on steady-state diffusion processes. In order to establish a relationship between cracks and diffusivity it is assumed that crack patterns can be approximated by a planar array of regularly spaced crack segments [16,17] (Fig. 10). The equivalent diffusivity of the cracked medium D is determined thanks to finite element simulations. Both sound matrix and cracks are meshed. Free water diffusion D_W is affected to cracks. For the uncracked material a coefficient of diffusion D_O is applied. For most cases, $D_W/D_O < 1000$. When $L_3=0$ it is found that $n = D_O/D$ is a function of $d = L_1/L_4$ (effective crack surface) : $n = d / (D_{fw}/D_o + d - 1)$. Fig. 11 illustrates the evolution of parameter n as function of d $L_3=0$. Assuming this law for a BIPEDE specimen, the main cracks detected by image-analysis have a crack width L_4 of about 0.02 mm and a crack spacing of about 50 mm (Fig. 9). Then, $d = 2500$ and $n = 0.71$. Thus, an increase of 40 % of the apparent diffusivity could be estimated. Now, if cracks are discontinuous, these simulations show that the crack density has to be extremely high to reach such a ratio. The main parameters are the crack width, the crack density and the ratio D_{fw}/D_0. Both because of the independence of the crack diffusivity with respect to its width, and of the low values of the ratio D_{fw}/D_0, the apparent diffusion coefficient can hardly overtake the sound diffusivity of the material. For mechanical loads a range of 10 % to 200 % can be expected.

However, in the local area surrounding a crack, the microstructure can be modified very quickly. In order to further investigate this point, the chemical model alone was implanted in a finite element code (section 4.3, non-linear equation of diffusion). The model simulates the calcium in ionic and solid phases distribution into a cement-based sample as a function of time. A cracked sample has then been simulated (Fig. 12). This shows the local influence of a crack on leaching by describing the evolution of Ca++ distribution.

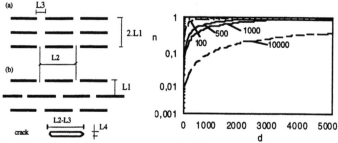

Fig. 10 : Parallel and periodical sets of cracks.

Fig. 11 : The factor n as a function of the d factor when $L_3=0$ (continuous cracks).

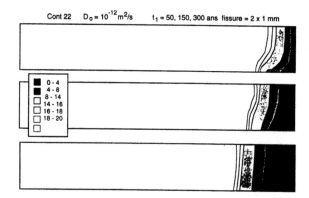

Fig. 12 : Evolution with time of the degradation of a cracked beam.
The iso-line indicates the calcium concentration (mmol/dm^3) in the pore water.

7 Current limitations of the model

All the developments undertaken in the course of this program were oriented towards the understanding of the physics involved in concrete ageing within a specific radioactive waste repository environment. Several assumptions were made which currently restricts the range of possible applications. The main assumptions are:

- The flow of fluids under pressure gradients is not considered in the model. Contrariwise to a porous media approach, the hydromechanical coupling presented follows only a "one-way path" (mechanics / hydraulics);
- All constitutive laws were developed assuming isothermal conditions. Nevertheless, the cement chemistry equilibria described are rather sensitive to temperature variations;
- No time-dependent effects such as creep of concrete were taken into account.

If assumptions 1 and 2 (isothermal conditions, no pressure gradient, tensile loading) seem justified under the degradation reference scenario attached to a low level radioactive waste repository, the third one should ideally be accounted for. This perspective will be further investigated in the future.

8 Extension to high level waste scenarios

Although this research was initiated and oriented towards a low-level radioactive waste surface repository environment (concrete containers, vault roofs and walls,...), the use of cement-based materials is currently being investigated for long term storage or high level waste geological disposal scenarios. A long term storage, either for spent fuel or glass canisters, could become an intermediate solution before geological. With a possible lifetime of 50 to 100 years, such a facility should combine the requirements from nuclear plants and existing low level waste surface repositories. Concrete would certainly be used extensively for building walls, vaults, or containers.

Cement-based materials could possibly be used in geological disposals, either in sedimentary clay layers or in crystalline bedrocks. Durability assessment in this context will first imply to consider the underground water content, which could lead to account for additionnal cement degradation mechanisms. Moreover, water pressures at depths such as -500 m down to -1000 m could involve hydro-mechanical couplings within concrete barriers such as sealings. At least, both heat generation and irradiation are likely to occur during the first hundred years. It is clear however that the cement-based materials are not expected to provide isolation over the total lifetime. Typically, a 1000 years durability should be considered as a satisfactory range for these structures.

Another range of applications lies in high integrity containers, designed to isolate the waste both under long term storage and deep disposal environments. The approach will consist in developing additional modules to the current model (temperature effects,etc.).

9 Conclusions and major perspectives

A numerical model has been developed to predict the long-term durability of concrete containment barriers within a radioactive waste disposal environment. Its main originality lies in the integration of chemical and mechanical loads and of their eventual interactions. Cracking, calcium leaching, diffusion and permeation processes and mechanical weakening are simultaneously considered.

The proposed model simulates the chemical attack penetration within a concrete structure taking into account the local effects of cracking in the kinetics. Moreover, thanks to a state coupling between mechanics and chemistry (evolution of elastic modulus as a function of calcium content), the crack developments due the chemical degradation can be simulated. The damage theory is applied to account for the evolution of cracking, whereas a non-linear diffusion model has been developed to simulate the underground water leaching. A so-called ageing function which affects the elastic modulus of degraded areas is introduced. Several numerical developments associated with the model are currently under progress to implement non linear diffusion and non-linear damage processes together.

Parallel to this approach of modelling, experimental developments have been realised. In order to study the chemical degradation in accelerated conditions, an electric field is applied to cement-based samples (LIFT procedure). The BIPEDE procedure allows the assessment of cracking mechanisms and permeability properties of concretes under tensile loads. Micro-hardness tests on LIFT specimens and image-analysis complete the investigation and collection of data on the several phenomena which are modelled. The data are used to identify the model parameters.

The limitations an the extensions of such a model based on a macroscopic approach have also been stated.

10 Acknowledgments

The authors would like to thank Mr. M. Lasne, head of the MISA research group at EDF for his initiative and support to carry this work. The author also thank A. Ammouche, J. Arsenault, L. Boisvert, X. Cespedes, O. Houdusse, J.C. Sartoris for their fine contribution in the development of the various experimental procedures.

REFERENCES

1 BOURNAZEL J. P., 1992,"Contribution à l'étude du caractère thermomécanique de la maturation du béton", PhD thesis, University Paris 6, France.

2 TORRENTI JM., PATIES C., PIAU J.M., ACKER P., DE LARRARD F., 1992,"La simulation numérique des effets de l'hydratation du béton", Proceedings of StruCo Me, Paris.

3 GERARD B., BREYSSE D., LASNE M., 1993, "Coupling between cracking and permeability, a model for structure service life prediction", Proceedings of SAFEWASTE'93, Avignon-FRANCE, Vol 3.

4 DIDRY O., 1993, "Conteneur de déchets FMA-Etat des contraintes résiduelles àl'issue de la période de maturation", Note EDF HTB2 93-O18A.

5 MAZARS J., 1984, "Application de la mécanique de l'endommagement au comportement non linéaire et à la rupture de béton de structure", ' Doctorat d'Etat ' thesis, Paris 6 University, France..

6 BREYSSE D., GERARD B., 1993, "Cracking of concrete, relevance and effects", Technical committe RILEM TC 146 Tightness of concrete with respect to fluids.

7 BERNER U.R., 1992, "Evolution of pore water chemistry during degradation of cement in a radioactive waste repository environment", Waste Management, vol. 12, pp 201-219.

8 GREENBERG S.A., CHANG T.N. ANDERSON E., 1960, "Investigation of colloidal hydrated calcium silicates, I. Solubility products", Journal of Physical chemistry, vol 64, 1151, 15 pages

9 SAITO H., NAKANE S., FUJIWARA A., 1994, "Experimental study of the deterioration of concrete by an electrical acceleration test method - Part 1: The characteristics of deterioration in various cement", Japan Architecture Congress, 6 September (in Japanese).

10 GERARD B., 1996," Vieillissement des structures de confinement en béton, modélisation du couplage endommagement-décalcification-diffusion", PhD thesis, ENS Cachan-France / Univ. Laval-Québec, to be submitted

11 PONCET B., 1993, "Etude de la durabilité du béton F44. Lixiviation du calcium sur 300 ans. Problème de Stéfan", Report EDF D585-SRE/GD 92-882.

12 HORNAIN H., REGOURD M., 1986, " Microcracking of concrete ", Proceedings of the Eighth International Congress on the chemistry of cement, Rio de Janeiro, Brazil, Vol. V, theme 4, pp 53-59.

13 GRAN H., 1995, " Fluorescent liquid replacement. A means of crack detection and water-binder ration determination in high strength concretes ", Cement and Concrete Research, Vol.25, pp 1063-1073.

14 AMMOUCHE A., HORNAIN H., 1995, " Caractérisation des matériaux vieillis et fissurés par analyse d'images ", LERM/Internal report for EDF-Research Division, MTC.

15 BEAUDOIN J.J., FELDMAN R.F., 1975, "A study of mechanical properties of autoclaved calcium silicate systems", Cement and concrete research, vol 5, 2, pp 103-118.

16 CHIRLIN G.R., 1985, "Flow through a porous medium with periodic barriers or fractures", Soc. of Petroleum Engineers journal, June 1985, pp 358-362.

17 SAYERS C. M., 1991, "Fluid flow in porous medium containing partially closed fractures", Transport in Porous Media 6: 331-336.

40 MODELLING OF THE CHEMICAL DEGRADATION OF A CEMENT PASTE

F. ADENOT and C. RICHET
Direction du Cycle du Combustible, DESC/SESD/SEM CEA
Saclay Commissariat à l'Énergie Atomique, Gif sur Yvette Cedex, France

Abstract
We describe a model (DIFFUZON) predicting the extent of degradation of a cement paste by pure water. DIFFUZON was validated by comparison with experimental results and its application has made it possible to show the importance of the diffusion coefficient in the degraded section and the surface gel on cement durability. DIFFUZON may be used to predict the degradation of other cement pastes corroded by other aggressive solutions where advective transport through the porous medium does not occur.
Keywords: Chemical degradation, model, diffusion, local equilibrium.

1 Introduction

Predicting the long-term behavior of concrete used in radioactive waste disposal requires knowledge of the effect of aggressive agents on its chemistry. Runoff (i.e., constantly renewed) water would appear to be one of these agents.

On contact with water, concrete undergoes chemical degradation which increases when the (chemical) composition of the aggressive solution is constant (with the progressive dissolution of cement constituents in the water, an equilibrium state between the solution and the concrete is reached, stopping the alteration).

To predict chemical degradation in this particularly unfavorable case, we have developed a model for calculating the variation of the thickness and composition of the degraded zone with time. This model which integrates the different physical-chemical phenomena involved has been established and validated for a cement paste for subsequent extension to concrete.

Mechanisms of Chemical Degradation of Cement-based Systems. Edited by K.L. Scrivener and J.F. Young. Published in 1997 by E & FN Spon, 2–6 Boundary Row, London SE1 8HN. ISBN: 0419215700.

2 Modelling

2.1 Physical-chemical phenomena

Little work has been reported on modelling the time dependence of chemical degradation of concrete by ground water. On the other hand, the phenomena involved in alteration of rocks by water and their modelling are well known to geochemists [1]-[5]. Thus, it appears that degradation of a saturated porous medium immersed in water depends on two main consecutive phenomena with different kinetics (as advective transport does not occur):

- *material transport by diffusion,* resulting from concentration gradients between the solid interstitial solution and the aggressive solution
- *dissolution-precipitation chemical reactions,* induced by the concentration variations brought about by diffusion.

The slowest process controls the overall velocity. In many geochemical systems, when the advective transport is very slow or neglected, the hypothesis of local equilibrium can be assumed. These are cases in which diffusion rates are much slower than those of the chemical reactions. Thus, the leaching fluxes are imposed by diffusion, i.e., they are proportional to the square root of the time as long as there is an unaltered zone and the aggressive solution (chemical) composition is constant. In the solid phase assemblage, at a given time, a region is observed where each zone with constant mineralogy is separated by a dissolution or precipitation front. These clearly defined fronts show that as soon as a solid phase is under- or supersaturated with respect to the contacting solution, it immediately precipitates or dissolves. Conversely, if the chemical reaction kinetics were slow with respect to diffusion, they would control the leaching flux. At a given time in the solid phase, this phenomenon is shown by a progressive spatial gradient of the concentrations of the different solid phases that dissolve or precipitate.

As hydrated cement is more reactive than natural rocks, it is probable that this assumption of local equilibrium is valid for the fluxes under consideration (diffusion).

Note that diffusion in the solid phase occurs, but it is very slow and can be ignored.

The degradation prediction model (DIFFUZON) was obtained from the simultaneous solution of transfer (by diffusion), conservation of matter and chemical equilibrium equations.

Model equations for unidirectional degradation

Fig. 1. Representation of the unidirectional zoning of the cement paste degraded section, made up of several constant mineralogy domains (zones) separated by moving fronts of dissolution or precipitation.

The degraded hardened cement paste can be divided into zones separated by dissolution or precipitation fronts (Figure 1)

By applying Fick's law for diffusion transport, a material balance over a volume element in a zone (*j*), gives:

$$\frac{\partial C_{i,i}}{\partial t} = D_j \frac{\partial^2 C_{i,i}}{\partial x^2} - \frac{1}{\phi_j} \frac{\partial S_{i,i}}{\partial t} \tag{1}$$

$C_{i,j}$: liquid phase concentration of constituent (*i*) in zone (*j*) (per unit volume of liquid)

$S_{i,j}$: solid phase concentration of constituent (*i*) in zone (*j*) (per unit volume of solid + liquid)

ϕ_j: porosity in zone (*j*)

x: distance from the surface (dm)

t: time (s)

D_j: diffusion coefficient in zone (*j*) in the liquid

Equation (1) has to be solved for the *n* diffusing species (*i*) simultaneously with the chemical equilibrium equations for the different solid phases at every point in each zone (*j*) with constant mineralogy.

Similarly, a material balance at each dissolution or precipitation front (*j*) gives:

$$(S_{i,j}^- - S_{i,j+1}^+)\frac{dl_j}{dt} = \phi_{s,j} D_j \frac{\partial C_{i,i}^-}{\partial x} - \phi_{s,j+1} D_{j+1} \frac{\partial C_{i,i+1}^+}{\partial x} \tag{2}$$

$C_{i,j}^-$ and $S_{i,j}^-$: constituent (*i*) concentrations in liquid and solid phases *left* of front

$C_{i,j+}$ and $S_{i,j+}$: constituent (*i*) concentrations in liquid and solid phases *right* of front

$\phi_{s,j}$ and $\phi_{s,j+1}$: surface porosities in zones (*j*) and (*j+1*)

lj: boundary (*j*) position at time (t)

D_j and D_{j+1}: diffusion coefficients in zones (*j*) and (*j+1*)

This equation also has to be solved for *n* diffusing species (*i*) simultaneously with chemical equilibrium equations for the different solid phases present.

Initial and limiting conditions

Our system consists of a semi-infinite media couple (maintenance of an unaltered zone during all of the attack and constant composition aggressive solution). It can be described mathematically by the following conditions:

- limiting condition (x = 0): $C_i = C_{i,0}$ for t > 0
- initial condition (t = 0): $C_i = C_{i,l}$ for x > 0

2.2 Numerical solution of model

The numerical solution of the model made up of several systems of equations is carried out using an algorithm associating input variables (or causes) with output variables (or consequences) and requires knowing a number of parameters. To assess the effects of aggressive conditions and concrete composition on durability, we took the aggressive solution composition (pH, the different aggressive ion concentrations) and material type (for now, DIFFUZON considers the cement composition (OPC, blended cement, etc.) and the water/cement ratio as the input variables; on extension to concrete, it will integrate sizes and amounts of aggregates, etc.). The output variables are the solid and liquid concentration profiles, porosity and diffusion coefficient profiles, the multi-mineral assembly and the degraded thickness at a given time. The parameters required for solving the model are the variables in the model equations: equilibrium constants and equations for determining the porosity and diffusion coefficient from the solid phase composition.

3 Model validation

3.1 Data used for solving model

Experimental observations
Experimental results [6], obtained for an OPC paste with a water cement ratio = 0.4, and immersed in a solution kept essentially deionized, showed that the hydroxyl, calcium, silicon and sulfate concentrations in the leaching fluxes were proportional to the square root of the time. The degradation was unidirectional and the altered section, which was 1.4 mm thick after 3 months, consisted of an assembly of several minerals, where each zone with a constant composition was separated by clearly defined dissolution fronts.

Diffusion rates are thus much slower than those of the chemical reactions and we can make the hypothesis of local chemical equilibrium between the solid phases and interstitial solutions.

Input variables
To validate the model, we applied it to a pure OPC paste with W/C = 0.4 immersed in water that was maintained essentially deionized. These are input conditions for which we have experimental degradation values.

Parameters required for solving the model
Equilibrium constants for the possible solid phases in OPC paste are taken from references [7]-[12].

The poorly crystallized components of the cement pastes with variable Ca/Si ratios [13]-[16] are modeled by a series of CSH phases with constant Ca/Si ratios each with an apparent solubility product:

$$(Ca^{2+})^{C/S}.(H_2SiO_4^{2-}).(OH^-)^{2C/S-2} = K_{Ca/Si}$$

There can be only two different CSH phases at a given point[1]. The corresponding Ca/Si ratio and $pK_{Ca/Si}$ values were interpolated from the experimental data [16], [17] (Table 1)

---

[1] This modelling of CSH chemical equilibria by a series of CSH phases with constant Ca/Si is a "discretization" of the solid solution

Tab. 1. Thermodynamic constants used for the CSH phases

C/S of CSH	1.65	1.45	1.30	1.15	1.05	0.95	0.90	0.85
pK of CSH	11.7	10.6	9.8	8.9	8.3	7.6	7.3	6.9

The porosity in each constant mineralogical zone is calculated from the average concentrations of the different solid phases in the considered zone, their molar volumes and densities. The solid phase concentration variations in a zone are neglected in this calculation. With the model solved, it is confirmed that these concentration variations have little effect on the porosity. The porosity increase resulting from CSH decalcification is also neglected: as this porosity is very fine (< 3 nm [18]), its effect on transfer and degradation are assumed to be negligible with respect to the existing porosity.

The diffusion coefficients, which depend essentially on the porosity and its spatial distribution, are determined from an empirical formula (extrapolation of a formula obtained from OPC pastes with different W/C giving diffusion coefficients in function of macroporosity).

3.2 Calculation results

The calculation results are shown in Figure 2. The calculated solid phase calcium, silicon and sulfate concentrations in the unaltered core are the initial cement paste concentrations (theoretical composition of this core). On the other hand, for aluminum, the solid phase concentration obtained by the model in the unaltered core is less than half that of the cement paste before degradation. This difference is explained by the simplifying assumptions in the model, i.e., the absence of aluminum in the CSH and surface gel and the fact that it is assumed that the Afm phase corresponds to monosulfoaluminate with Al/S = 2, while according to some reports [20], this is a solid solution between the monosulfoaluminate and C_4AH_{13} with Al/S < 2.

The solid phase calcium concentration from the specimen surface (x =0) up to the unaltered core (x_{in} = 1.2 mm) increases while that of silicon is practically constant: the Ca/Si ratio in the CSH increases from 0 (silica gel; zone I) up to 1.65 (unaltered core, zone V).

The boundary between the unaltered zone (zone V) and zone IV is characterized by an abrupt decrease in the calcium concentration which corresponds to complete dissolution of the portlandite.

The solid phase aluminum and sulfate concentrations vary. Going from the unaltered core in the direction of decreasing x there is an increase in aluminum concentration in zone IV, while the sulfate concentration is constant: the Al forms monosulfoaluminate at the expense of ettringite. On the other hand, at the front separating zones III and IV, the monosulfoaluminate completely dissolves yielding ettringite. We have precipitation of ettringite in zone III then complete dissolution of this phase at the front separating zones II and III. Precipitation of ettringite in the altered section is due to diffusion of part of the surface aluminum and sulfate to the core.

The precipitation of monosulfoaluminate and ettringite results in a porosity decrease which may be the source of stresses in the specimen. These stresses will increase with the amount precipitated (this is dependent on the initial cement paste composition).

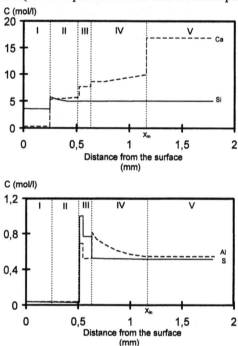

Zone	I	II	III	IV	V
Mineralogical composition	Silica gel	CSH	Ettringite CSH	Monosulfoalu. Ettringite CSH	Unaltered core Portlandite Monosulfoalu. Ettringite CSH

Fig. 2. Modeled solid phase calcium, silicon, aluminum and sulfate profiles in an OPC paste with W/C = 0.4 after 3 months degradation in essentially deionized water at pH 7.

It is also observed that the solid phase concentration variations in each zone are very low (the significant solid phase concentrations are at the fronts). The assumption of constant porosities and diffusion coefficients in each zone is thus justified.

Tab. 2. Comparison of calculated and experimental values for leached amounts

		Calculated leached amounts	Experimental leached amounts
Calcium	$(mol.d^{-1/2}.dm^{-2})$	$2.8.10^{-2}$	$2.8.10^{-2}$
Silicone	$(mol.d^{-1/2}.dm^{-2})$	$5.6.10^{-4}$	$5.7.10^{-4}$
Aluminum	$(mol.d^{-1/2}.dm^{-2})$	$4.3.10^{-4}$	$< 2.8.10^{-4}$
Sulfate	$(mol.d^{-1/2}.dm^{-2})$	$4.3.10^{-4}$	$4.3.10^{-4}$

3.3. Comparison with experimental results

The model has allowed us to predict the degraded thickness (1.2 mm) after three months, the successive zones, their mineralogical composition and the C/S ratio in the CSH. These values were consistent with our experimental results.

Table 2 shows that leached amounts of calcium, silica and sulfate agree well with the experimental values.

Nevertheless, the calculated amount of leached aluminum differs from the experimental value (Table 2). We have in fact been led to consider a relatively high leaching rate for aluminum in order to obtain the appearance of monosulfoaluminate in the degraded zone. This constraint results from modelling the ettringite chemical equilibrium by a constant solubility product. However, an increase in the apparent solubility product with the calcium concentration is observed experimentally [9].

The secondary precipitation of ettringite in the degraded section predicted by the model was verified experimentally a posteriori [21].

4 Long-term prediction and perspectives

The established model (DIFFUZON) is based on the assumptions of local equilibrium and diffusion transport according to Fick's law. It is made up of systems of non-linear equations and partial differential equations. We have established an algorithm providing solutions of these equations in each zone and at each front. Knowing the composition of the aggressive solution and the non-degraded cement paste composition (input variables), the equilibrium constants, the solid phase molar volumes and the function for determining diffusion coefficients from the porosity (indispensable parameters for the calculation), we can determine the zoning of the degraded section, the solid and liquid phase concentration profiles as well as the porosity and diffusion coefficient in each zone (output values).

For now, DIFFUZON does not take precipitation of phases, which do not exist initially in the unaltered zone, into account.

Using DIFFUZON, by taking the principal constituents of a cement phase into consideration, has made possible to validate short-term (three months) predictions by comparison with the experimental results. The small differences between the calculated and theoretical values are explained by the simplifying assumptions made: we neglected aluminum in the CSH and surface gel and assumed that the apparent solubility product of ettringite was constant.

The degraded thickness increases with the square root of time. We thus have a degraded thickness of 1.2 mm at 3 months and 4 cm at 300 years. This long-term estimation is reliable to the extent that DIFFUZON has been validated by comparison with experimental results and that no other process than those taken into account is involved.

Solving our model depends on the existence of a stable surface gel. However, if it is assumed that it disappears by erosion or dissolution (due to chemical instability), the degraded thickness at 300 years will be much greater: more than 8 cm. This significant difference induced us to characterize it precisely (determine its composition, chemical stability, porosity, porosity form factor, mechanical strength) and to be certain that it will remain in the long term [22]. A study in collaboration with ANDRA[2] is now in progress in the CEA.

In addition, it also appears that the diffusion coefficients in the altered section govern the degraded thickness.

The model bases have been established and verified, it is reasonable to envisage its medium- term extension to:

• different (chemical) compositions of the aggressive solution, making allowance for precipitation of phases not present in the initial paste,
• different types of cement (study of chemical and textural properties of each cement) (DIFFUZON has already been used and validated for predicting the alteration of a blended cement paste with W/C = 0.38 cured in a saturated portlandite solution for 3 years. For this material, we observed very good agreement between DIFFUZON and the experiments. The main difference between the long-term behavior of the OPC and this cement is the alteration rate which is much slower for the latter (2 cm degraded thickness after 300 years)),
• different W/C (behavior of anhydrous during degradation),
• concretes (effects of aggregates and the high porosity of their interfacial transition zone on degradation). Bourdette has already partially studied this point by validating the model with a mortar [23] and [24].

5 Acknowledgements

This work has been supported by the Agence Nationale pour la gestion des Déchets Radioactifs (ANDRA) and the Commission of the European Communities.

6 References

1.   Weare, J.H., Stephens, J.R., Eugster, H.P. (1976) Diffusion metasomatism and mineral reaction zones: general principles and application to feldspar alteration, *American Journal of Science*, Vol. 276, pp. 767-816.

2.   Frantz, J.D., Mao, H.K. (1976) Bimetasomatism resulting from intergranular diffusion: I. a theorical model for monomineralic reaction zone sequences, *American Journal of Science*, Vol. 276, pp. 817-840.

3.   Frantz, J.D., Mao, H.K. Bimetasomatism resulting from intergranular diffusion: II prediction of multimineralic zone sequences, *American Journal of Science*, Vol 279, pp. 302-323.

4.   Lichtner, P.C., Oelkers, E.H., Helgeson, H.C. (1986) Exact and numerical solutions to moving boundary problem resulting from reversible heterogeneous reactions and aqueous diffusion in a porous medium, *Journal of Geophysical Research*, Vol. 91, N°B7. pp. 7531-7544.

---

[2]Agence Nationale pour la gestion des Déchets Radioactifs - National Radioactive Waste Management Agency)

5. Lichtner, P.C. (1993) Scaling properties of time-space kinetics mass transport equations and the local equilibrium limit, *American Journal of Science*, Vol. 293, pp. 257-296.

6. Adenot, F., Buil, M. (1992) Modelling of the corrosion of cement paste by deionized water, *Cement and Concrete Research*, 22, pp. 489-495.

7. Zhang, F., Zhou, Z., Lou, Z. (1980) Solubility product and stability of ettringite, 7th *International Congress on the Chemistry of Cement*, II-88.

8. Damidot, D., Glasser, F.P. (1993) Thermodynamic investigation of $CaO-Al_2O_3-CaSO_4-H_2O$ system at 25°C and the influence of Na_2O, *Cement and Concrete Research*, 23, pp. 221-238.

9. Jones, F.E., Roberts M.H. (1959) The System $CaO-Al_2O_3-H_2O$ at 25°C, *DSIR Building Research Station*, Note E 965. HMSO, London.

10 Atkins M., Glasser F.P., Kindness A., Macphee D.E. (1991) Solubility data for cement hydrated phases (25°C), DOE Report n° DOE/HMIP/RR/91/032.

11. HATCHES (1991) database NEA release 4.

12. Lea, F.M. (1971), *The Chemistry of Cement and Concrete*, 3ème édition-Arnold.

13. Brunauer, S., Greenberg, S.A. (1960) The hydratation of tricalcium silicate and b-dicalcium silicate at room temperature, 4th *International Symposium on the Chemistry of Cement*, III.1, pp. 135-165.

14. Fujii, K., Kondo, W. (1981) Heterogeneous Equilibrium of Calcium Silicate Hydrate in Water at 30°C, *Journal of Chemistry Society.*, Dalton Trans 2, pp. 645-651.

15. Greenberg, T.N., Chang, S.A. (1965) Investigation of the Colloidal Hydrated Silicates II. Solubility Relationships in the Calcium Oxide-Silica-Water System at 25°C, *Journal of Physical Chemistry*, 69, pp.182-188.

16. Jennings, H.M. (1986) Aqueous Solubility Relationships for two types of Calcium Silicate Hydrates, *Journal of the Ceramic Society*, 69, pp. 614-618.

17. Berner, U.R. (1988) Modelling the Incongruent Dissolution of Hydrated Cement Minerals, *Radiochimica Acta*, 44/45, pp. 387-393.

18. Adenot, F., Auvray, L., Touray, J.C. (1993) Analyse de la dimension fractale d'agrégats de C-S-H (silicates de calcium hydratés) de composition et de mode de dégradation différents. Implications pour les études de durabilité des pâtes de cimen, . *Compte Rendu de l' Académie des Sciences*, t.317, Série II, pp. 185-189, Paris.

19. Adenot, F. (1992) *Durabilité du béton : caractérisation et modélisation des processus physiques et chimiques de dégradation du ciment*, Thèse de l'Université d'Orléans (Fr).

20. Roberts, M. H. (1969) Calcium Aluminate Hydrates and Related Basic Salt Solid Solution, 5th *International Congress on the Chemistry of Cement*, II-29, pp.104-117.

21. Adenot, F., Faucon, P. (to be published) Précipitation d'ettringite secondaire au cours de l'altération d'une pâte pure par de l'eau déminéralisée.

22. Faucon, P., Adenot, F., Jacquinot, J.F., Virlet, J. (1995) Dégradation des pâtes de ciment : Etude de la couche superficielle, *Colloque Géomatériaux Environnement Ouvrages*, Aussois (Fr).

23. Bourdette, B. (1994) *Durabilité du mortier: prise en compte des auréoles de transition dans la caractérisation et la modélisation des processus physiques et chimiques d'altération*, Thèse de l'Institut National des Sciences Appliquées de Toulouse (Fr).

24. Bourdette, B., Adenot, F. (to be published) Modélisation de la dégradation chimique d'un mortier : Influence des auréoles de transition.

41 DURABILITY OF CEMENT STABILISED LOW LEVEL WASTES

T. BAKHAREV, A.R. BROUGH,
R.J. KIRKPATRICK, L.J. STRUBLE and
J.F. YOUNG
University of Illinois, Urbana, IL, USA
R. OLSON, P.D. TENNIS, D. BONEN, H. JENNINGS and
T. MASON
Northwestern University, Evanston, IL, USA
S. SAHU and S. DIAMOND
Purdue University, West Lafayette, IN, USA

Abstract
Cementitious materials containing high proportions of slag and fly ash have been tested for suitability to immobilize simulated alkaline and carbonated off–gas waste solutions after vitrification of low–level tank wastes stored at Hanford. To assess their performance, long–term durability was assessed by measuring stability of compressive strength and weight during leaching and exposure to sulfate and carbonate solutions. The important parameter controlling the durability is pore structure, because it affects both compressive strength and susceptibility to different kinds of chemical attack. Impedance spectroscopy was utilized to assess the connectivity of the pore system at early ages. Mercury intrusion porosimetry (MIP) and SEM were utilized to assess development of porosity at later ages. Phase alterations in the matrix exposed to aging and leaching in different media were followed using XRD.

Mixtures were resistant to deterioration during immersion in solutions containing high concentrations of sulfate or carbonate ions. Mixtures were also resistant to leaching. These results are consistent with microstructural observations, which showed development of a finer pore structure and reduction in diffusivity over days or months of hydration.
Keywords: Durability, permeability, diffusion, deterioration, cement, wastes.

1 Introduction

Immobilization of tank wastes at Hanford is expected to utilize a vitrification process. It is assumed that there will be an off–gas waste produced during vitrification, and the objective of our studies is to explore stabilization of this off–gas waste using cementitious materials. For these studies, two off–gas waste compositions were used, one with a composition similar to the tank waste but more dilute, and the other with added carbo–

Mechanisms of Chemical Degradation of Cement-based Systems. Edited by K.L. Scrivener and J.F. Young. Published in 1997 by E & FN Spon, 2-6 Boundary Row, London SE1 8HN. ISBN: 0419215700.

nate CO_2 to simulate waste from a gas–fired melter, where CO_2 levels will be high.

There are two important issues to consider in designing a durable cement–stabilized waste form. One is microcracking due to thermal stresses resulting from the temperature rise during early hydration. Therefore it is important to design the material such that excessive temperature rise does not occur. The other important issue is that the material be resistant to leaching and chemical attack. If the material is free of microcracks and low in permeability, it is likely to resist these processes. Because the waste form is to be buried underground, any chemical attack would likely involve sulfate or carbonate solutions.

It is known that pozzolans improve the durability of cementitious materials. They reduce heat evolution, and they increase the resistance of concrete to deterioration by aggressive chemicals such as chlorides, sulfates, etc. [1]. Much of this increase in durability is attributed to decreased permeability and reduced ion diffusivity resulting from a finer pore structure. Therefore the waste materials in this study contained ~60 weight % of fly ash and slag. Attapulgite clay (~10% by weight) was used to prevent excessive bleeding.

In order to assess the durability of waste materials, they have been exposed to leaching solutions and to solutions containing high concentrations of sulfate or carbonate ions, and their pore system development has been studied.

This paper describes the durability of the off–gas wasteforms developed for Hanford. A companion paper [2] deals with the chemical evolution of these materials.

2 Experimental procedure

To prevent excessive heat evolution during hydration of cementitious mixtures combined with a highly alkaline liquid waste, high proportions of slag and fly ash were utilized. The mix design was adjusted to meet the requirement that temperature on adiabatic curing not exceed 90°C. The temperature rise was measured using an adiabatic calorimeter constructed in our laboratory. The mixtures tested are designated by their percent (by weight) of cement[1], fly ash[2], slag[3] and clay[4]. For example, the material 3:3:3:1 contains 30% cement, 30% fly ash, 30% slag and 10% clay. Two types of waste solutions were tested: alkaline (A) and carbonated (C). Table 1 shows their chemical compositions. The proportion of solid to liquid was 1 kg to 1 liter. The materials were cured adiabatically for 4 days (reaching the temperatures shown in Table 2); after that they were sealed in tubes and cured isothermally at 70°C for 10 months. Additional details are provided in Ref. [2].

Impedance spectroscopy was utilized to observe changes in electrical conductivity of the tested materials during the first few days of hydration. As described elsewhere [3], the measured sample conductivity (σ) was normalized by the measured conductivity of the pore solution (σ_0). The resulting normalized conductivity (σ/σ_0) is a microstructural parameter that characterizes the capillary pore network. It is directly related to the diffusivity through the Nernst–Einstein equation, and can be used to compute permeability through the Katz–Thompson equation,

1. Type I/II, Ash Grove,Durkee, OR
2. Type F, Centralia, Ross Sand and Gravel Co., Portland, OR
3. Koch Minerals Co., Chicago, IL
4. Engelhard Co., Iselin, NJ

Table 1. Composition of solutions

	Alkaline	Carbonated
$Al(NO_3)_3 \cdot 9\,H_2O$	7.03 g/l	7.03 g/l
$Na_3(PO_4) \cdot 12\,H_2O$	24.53 g/l	24.53 g/l
$NaNO_2$	12.17 g/l	12.17 g/l
NaOH	53.76 g/l	——
$NaNO_3$	32.73 g/l	32.73 g/l
Na_2CO_3	——	39.97 g/l

$$k = \frac{1}{226} d_c^2 \frac{\sigma}{\sigma_0} \tag{1}$$

where k is the coefficient of permeability and d_c is the threshold pore diameter (obtained by mercury intrusion porosimetry).

Mercury intrusion porosimetry was utilized to assess the changes in pore size distribution and porosity at ages up to 10 months. This method was also used to measure d_c used in equation (1). Prior to measurement the samples were immersed in methanol for 3 days and subjected to drying at 105°C to reduce damage to microporosity during drying. The minimum intruded pore diameter was 3 nm.

The leaching test was designed to simulate leaching by ground water percolating through the burial vault. The test was based on the ANSI/ANS–16.1–1986 test. Four leaching agents were used: deionized water, simulated ground water, 5.3 mmole/L solution of magnesium sulfate, and 5.3 mmole/L solution of sodium bicarbonate. Table 3 shows the composition of the simulated ground water, which was based on analyses of the ground water at Hanford [4]. The experiment was performed on cylinders (50 mm

Table 2. Adiabatic curing temperatures

Materials	Temperature, °C
3:3:3:1C	60
3:0:6:1C	65
3:4:2:1A	54
3:5:1:1A	52

Table 3. Simulated ground water composition

Constituent	Quantity, µg/mL
Calcium	50.7
Magnesium	15.2
Sodium	28.4
Sulfate	51.2
Total Carbon	32.0

in length, 26 mm in diameter) at 45°C. All samples were immersed in 250 mL bottles containing the leach medium. The leach medium was changed at 1, 2, 3, 7, 14, 28, 56, 84, and 112 days. Compressive strength and weight were monitored periodically.

The test for sulfate deterioration was based on ASTM C1012, "Standard Test Method for Length Change of Hydraulic–Cement Mortars Exposed to a Sulfate Solution". Beginning at the age of 1 month, cylindrical samples were immersed in a 0.352 mol/L solution of magnesium sulfate and stored at 45°C for 10 months. ASTM C1012 calls for measurement of length change of bars. We were unable to make such measurements due to the low strength of these materials. Instead, weight change and compressive strength were measured periodically.

An analogous test was used for carbonate deterioration. The cylindrical samples at the age of 1 month were immersed in a 0.352 mole/L solution of sodium bicarbonate and stored at 45°C for 4 months. Weight change and compressive strength were measured periodically.

3 Results

Figure 1 shows the normalized conductivity of 3:0:6:1C. The value of normalized conductivity dropped rapidly after 24 hours, reaching 0.001 by 3–4 days. This value is more than an order of magnitude lower than a conventional Portland cement paste with w/c of 0.5 [3].

Table 4 summarizes the MIP measurements at 1 month and 10 months. All samples showed an increase in pore surface area and a decrease in pore size with time. Most samples also showed a small increase in pore volume during this time interval.

Figure 2 presents the evolution of compressive strength during the leaching experiments. The strength values have a scatter due to the stochastic nature of the test and the heterogeneity of the samples. An increase in compressive strength after 1 month was ob-

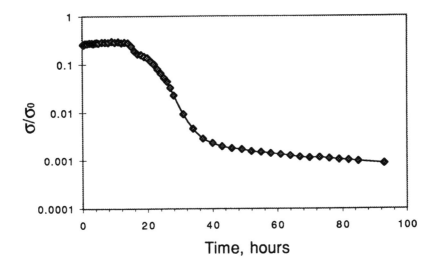

Figure 1. Normalized conductivity of 3:0:6:1C.

Table 4. Summary of MIP measurements for off–gas materials

Material	Age, month	Porosity, vol.%	Pore area, m^2/g	Threshold diameter, μm	Median pore diameter, μm
3:3:3:1C	1	50.6	62.5	1.0	0.009
	10	52.4	124.8	0.1	0.007
3:0:6:1C	1	53.7	94.24	0.3	0.009
	10	49.9	134.7	0.15	0.006
3:4:2:1A	1	58.6	84.6	0.2	0.018
	9	63.8	101.6	0.15	0.009
3:5:1:1A	1	31.8	35.5	1.0	0.18
	9	52.2	92.0	0.1	0.01

served in 3:4:2:1A, attributed to continuing hydration. All the other materials were reasonably stable in weight and compressive strength when subjected to leaching.

Table 5 presents results of the sulfate test. Samples with carbonate waste (3:3:3:1C and 3:0:6:1C) had a very small increase in weight (2–3%) and small decrease in strength during 10 months. Samples with alkaline waste (3:5:1:1A and 3:4:2:1A) had lower initial compressive strengths, but little strength loss or weight change. In the absence of a predetermined failure criterion for strength loss or weight change on exposure to sulfate solution, we can only conclude that the low changes observed appear to reflect satisfactory performance, especially taking into account the very aggressive, highly concentrated solutions (10^4 times the level of sulfate in the ground water) and the low strength of these materials at the start of exposure. Examination of materials exposed to sulfate attack using SEM showed a dense rim 1–2 mm thick rich in calcium and sulfur; by XRD the prevalent phase was gypsum.

Table 6 presents the results of the carbonate test. Only small changes in weight (1–2%) and strength were observed. Examination of the material using SEM showed that a calcium rich phase was uniformly distributed in pores of the matrix; by XRD the prevalent phase was calcite.

4 Discussion

Permeability to water and diffusion of ionic species in cementitious materials are important keys to durability. Therefore the development of porosity in these waste materials was assessed using different techniques. The measurements of electrical conductivity in the first few days of hydration showed a rapid drop in diffusivity of the samples. The estimated permeability of materials at the age of 1 month using eq. (1) was in the range of 10^{-16} m/s. It is known that addition of mineral admixtures such as slag or fly ash leads

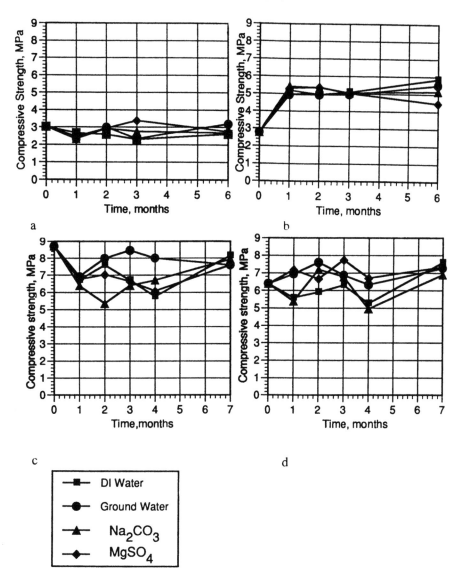

Figure 2(a–d). Compressive strength evolution in the leaching experiment:
(a) 3:5:1:1A, (b) 3:4:2:1A, (c) 3:0:6:1C, (d) 3:3:3:1C.

Table 5. Results of immersion in sulfate solution

Material	Initial compressive strength, MPa	Compressive strength at 10 months, MPa	Increase in Weight , %
3:3:3:1C	6.2	6.2	3.2
3:0:6:1C	8.3	5.8	2
3:5:1:1A	3.0	2.5	7.5
3:4:2:1A	2.8	2.5	7.8

Table 6. Results of immersion in the carbonation solution

Material	Initial compressive strength, MPa	Compressive strength at 10 months, MPa	Increase in Weight , %
3:3:3:1C	6.2	5.8	2.4
3:0:6:1C	8.3	6.2	2
3:5:1:1A	3.0	2.9	1
3:4:2:1A	2.8	5.1	2

to a decrease in pore size and an increase in the fraction of pores in the finer size range of 15 nm or less [5], and the MIP measurements showed this trend. At later ages the pores became smaller in size and formed a pore structure of greater surface area, changes that are expected to reduce permeability.

All the materials showed a resistance to deterioration in concentrated sulfate solution (Table 5). Ingress of sulfate ions is presumed to be controlled by the rate at which sulfate diffuses through the pore structure. Although the direct measurements of diffusivity have not been conducted, the results of the sulfate test suggest that the materials have quite low diffusivities. The diffusivity was likely also reduced by precipitation of calcium sulfate.

Leaching, like chemical attack, involves movement of water and dissolved species due to a concentration gradient. So leaching also depends on the pore structure and permeability of the paste. All the materials showed resistance to leaching. These results correlate quite well with the results of chemical attack, and can be explained by low diffusivities and low permeabilities of the materials.

Phase alterations observed by XRD suggest that a further increase in strength may be expected. In a companion paper [2] it is suggested that calcium silicate hydrate has a tendency to increase polymerization and crystallization, leading to improved strength and decreased permeability. The observed low content of calcium hydroxide in these materials is also favorable for durability.

5 Conclusions

It has been the experience that well cured cementitious materials with low water–solid ratios have low permeability and resist deterioration. This study showed that cementitious materials containing a high proportion of slag, fly ash, and alkaline solution and cured at elevated temperatures, despite having very high water–solid ratios, may also have low permeability and resist deterioration.

6 Acknowledgement

The research was funded by grant MTS–SVV–097600 from the Westinghouse Hanford Company. We thank the Ross Sand and Gravel Co., Koch Mineral Co., Engelhard Co., and the Ash Grove Cement Co. for materials.

7 References

1. Hooton, R.D. (1986)Permeability and pore structure of cement pastes containing fly ash, slag and silica fume, in *Blended Cements,* STP 897, (ed. G. Frohnsdorff), American Society for Testing and Materials, Philadelphia, pp. 128–143.
2. Bakharev, T., Brough, A.R., Kirkpatrick, R. J., Struble,L. J. and Young, J. F. (1996) Chemical evolution of cementitious materials with high proportion of fly ash and slag, in *Mechanisms of Chemical Degradation of Cement – Based Systems*, (ed. K.L. Scrivener and J.F. Young), E & FN Spon, London, current volume.
3. Christensen, B. J., Coverdale, R. T., Olson, R. A., Ford,S. J., Garboczi,E. J., Jennings, H. M., and Mason, T. O. (1994) Impedance spectroscopy of hydrating cement–based materials: measurement, interpretation, and application. *J. Amer. Ceram. Soc.*, Vol.77, No. 11. pp. 2789–804.
4. Johnson,V. G. (1993) *Westinghouse Hanford Company Operational Groundwater Status Report, 1990–1992*, U.S. Department of Energy, Hanford.
5. Roy, D.M. (1989) Relationships between permeability, porosity, diffusion and microstructure of cement pastes, mortar, and concrete at different temperatures. *Mater. Res. Soc. Proc*, (ed. L.R. Roberts and J.P. Skalny),Vol. 137, Pittsburgh, PA, pp. 179–189.

42 C's RETENTION IN ZEOLITES IMMOBILISED BY PORTLAND CEMENT

L.P. ALDRIDGE, R.A. DAY, S. LEUNG
Australian Nuclear Science and Technology Organisation, Menai, NSW, Australia

A.S. RAY, M.G. STEVENS, R.S. KNIGHT and C.F. MAPSON
Department of Materials Science, University of Technology, Sydney, NSW, Australia

Abstract
Three zeolites were exchanged with Cs cations and immobilised in Portland cement at a waste to cement loading ratio of 1 (by weight). The stability of the zeolites in the cement paste was monitored over 1.5 years. Two zeolites reacted within a week to consume all of the calcium hydroxide in the cement paste. In contrast the other cemented zeolite still contained calcium hydroxide after 1.5 years. When this zeolite was ground finer it did react with calcium hydroxide. Leach tests in distilled water at 40 and 70 °C measured the retention of Cs by the three cemented zeolites.
Keywords: Nuclear waste, immobilisation, stabilisation, Portland cement, zeolite

1 Introduction

Zeolites are used to remove radioactive cations from contaminated waters. The contaminated zeolites must themselves be conditioned before they can be disposed of in a near-surface repository. Portland cement is commonly used to immobilise radioactive wastes. However portlandite (calcium hydroxide), formed in hydrating cement, reacts with many zeolites to form calcium silicate hydrate. This pozzolanic reaction destroys the zeolite, freeing mobile cations such as caesium [1] [2].

An earlier report [3] contrasted the retention of caesium in three different zeolites immobilised in hydrating cement. An Australian natural zeolite from the Werris Creek deposit had less pozzolanic activity than either a zeolite from the Teague deposit (USA) or zeolite A. In this work the stability of the three cemented zeolites cured for one and a half years was compared and the Werris Creek zeolite was characterised in order to study its enhanced retention of caesium.

Mechanisms of Chemical Degradation of Cement-based Systems. Edited by K.L. Scrivener and J.F. Young.
Published in 1997 by E & FN Spon, 2-6 Boundary Row, London SE1 8HN. ISBN: 0419215700.

2 Experimental

Three different zeolites were used in this work: a sample from Werris Creek (Australia), zeolite XY from Teague (USA), and an A type. The respective suppliers stated that; Werris Creek zeolite was at least 65% pure clinoptilolite, Teague zeolite was over 90% clinoptilolite with small amounts of cristobalite calcite and montmorillonite, and Zeolite A was essentially pure crystals of less than 1 μm in size.

The cation exchange capacity of the Werris Creek and the Teague zeolites were 114 and 163 meq/100g respectively (A. Aharon of the Physical Chemistry Department University of New South Wales Kensington, NSW, private communication). The Werris Creek zeolite had a mean particle size of 73 μm (80% of the particles being between 6 and 160 μm). The Teague zeolite had a mean particle size of 35 μm (80% of the particles were between 1 and 58 μm). The particle sizes were determined using a Malvern Mastersizer after the sample was dispersed by ultrasonic agitation for 10 minutes

The cement used in this study was Aalborg Lion brand. Cemented zeolite samples were mixed by a Hobart mixer. The zeolite (200g) and the cement (200g) were dry mixed for 2 minutes, then 84 mls of water were added. The sample was then mixed for three minutes. The mortar was cast into small cylinders (2 cm in diameter and 3 cm in height) which were sealed and cured at room temperature.

Two types of caesium ion exchange treatment were used and these are designated as MEDIUM Cs exchange or LOW Cs exchange. The MEDIUM Cs exchanged form was prepared by stirring 40.95 g of $CsNO_3$ in 1 litre of water and adding 300g of zeolite to the solution which was stirred for one hour before the zeolite was filtered off and dried. It was calculated that this procedure resulted in the replacement by caesium of 50% of the ion exchange sites on the Werris Creek Zeolite. The LOW Cs exchanged form was prepared by a similar procedure except that only one fifth the amount of Cs was exchanged.

3 RESULTS

3.1 Characterisation of the Werris Creek Zeolite

The Scanning Electron Microscope (SEM) image of the phases in un-exchanged Werris Creek zeolite is shown in Fig 1. Here clinoptilolite, quartz, and sanidine are differentiated by backscattered electron contrast. A phase count made from 100 points separated by 200 μm indicated that the zeolite consisted of 57% clinoptilolite, 22% quartz, and 11% sanidine by weight (after correction for density). Additional minerals identified included apatite, calcite and rutile.

In order to check the homogeneity of the sample forty separate grains were selected with each grain being 200 μm apart. Of these grains, nine had one dimension greater than 100 μm and four were less than 5 μm in size. Only eight of the forty consisted of a single mineral crystal - each less than 10 μm in diameter.

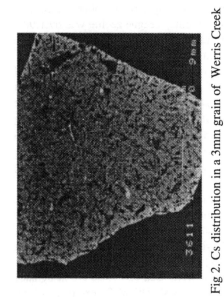

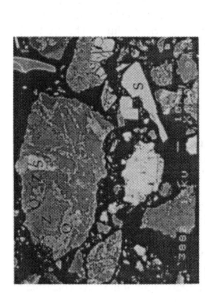

Fig 1. A typical area of Werris Creek zeolite. Z=Zeolite, S=Sanidine, Q=Quartz.

Fig 3. Detail of area in fig 2 showing Cs distribution. (The bright areas are generally high in Cs).

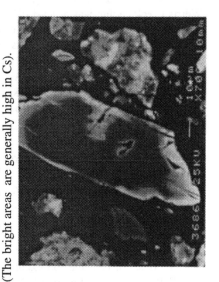

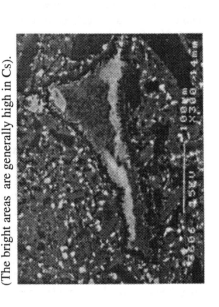

Fig 2. Cs distribution in a 3mm grain of Werris Creek zeolite. (The bright areas are generally high in Cs).

Fig 4. Single large crystal showing Cs distribution in zeolite. (The bright areas are generally high in Cs).

Except in very rare cases it was difficult to carry out representative EDAX analyses of the separate minerals because of their fine grain size. Most minerals were less than 2 μm from a grain of a different mineral. The best estimates of the zeolite compositions lie between $Ca_{3.4}Na_{0.8}Al_{7.6}Si_{24.3}O_{72}$ and $Ca_{2.0}Mg_{1.0}Na_{0.6}Al_{6.7}Si_{24.8}O_{72}$. The sanidine composition is approximately $K_{0.4}Na_{0.6}Al_{1.1}Si_{2.9}O_{8}$.

Fig. 2 shows a 3 mm-sized grain of Werris Creek zeolite which had been soaked in caesium nitrate solution. The image was taken with backscattered electrons and so the bright areas are mostly caesium rich. Caesium concentrations are higher on the outside of the crystal and in some areas inside the porous grain (Fig 3).

Fig. 4 shows the backscattered image of a single crystal in a Werris Creek zeolite which had been exchanged at LOW Cs exchange. The size of this crystal is unusual. The backscatter image indicates that caesium has concentrated; on the surface, around the pores, and along the cleavage planes. EDAX analysis indicates that the unexchanged zeolite has a composition of $Ca_{3.8}Al_{8}Si_{24}O_{72}$ while the Cs-rich area has a composition of $Cs_{2.2}Ca_{2.5}Al_{8}Si_{24}O_{72}$.

Quantitative x-ray analysis (SIROQUANT [4]) indicated that the Werris Creek zeolite consists of 62% clinoptilolite, 2% mordenite, 29% quartz, and 7% sanidine. This estimate is likely to be more accurate than that derived from the point count where only 100 points were studied.

3.2 Pozzolanic Reactions of the Cemented Zeolites

X-ray diffraction patterns (XRD) of the cemented zeolites, taken with Co Kα radiation, are shown in Fig 5. The zeolites had been allowed to cure for 1.5 years before they were crushed and ground. It is clear from the patterns that in both the zeolite A and the Teague zeolite almost all of the portlandite had reacted with the zeolite. In contrast the XRD from cemented Werris Creek zeolite showed that a significant amount of portlandite was still present. The intensity of the portlandite peaks is similar in cemented Werris Creek zeolite cured for 1, 3 and 18 months.

It was considered that the reduced pozzolanic activity of the Werris Creek zeolite could have been related to the larger particle sizes of this zeolite. To test this hypothesis Werris Creek zeolites was first ground for three hours and then cemented (using the standard procedure already described), and finally cured at room temperature for 7 days. The x-ray diffraction pattern of this sample showed a only small intensity in the calcium hydroxide peaks.

XRD from the cemented Teague zeolite cured for 3 days showed significant portlandite peaks. However the XRD of cemented Teague zeolites cured for 7 days or over showed that only insignificant amounts of portlandite remain. There were no significant amounts of portlandite peaks in any XRD of the cemented A zeolites.

3.3 Leaching

The leaching test followed the procedure of Zamorani and Serrini [5]. The cemented zeolite cylinders cured for one and a half years were suspended in a sealed plastic container containing 250 ml of distilled water. The containers were kept in a constant temperature bath at 40°C and 70°C. At 3, 7, 14, and 21 days a 5 ml aliquot of the

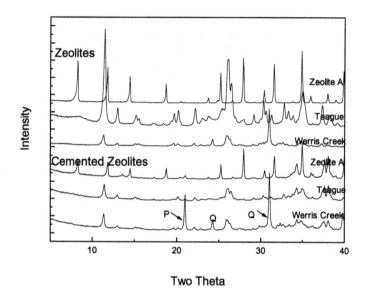

Fig. 5: XRD of zeolites and cemented zeolites cured for 1.5 years (Co Kα radiation).

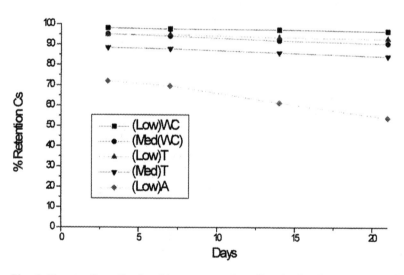

Fig. 6: Cs retention after leaching cemented zeolites in distilled water at 70°C.

solution was taken. The clear solution was acidified by HNO_3 and analysed by ICP. The amount of caesium retained in the zeolite as a function of time after leaching at 70° is shown in Fig 6. The analysis of the 21 days leachate is shown in Table 1. In both the table and the figure the letters WC, T, and A designate Werris Creek, Teague and A zeolites receptively. (Low) and (Med) designate the amount of caesium exchange as was defined in the experimental section. The results from the leach test are similar to those reported after one month's leaching at 50°C [3]. The amount of caesium removed by leaching has increased slightly. Clearly cemented zeolite A is the least efficient and cemented Werris Creek zeolite is the most efficient system for caesium retention.

The relatively high amount of calcium in the Werris Creek leachate is an indication of the presence of portlandite. The relatively higher amounts of aluminium in the A leachate is an indication that the alumina rich A zeolite has broken down. It is noted that there is slightly greater amounts of Si and S in the leachate for the cemented zeolites which have the least amount of portlandite (Teague and A).

Table 1: Results of Leachate Analysis after leaching at 40 and 70° C

	40°C					70°C				
	Ca (ppm)	Al (ppm)	S (ppm)	Si (ppm)	Cs (ppm)	Ca (ppm)	Al (ppm)	S (ppm)	Si (ppm)	Cs (ppm)
WC(Low)	451	2	9	1	7	363	3	18	1	13
WC(Med)	402	2	8	1	22	388	3	26	1	47
T(Low)	108	3	24	8	12	75	3	38	10	21
T(Med)	121	3	18	7	115	90	3	30	8	163
A(Low)	11	70	67	22	121	8	100	83	35	137

4 Discussion

The design of a good ion exchange system for the removal of radioactive cations from contaminated waters would need to consider; (a) the flow through the exchange column, (b) the efficiency of the exchange process, (c) the limit to the level of exchange and (d) the disposal of the contaminated zeolites.

Although zeolites, such as clinoptilolite, have a substantial ion exchange capacity it is impossible to use all of the capacity in cemented zeolites because (a) radioactive zeolites generate substantial amount of heat, and (b) calcium from the cement's pore water exchanges into the zeolite to free caesium [1]. (10g of fully exchanged caesium clinoptilolite released 66% of its caesium on exposure to 500 ml of saturated calcium hydroxide solutions[1] [2].)

The capacity of a zeolite and the rate at which the zeolite will exchange will control the efficiency of an ion exchange column to remove caesium from a solution. Ion selective electrodes can continuously monitor the rate at which some cations are removed from solution. There is no ion selective electrode for caesium but the clinoptilolite selectivity for caesium is similar to ammonium ions. The rate at which three zeolites (8g) remove ammonium ions (original concentration about 0.001M) from 150 mls of ammonium chloride solution was compared by measuring the ammonium concentration every 20 seconds by an ion specific electrode. Zeolite A removed most and the Werris Creek zeolite removed the least ammonium from solution. The Werris Creek zeolite was the slowest to reach equilibrium and ammonium was still being removed from the solution after 20 minutes. The rate of caesium exchange improves if the Werris Creek zeolite is ground but, as was shown before, grinding decreases the stability of the cemented zeolite.

The Werris Creek zeolite is harder to grind that the Teague zeolite. This makes it easier for the Werris Creek zeolite to be supplied in a coarse, attrition-resistant powder: that resists pozzolanic activity. However the Werris Creek Zeolite, because of its low clinoptilolite content, is the least efficient for extraction of caesium: this could counterbalance its improved retention properties on cementation.

From Fig 6 it can be seen that the amount of caesium leached from the cemented Werris Creek or Teague zeolites by distilled water seem to be in equilibrium with the leach solution after three weeks. This was checked by running the leach tests for an extra week and no significant extra caesium appeared in the leach solution.

The percentage of caesium retained in the cemented zeolites after leaching for three weeks at 70°C is shown in Table 2. Under similar leaching conditions no caesium would remain in zeolite-free cement pastes [5].

Table 2: The percentage of caesium retained in cemented zeolite leached for three weeks at 70°C

	LOW LEVEL % Cs Retained	MEDIUM LEVEL % Cs Retained
Cemented A Zeolite	54	-
Cemented Teague	93	84
Cemented Werris Creek	96	90

The level of caesium retention by cemented zeolite that would be deemed satisfactory must depend on the waste classification and the regulations governing the waste. Hence no absolute recommendation can be made as to the best caesium loading in zeolite or which zeolite can would best meet the immobilisation criteria for a particular waste stream.

It is now considered that the curing period of 1.5 years was excessive and that a curing period of 3 months would probably have given just an good as estimate of the caesium retention in cemented zeolites over the life time of the waste form because (a) The current leach results differ only by 10% from leach results carried out carried out after one month's curing [3] and (b) from the stoichiometry of the pozzolanic reaction. It is possible to calculate that at complete hydration only half of the zeolite can be destroyed, at a zeolite to cement ratio of 1:1, assuming that the reaction product of the calcium in the cement is $C_{1.7}$-S-H_4. After three months it can be assumed that only small amounts of extra portlandite will be produced. The XRD seems to show that there is not significantly different amounts of portlandite between the cemented Werris Creek zeolite cured for 2 and 18 months. Hence it could be assumed that little extra pozzolanic activity occurs after 3 months. Thus it is unlikely that longer time trials would show any significant difference to the stability of the cemented Werris Creek zeolite system.

5 Conclusions

1. The addition of clinoptilolite significantly increased the caesium retention of cemented pastes.
2. The caesium retention in cemented zeolite depended on the caesium loading of the zeolite. A lower loading increased the retention.
3. Werris Creek clinoptilolite resisted the pozzolanic reaction of cement paste better that the Teague clinoptilolite
4. The resistance of the Werris Creek zeolite to the pozzolanic activity depends, on a large measure, to the greater particle size of the zeolite.

6 Acknowledgments

We would like to thank, ANISE for funding the work (Grant 95/103Eng Mat), Zeolite Australia, Teague, and Degaussa for generously suppling the zeolites and help on product data.
We thank Liz Keegan and Elaine Loi for the ICP analysis, Tim Nicholls for preparation of the SEM samples, Yvonne Farrar for the NAA of the exchanged zeolites and Ross Campbell for the particle size analysis.

7 References

1   Angus, M.J., McCulloch, C.E., Crawford, R.W., Glasser, F.P. and Rahman, A.A. (1984) Kinetics and mechanism of the reaction between portland cement and clinoptilolite, in *Advances in Ceramics: Nuclear Waste Management,* Vol 8. (ed G.G. Wicks and W.A. Ross), American Ceramic Society, pp 429-440.
2   Glasser, F.P., Rahman, A.A., Crawford, R.W., McCulloch, C.E. and Angus, M.J. (1982) *Immobilisation and leaching mechanisms of Radwaste in cement-Based Matrices*, Report DOE/RW/82.108.
3   Aldridge, L. P., Ray, A., Stevens, M.G., Day, R.A., Leung, S., Morassut, P. and. Roukis, G. (1994) Immobilisation Of Caesium Contaminated Zeolite By Portland Cement, in *Proceedings of the International Ceramics Conference Austceram 94,* (ed. C.C. Sorrell and A.J. Ruys), Australian Ceramic Society Sydney, pp. 1340-1345.
4   Taylor, J.C. (1991) Computer Programs for Standardless Quantitative Analysis of Minerals Using the Full Powder Diffraction Profile. *Powder Diffraction,* Vol 6 No. 1. pp. 2-10.
5   Zamorani, E. and Serrini, G: (1990) Evaluation of Cement Matrix preparation and leaching procedures used in radioactive waste management. *Radioactive Waste Management and the Nuclear Fuel Cycle,* Vol. 14 No. 3. pp. 239-251.

43 FORMATION OF ZEOLITE IN MIXES OF CALCIUM ALUMINATE CEMENT AND SILICA FUME USED FOR CESIUM IMMOBILIZATION

H. FRYDA and P. BOCH
Ecole Supérieure de Physique et Chimie Industrielle, Paris, France
K. SCRIVENER
Imperial College, London, UK

Abstract

The immobilization of cesium with a mix of calcium aluminate cement (CAC) and silica fume (SF) was studied. The presence of cesium within the solution of hydration leads to the formation of a zeolitic phase of the chabazite family. Cesium is immobilized within the structure of the zeolite during its crystallization. The efficiency of the immobilization is demonstrated by a severe leach test. The kinetic of crystallization depends on the temperature. The same zeolite crystallizes when a prehydrated mix of CAC-SF is in contact with a Cs-solution. Thermal stability of the zeolite has been assessed by simultaneous TGA-TDA analysis. The full destruction of the structure requires temperatures higher than 750°C.
Keywords: calcium aluminate cement, silica fume, zeolite, chabazite, cesium immobilization, leaching, thermal stability.

1 Introduction

[137]Cesium is a dangerous radionuclide found in nuclear wastes. Ordinary Portland cement (OPC) is commonly used for treatment of low- and medium-activity wastes and, therefore, its related properties have been widely investigated. It was found that cesium cannot be immobilized within the solid phase during the hydration of OPC [1]. The large diameter of Cs^+ (0.33 nm) prevents this ion from entering the structure of OPC hydrates and the adsorption onto the C-S-H is very limited. Moreover, cesium salts are very soluble in water (2.6 Kg.l^{-1} for Cs_2CO_3 at 20°C), whatever the pH. Consequently, OPC cementitious compositions are not well adapted to the immobilization of cesium within a solid phase.

Mechanisms of Chemical Degradation of Cement-based Systems. Edited by K.L. Scrivener and J.F. Young. Published in 1997 by E & FN Spon, 2–6 Boundary Row, London SE1 8HN. ISBN: 0419215700.

An alternative is to use natural clays, as montmorillonite, or zeolites, as clinoptilolite, which can provide a selective ionic exchange with cesium. However, storage as solid block imposes such treated wastes to be embedded in a cementitious matrix. If the cement is rich in calcium, the reaction between $Ca(OH)_2$ and the zeolite or clay, a pozzolanic reaction [2], leads to the destructin of the ion exchanger and to the release of cesium.

Hoyle et al [3] have studied materials made of OPC, CAC, and SF, hydrated with a solution of cesium hydroxide. They found that these compositions behave much better than simple OPC compositions regarding the retention of cesium, and that the retention is improved in materials with high contents of equivalent alumina and silica.

The present paper reports on a study devoted to the immobilization of cesium by reaction of a CAC+SF mix with aqueous solutions of cesium carbonate and cesium hydroxide. It was focused on the ability of the mix to immobilize cesium within a solid phase, and the possible formation of zeolite. This study was part of a larger study, dedicated to the trapping of cesium within a stable mineral structure [4-6].

2 Experiments

For all experiments, we have used the same mix of 60 wt.% of CAC (SECAR71 from *Lafarge Aluminate*) and 40 wt.% of SF (from *ELKEM*), whose characteristics are given in table 1. The mix was reacted with deionized water or with solutions of Cs_2CO_3 or CsOH (inactive [133]Cs was used). The following curing conditions were studied :

Paste : the ratio solution/solid was 1. The anhydrous solid was mixed with the solution in a beacher. Tripolyphosphate of sodium was used as a deffloculating aggent. After mixing to homogeneity, the paste was poured in a glass tube which was sealed and stored at a controled temperature. Initial cesium concentration was 0.75 mol.l^{-1}. Temperature was 50°C and 90°C.

Suspension : the ratio solution/solid was 10. The anhydrous solid was mixed with the solution in a reactor whose temperature was controlled. Curing was performed in the same reactor which was kept sealed to avoid any evaporation. Temperature was 20°C and 50°C. Initial cesium concentration was 0.1 mol.l^{-1}.

Suspension with prehydration : conditions were the same as suspension, but the starting material was the mix CAC-SF already hydrated with deionized water at 50°C with the suspension procedure for 7 days. Temperature of curing was 50°C. Initial cesium concentration was 0.1mol.l^{-1}.

For each type of curing, liquid and solid were separated by filtration. Solid was dryed by washing with acetone and ether to stop the cure and to eliminate the exces solution.

Cesium concentration : for the suspensions, the Cs concentration of the solution was determined by atomic emission spectroscopy. For the pastes, the amount of bonded water of the solid was determined by TGA, and the amount of cesium within the solid was determined using atomic emission spectroscopy by analysing the solution resulting from the total dissolution of the solid.

Solid characterization : solid was analysed by powder X-Ray diffraction and simultaneous thermogravimetric and thermodifferential analysis (TGA-TDA).

Leaching tests : we used a leaching test where 0.6 g of dried solid were ground (particle size < 80μm), then mixed with 20 g of pure, deionized water heated to 50°C, then stored at 50°C in a sealed plastic bottle. After a first period of leaching, the concentration of cesium in the leaching solution was measured and the leaching solution was renewed by a fresh, Cs-free solution. The same operation was repeated at the end of each period, whose duration increased throughout the test. This test was performed on solid cured as paste.

Table 1: Starting materials.

	SECAR71	Silica fume
Composition (wt.%)	$Al_2O_3 \sim 71\%$ $CaO \sim 28\%$ $SiO_2 < 0.7\%$ $Na_2O < 0.3\%$ $MgO < 0.25\%$ $Fe_2O_3 < 0.11\%$ $TiO_2 < 0.1\%$ $SO_3 < 0.3\%$	$SiO_2 \sim 93\%$ $Al_2O_3 \sim 3.1\%$ $ZrO_2 \sim 2.5\%$ $CaO \sim 0.4\%$ $Fe_2O_3 \sim 0.15\%$ $Na_2O \sim 0.06\%$
Characteristics	Crystallized material, major phases: CA and CA_2	Amorphous material, spherical particles: $BET \approx 14 \, m^2 . g^{-1}$, $\Phi \approx 1\mu m$.

3 Results

3.1 Powder X-Ray diffraction of the solid phase

Table 2 relates the crystalline phases detected by diffraction. The stable phase assemblage without cesium is $C_3AS_nH_{6-2n}$ (silicon hydrogarnet), AH_3 (gibbsite) and C_2ASH_8 (strätlingite).

The presence of cesium, introduced as cesium carbonate or cesium hydroxyde, changes the pattern of the reaction and leads to the occurence of new peaks on the XRD pattern. In all materials where they were detected, these peaks were located at the same angular position and exhibited the same relative intensities. Consequently, they were hypothesized to be related to a single phase that is noted Z in table 2.

Identification of the zeolite : the XRD pattern of phase Z does not agree with any JCPDS pattern, but it is close to that of a zeolite, calcium-chabazite ($CaAl_2Si_4O_{12}.6H_2O$). The XRD pattern of phase Z can be indexed on an hexagonal unit cell, with $a = 1.382$ nm and $c = 1.525$ nm. Chabazite belongs to space group R m, with $a = 1.378$ nm and $c = 1.499$ nm. Hoyle et al. [3] have already assumed that the reaction between CAC, SF and cesium leads to chabazite with cesium trapped within the structure. To confirm this assumtion, the XRD pattern of phase Z was compared with the pattern of a pure Cs-chabazite. The sample of experiment C2 cured for 1 year has been heated at 500°C in air. This treatment leads to the destruction of all phases, except phase Z and $CaCO_3$. Peaks of phase Z were not changed by the treatment. Therefore, the resulting XRD pattern presents contributions of phase Z, $CaCO_3$ and amorphous phases. The comparison with a pure Cs-chabazite (see figure 1) shows a good agreement of the distances and relative intensities which confirms that the structure phase Z is the same than that of chabazite and that it contains some Cs cations.

Table 2 : crystalline phases detected by powder X-ray diffraction. Results of 7 experiments, reported in row-column A2, B2, C2, C3, D1, D2, D3 and E2.

	[1] T=20°C	[2] T=50°C	[3] T=95°C
[A] P, no cesium		1 year : h, s, g	
[B] P,[Cs]=0.7mol.l⁻¹ as CsOH		7 days : h, g 28 days : h, g, c, Z***	
[C] P,[Cs]=0.7mol.l⁻¹ as Cs₂CO₃.		7 days: m, h, g 28 days: m, h, g, Z*** 1 year: m, h, g, Z****	1 day : h, g, Z* 8 days : h, g, Z****
[D] S,[Cs]=0.1mol.l⁻¹ as Cs₂CO₃.	32 days: m, s, g 60 days: m, s, g, Z*** 118 days: m, h, g, Z****	7 days: m, h, g, Z* 28 days: m, h, g, Z*** 124 days: m, h, g, Z****	
[E] SPH,[Cs]=0.1mol.l⁻¹ as CsOH		0 days: h, s, g 9 days: h, s, g, Z** 28 days: h, s, g, c, Z****	

P = paste   S = suspension   SPH = suspension with prehydrated solution
m = $C_4A(CO_3)H_{11}$ h = $C_3AS_nH_{6-2n}$ g = AH_3 c = $CaCO_3$ s = C_2ASH_8
Z = zeolite, * to **** = trace to major phase.

3.2 Cs immobilization

Figure 2 shows the evolution of Cs concentration within the solution during the experiment D2. There is a clear correlation between the crystallisation of the zeolite and the decrease of the Cs concentration, which was 105 mmol.l⁻¹ at the begining of the cure, and dropped at less than 0.14 mmol.l⁻¹ after 28 days, and remained at this level until the end of the experiment which was 124 days. Samples from the experiment C2 were leached after 7, 10 and 28 days of cure (see figure 3). The same correlation can be made between immobilization of Cs and crystallisation of chabazite. These results confirm that the crystallisation of the zeolite leads to an efficient immobilisation of Cs within the solid phase.

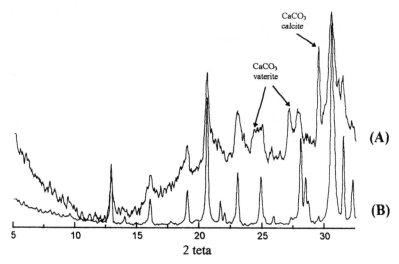

Fig. 1. Comparaison between XRD pattern of phase Z and pure Cs-chabazite
A : XRD pattern of sample from experiment C2 after 1 year of cure and heating at
500°C.
B : XRD pattern of pure Cs-chabazite.

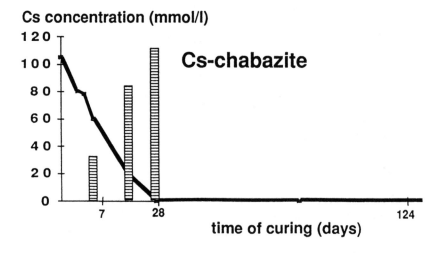

Fig. 2. Cs immobilisation, experiment D2
curve = Cs-concentration of the solution of hydration VS time of curing.
bars = intensity of the main XRD peak of chabazite.

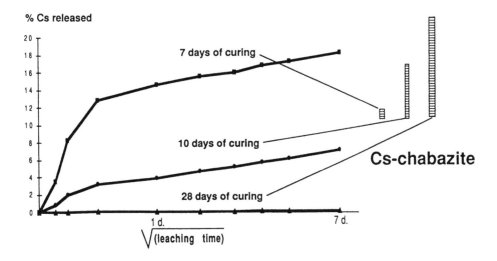

Fig. 3. Leaching test, experiment C2 after 7, 10 and 28 days of curing.
IM = coefficient of immobilisation of Cs within the solid after the cure
curves = cumulative % of Cs released VS time of leaching.
bars = intensity of the main XRD peak of chabazite.

3.3 Thermal stability

Samples of experiment C2 cured for 7 days and 1 year of cure were analysed by
TGA-TDA. For the 7 day cured sample, 3 main endothermic peaks, each associated
with weight loss, occur at 160°C, 285°C, and 315°C. XRD patterns of materials heat
treated at the corresponding temperatures showed that these peaks relate to the
dehydration of $C_3A(CaCO_3)H_{11}$, AH_3 and $C_3AS_nH_{6-2n}$.

For the 1 year cured sample, the TGA-DTA diagrams show two more
endothermic peaks at 200°C and 700°C, associated with weight losses. These
decompositions are respectively attributed to the dehydration of the chabazite and
the decomposition of $CaCO_3$.

As all zeolites, the water contained within the chabazite is not involved in the
chemical bonds of the structutre, but is present as water mollecule within the cavities
of the structure. This zeolitic water is volatilized at 200°C [7], but the structure of
the crystal is not destroyed. Powder XRD experiments carried out on materials heat
treated at increasing temperatures showed that the full destruction of chabazite
requires temperatures higher than 750°C.

Chabazite, as all zeolites, is a ceramic-like phase. Therefore, chemical bond is
more stable to heat treatment than hydrates.

3.4 Conditions of formation

It can be seen form table 2 that chabazite crystallizes with either Cs_2CO_3 or CsOH. Although anions and pH (10 for Cs_2CO_3 solution, 14 for CsOH solution) certainly influence the formation of the hydrates, there is no important difference relating to the crystallization of chabazite between Cs_2CO_3 or CsOH solutions. The primary factor for chabazite crystallization seems to be the presence of Cs within the solution. Crystallization of chabazite is rather slow in comparaison with that of hydrates. At 50°C, the latter has been completed after 1 day whereas XRD peaks of chabazite occur only after 7 days. However, temperature greatly influences the kinetic of crystallisation of the zeolite. At 20°C, chabazite has not been formed after 30 days and is detected after 60 days, whereas at 95°C, XRD peaks are detected only after 1 day. Finally, it can be seen from experiment E2 that the presence of anhydrous material at the begining of the cure is not necessary for chabazite to be formed. After 7 days of hydration with deionized water at 50°C, the replacement of the initial solution by a solution of CsOH leads to the formation of chabazite. In each case, crystallization of chabazite is accompagnied by a decrease of the diffracted intensities of hydrates.

4 Discussion and conclusion

The crystal structure of chabazite is able to accomodate different type of cations such as Na^+, K^+, Rb^+, Cs^+, Ca^{2+}, Sr^{2+} or Ag^+. It is composed of a negatively charged 3D framework of TO_4 tetrahedra, T=Si or Al, and of cavities which contain the cations. Cavities can communicate each other by 8-ring windows formed of 8 tetrahedra TO_4, as schematised on figure 4. Calligaris et al. [8] have studied the structure of a chabazite containing cesium. Its location was found to be at the centre of the 8-ring window.

Crystallisation of zeolites proceeds by precipitation of precursor species from the solution. These species are aluminosilicate polyanions whose condensation leads to the formation of the rings of the framework. Cations act as a template [9] by stabilizing such structures. In our experiments, the crystallization of chabazite requires the presence of cesium within the solution, and the crystallization leads to a decrease of the Cs concentration. Therefore, we assume that the presence of Cs^+ within the solution allows the stabilisation of the 8-ring, and that the crystallisation proceeds by precipitation and lead to the trapping of Cs+ within the 8-ring. The apperture of the 8-ring is 0.38 nm whereas the Cs^+ diameter is 0.33 nm. Such a fit allows strong electrostatic interactions between O^{2-} (which bear the negative charge of the framework) of the 8-ring, , and Cs^+, and leads to stabilization of the structure.

The good fit between the 8-ring and the diameter of Cs^+ can be correlated to the following features (i) Thermal stability of chabazite increases with the size of the cation, from Na^+ to Cs^+ [7]. Upon heating, open framework of zeolites tend to collapse in dense structures. As the 8-ring is filled with large cation, the collapse is shifted to higer temperature. (ii) Upon ionic exchange, chabazite shows the higher selectivity for Cs^+ [7, p.557]. This is a result of the energy balance involved in the ionic exchange, which is in favour of Cs^+.

sites I, II, III   small cations : Ca^{2+}, Na^+
site IV   large cations : K^+, Cs^+

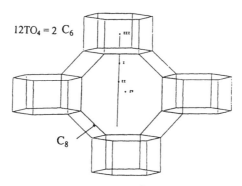

Fig. 4. Structure of chabazite. 4 cation sites. Cs^+ is located on site IV, at the center of the 8-ring window.

5 Reference

1.   R.W. Crawford et al. (1984) *Intrinsic sorption potential of cement component for* ^{134}Cs, Cement and Concrete Research, Vol 14, pp 595-99.
2.   H.F.W. Taylor (1990) Cement Chemistry, p. 304, Academic Press, San Diego.
3.   S.L. Hoyle, M.W. Grutzeck *Incorporation of cesium by hydrating calcium aluminosilicates*, J. Am. Ceram. Soc., 1989, Vol 72, No 10, pp. 938-47.
4.   P. Boch, M. Seiss, G. Vetter, M. Jacquin (1992) *High-alumina cements for cesium trapping*, Chemistry of Cements for Nuclear Applications, Vol 22, p-369-74, Pergamon, Oxford.
5.   H. Fryda, G. Vetter, P. Boch. (1993) *High-alumina cement/silica fume material for caesium trapping*, Proceedings of Third Euro-Ceramics, Vol 3, p-213, P.Duran & J.F.Fernandez Edit., Faenza Editrice Publ.,San Vicente, Espagne.
6.   H. Fryda, F. Babonneau, G. Vetter, P. Boch. (1994) *Capture of cesium into ceramizable cementitious materials*. In PressProceedings of 8th CIMTEC, Firenze, Italy.
7.   D.W.Breck (1974) *Zeolite Molecular Sieve : Structure, Chemistry and Use*, p. 446, John Wiley & sons, New York.
8.   M.Calligaris et al. (1986) *Crystal structure of the hydrated and dehydrated forms of a partially cesium-exchanged chabazite*. Zeolites, Vol 6, pp. 137-141.
9.   R.M.Barrer (1982) *Hydrothermal chemistry of zeolites*, p.166, Academic Press, London.

44 STABILIZATION OF SIMULATED ALKALINE NON-VITRIFIABLE LOW-LEVEL RADIOACTIVE WASTE BY CARBONATE-BEARING AFm AND AFt PHASES

D. BONEN, P.D. TENNIS, R.A. OLSON, H.M. JENNINGS and T.O. MASON
Center for Advanced Cement-based Materials, Northwestern University, Evanston, IL, USA

Abstract
Studies on simulated non-vitrifiable alkaline liquid waste from an off-gas system of the Low-Level Waste Disposal Vitrification Facility have shown that incorporation of this wasteform into high liquid-solid cementitious pozzolanic mixtures hinders hydration that results in excessive bleed water and delays setting time for days. In order to offset these adverse effects, a bleed-free system composed of cement, metakaolin, and gypsum was fabricated. While maintaining the bleed-free characteristics, fluidity and heat of evolution can be controlled by substituting slag for cement down to about 40%. This system is identifiable by a rapid build up of sulfate and calcium in the pore fluid that gives rise to an early precipitation of AFm phases and portlandite followed by ettringite and probably tobermorite. Solvent exchange and impedance spectroscopy techniques show a reduction of the transport properties over time that are comparable to cement paste prepared at a considerably lower water-to-cement ratio.
Keywords: Stabilization, alkaline low-level radioactive waste, AFm, AFt

1. Introduction

In a previous paper, we described the characteristics of a cementitious blend made of portland cement, fly ash, and attapulgite clay that was mixed with high-alkaline solution simulating the immobilization of Hanford low-level radioactive waste stream[1,2,3]. It has been shown that the blend displayed a rapid flow loss as a result of an early precipitation of hemicarbonate, calcium phosphate and zeolite.

A second set of studies were initiated to investigate the immobilization of simulated non-vitrifiable off-gas alkaline liquid waste recovered from a Low-Level Waste Disposal Vitrification Facility at the Hanford site. Preliminary tests showed that mixtures with compositions similar to that reported above yielded low reactivity rates when cured at room temperature. This was best manifested by XRD patterns showing an abundance of unreacted

Mechanisms of Chemical Degradation of Cement-based Systems. Edited by K.L. Scrivener and J.F. Young.
Published in 1997 by E & FN Spon, 2–6 Boundary Row, London SE1 8HN. ISBN: 0419215700.

clinker phases even after a 14-day hydration period. In turn, the low hydration rates resulted in excessive bleed water that was as high as about 20%. The amount of bleed water, however, could be substantially reduced by heat curing the blends at temperatures ranging from 55 to 70 C. At about 70 C, the bleed water was about 1 to 2%. Properties of mixtures that were heat cured under simulated adiabatic conditions are reported in this volume[4].

In order to eliminate the bleed water at ambient temperature, a new mix design was engineered. The incentive to do so becomes evident if one considers the adverse effects of high-temperature curing on the long-term volume stability of the solidified wasteform. Low-temperature curing reduces thermal stresses and shrinkage. In addition, such blends can be suitable for small and medium scale fixation operations as well as other applications.

As previously indicated, the bleed water is the result of high sedimentation due to a low reactivity rate. This is in agreement with Sahu's[5] observations which demonstrated that the initial calcium content of the pore solution of these blends is low and stays low for a long period of time.

Pore solution analyses have been carried out for a number of years by various workers. It is widely accepted that the pore solution of ordinary portland cement paste is saturated with respect to $Ca(OH)_2$ [6]. It has been demonstrated that the solubility of $Ca(OH)_2$ is substantially decreased at high Na^+ and OH^- contents[7]. Despite the difficulties involved in the calculation of the ion concentrations from the solubility products of solutions with high ionic strength[8-9], it is recognized that the levels of ions in solution are influenced in a complex way by the presence of other species. For example, Way and Shayan[10] noted that an increase in NaOH from 0 to 0.5N was associated with a rapid increase of SO_4^{2-}. However, further increase in NaOH did not significantly change the sulfate concentration. Rechenberg and Sprung[11] have shown that sulfate ions influence the Ca ion solubility, and the Ca concentration is higher than the concentration that would be predicted from Ca-OH relations [12].

The philosophy that has been undertaken was to engineer a mixture design by facilitating early precipitation. This was accomplished by increasing the calcium content of the pore solution by addition of gypsum, and substituting metakaolin (reactive calcined kaolin) for attapulgite. The objectives of this investigation were to formulate a bleed-free wasteform composed of cement, metakaolin, gypsum, slag, and fly ash, and to characterize its course of hydration and transport properties over time.

2 Materials and methods

Mixtures composed of cement, metakaolin, gypsum, slag, and fly ash and were prepared with high alkaline simulated liquid waste at a solution:solid ratio of 1 liter:1 kg. Table 1 shows the chemical composition of the simulated waste solution. For convenience, the specimens are grouped in two series: CGK series that includes specimens CK, CGK, CGKF and CGKS and GS series that includes specimens GS 30, GS 40, and GS 50 (See Table 2).

Mixtures for pore fluid analyses and X-ray diffraction (XRD) examinations were cast in 2x3" plastic cylinders and sealed to avoid evaporation. Companion mixtures for bleed water determination were cast in graduated 50 ml test tube. Specimens for compressive strength were cast in 2x4" cylinders. Impedance spectroscopy measurements were carried out using Hewlett-Packard Model 4192A impedance

Table 1. Solution Content

Compound	g/l
$Al(NO_3)_3 \cdot 9H_2O$	7.03
$Na_3(PO_4) \cdot 12H_2O$	24.53
$NaNO_2$	12.17
Na_2CO_3	39.97
$NaNO_3$	32.73
Na_2SO_4	0.21
$NaCl$	0.28
KCl	0.20
$Ca(NO_3)_2 \cdot 4H_2O$	0.02
$Ni(NO_3)_2 \cdot 6H_2O$	0.02
$Na_2B_4O_7$	0.01

Table 2: Mixture proportions

Mix	Composition					Bleed water
	cement	gypsum	meta-kaolin	fly ash	slag	
CK	80	-	20	-	-	< 5 %
CGK	68	12	20	-	-	none
CGKF	51	9	20	20	-	none
CGKS	51	9	20	-	20	none
GS 30	42.5	7.5	20	-	30	< 1 %[1]
GS 40	34	6	20	-	40	< 3 %[1]
GS 50	30	5.3	14.7	-	50	~ 5 %[2]

[1] At 2 days decreases to less than 0.5 % and completely absorbed at 3 days.
[2] Absorbed after a few days.

analyzer on specimens cast in 3-mm polycarbonate mould with inner dimensions of 1x1x4". Specimens for solvent replacement determination were carried out on sliced discs with 25.4 mm diameter and 3 mm thick. Methanol was found to be a reproducible and relatively fast replacement medium. Pore solutions were extracted from companion specimens using a steel die press and tested for calcium, aluminum, sulfur, sodium, and potassium by ICP (Plasma 40, Perkin-Elmer). X-ray diffraction analyses were carried out in a Phillips diffractometer equipped with a graphite monochromator. All specimens were cured at room temperature until testing.

3 Results and Discussion

3.1 Bleed Water and Compressive Strength
Table 2 shows the mixture proportions of the solid ingredients and the amount of bleed water registered. All of the specimens were prepared at solution:solid weight ratio of 1.07:1 and cured at room temperature until testing. Substituting metakaolin for attapulgite had a very favorable effect on the amount of bleed water, which was reduced from about 20% down to about 5%. Further reduction was obtained by incorporation of gypsum. The gypsum replaced 15% of the cement by mass. In CGK, CGKF, and CGKS samples addition of gypsum eliminated bleeding. In fact, the latter three mixtures were characterized by a rapid loss of fluidity, and although they were castable and workable, they lost their fluidity within the first 30 minutes after mixing.

Proper fluidity could be maintained by increasing the slag content. Accordingly, no observable fluid loss was registered during the first 30 minutes after mixing in batches containing more than 30% slag. In turn, successive additions of slag brought about bleeding. In GS 30 the bleed water was less than 1%, in GS 40, about 3%, and in GS 50 about 5%. The bleed water, however, reabsorbed after a few days after mixing.

Following hardening, the CGK series (i.e. CK, CGK, CGKF and CGKS specimens) developed considerable strength. At 120 days the compressive strength of CK, CGK, and CGKS was about the same and averaged 11 at MPa (~1600 psi), whereas that of the fly ash specimen was much lower and averaged 7 MPa (~1000 psi).

As the cement-based waste is supposed to be placed in large near-subsurface vaults, high-volume pozzolanic mixtures are favorable for controlling the heat of evolution and keeping the corresponding temperature below $90^{\circ}C$. Preliminary results on the temperature profile of the GS 30 blend that was mixed at room temperature and cured under adiabatic condition, has shown that the maximum temperature was reached after 3 days and levelled off at $72^{\circ}C$ [13].

3.2 Pore Solution Chemistry

Figure 1 shows the evolution of the pore fluid chemistry in the CGK series up to 180 days. Evidently, gypsum addition has favorable effects on the initial concentration of calcium in the pore solutions. The initial calcium concentration in the pore solution of CK is relatively low, about 1.2 milli molar (mM). At 1 day it increases to about 4 mM and then gradually decreases to a fraction of 1 mM over an additional 6-day period. In contrast, in the presence of gypsum the calcium solubility is increased, and the corresponding values for CGK and CGKS specimens are about 16 mM. The calcium content of the pore fluids of CGKF reaches about the same level at 1 day. Subsequently, from 1 to 3 days the calcium content of the gypsum-bearing specimens is strongly depleted and ranges from about 0.3 to 0.7 mM. Interestingly, from 7 days and on a slight increase in the calcium content is observed, and at 180 days it ranges from a low of 0.4 to a high of 2.0 mM for CK and CGKF, respectively.

Aluminum displays a somewhat complicated pattern. In the absence of gypsum (i.e. CK) it is depleted in the pore solution and drops from 5.6 mM after mixing to 0.22 mM at three days. Then, it is enriched and gains about the same level as its initial concentration. Gypsum addition suppresses the initial aluminum concentration to about 0.3 mM. Upon removal of much of the sulfate ions from the pore solution from 1 to 3 days, the aluminum concentration is sharply increased to 4.4-5.7 mM. In the fly ash and slag specimens (i.e. CGKF and CGKS) the aluminum reaches a maximum at about 7 days and then is progressively removed from the pore fluids.

All the samples display similar patterns for sulfur, sodium, and to a lesser extent potassium. Gypsum addition increases the sulfate content 3.5 times from about 130 mM in CK to about 460 mM in the gypsum-bearing specimens. The initial sulfur content of all the gypsum-bearing specimens is about the same indicating that the pore solutions are saturated with respect to sulfate, irrespective to the gypsum content added. The major removal period of sulfate from the pore fluid takes place from 1 to 3 days, where its content drops to 56 mM in CK and ranges from 104 to 130 mM in the gypsum-bearing specimens. These differences become smaller over time and at 180 days the sulfur content of CK is about 60 mM and that of the gypsum-bearing specimens around 84 mM.

In all the specimens the sodium content in the pore solutions remains largely unchanged during the first day and ranges from 1540 to 1620 mM. From 1 to 7 days it is removed from the pore solution and drops to around 1050 mM. Thereafter, the sodium concentration tends to slightly increase over time. Potassium is mainly concentrated in the pore solution. Nonetheless, it is differentiated according to the blending ingredients. Accordingly, its levels in the fly ash and slag specimens are notably lower than in CK and CGK illustrating the potassium uptake capacity of the former blends. Furthermore, in CGKS potassium reaches a maximum at 28 days and then is progressively removed from the solution. Similar results were reported by Glasser et al.[14].

Successive additions of slag in GS series yield almost identical patterns for sulfur and sodium, and some changes for aluminum and potassium. Of interest are the effects on the calcium content. The initial concentration of calcium in GS 30 (30% slag) is about the same as that for CGKS (20% slag). However, further increase of the slag content is followed by a sharp drop of its initial content to less than 2.6 mM. It appears that the low calcium concentration in GS 40 and GS 50 (40 and 50% slag, respectively) is associated with the appearance of bleed water and a low hydration rate that has been observed by XRD examination.

3.3 Course of Hydration

The amount of unevaporable water (chemically bound water as determined by the weight loss on ignition from 100 to 1000°C) is shown in Figure 2. The weight loss is a measure of the

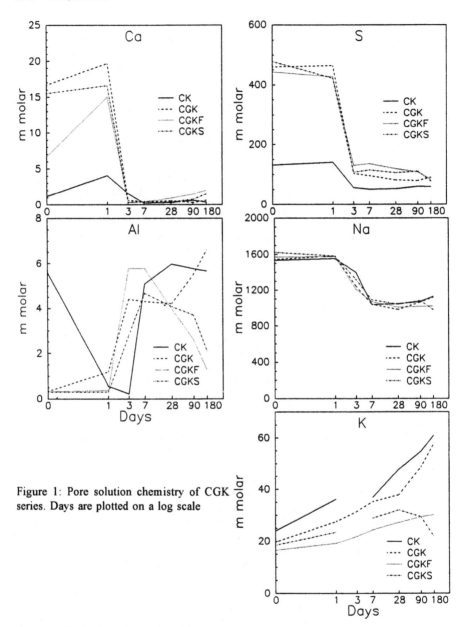

Figure 1: Pore solution chemistry of CGK series. Days are plotted on a log scale

degree of hydration; the higher the weight loss the greater the latter. All the specimens display sub-parallel patterns. Evidently, most of the hydration reactions appear to be completed by 28 days. However, the hydration continues up to 180 days, although at a slow pace. The blends with the highest proportion of cement, CK and CGK, have the highest amounts of unevaporable water totalling 17.8 and 19.5 %, respectively, at 180 days. For comparison, the weight loss of plain ASTM Type I portland cement at complete hydration is about 23%. However, such comparison should be made with caution because of the uncommon nature of the hydrated products.

Osbaeck and Jons[15] noted that alkali sulfate enhances the rate of hydration of cements that are otherwise similar. In accordance with this observation, addition of gypsum to CK increases the bound water, especially at early age by about 3%. This illustrates the favorable effect of gypsum on facilitating early reactions that bind water and suppress bleeding. Substituting slag or fly ash for cement (i.e, CGKS and CGKF) reduces the unevaporable water to about 16.2% at 180 days, but increases the weight loss per gram of portland cement by about 10%. In the GS series, successive addition of slag further decreases the weight loss as well as the rate of reactions, especially for GS 40 and GS 50 specimens.

Figure 3 a-c shows what we consider as the main region of interest of the XRD patterns. The low calcium and sulfate content in CK pore fluid brings about a relatively low hydration rate. Up to 7 days, the XRD pattern is dominated by conspicuous peaks of clinker minerals. Nonetheless, at 1 day Fig. 3a displays distinct peaks of CH [$Ca(OH_2)$] and hemicarbonate ($Ca_4Al_2C_{0.5}O_8 \cdot 12H_2O$) that at 7 days are followed by monocarbonate ($Ca_4Al_2CO_9 \cdot 12H_2O$) and probably trona [$Na_3H(CO_3)_2 \cdot 2H_2O$]. At 28 days, CH content reaches a maximum and clinker minerals cannot be unequivocally identified. Through aging, both the CH and hemicarbonate peaks decreased and the latter is probably converted to monocarbonate. The 180-day XRD pattern, therefore, is mainly composed of well-defined peaks of C-S-H at 3.06, 2.78, and 1.83 Å, along with peaks of monocarbonate, trona, CH, and hemicarbonate.

Gypsum addition notably changes both, the rate of hydration and phase composition. In CGK, (Fig 3b) the XRD pattern of clinker mineral is shortened to three days. A rather large amount of CH is formed along with hemicarbonate and a small amount of monocarbonate. CH peaks at three days and then is progressively taken up and it constituents only a minor amount at 180 days. The high content of CH observed in Fig. 3b is due to the reaction of sulfate ion with the sodium hydroxide in the pore solution according to the reaction:

$$2Na(OH)_2 + CaSO_4 \cdot 2H_2O \rightarrow Ca(OH)_2 + Na_2SO_{4\,(aq.)} + 2H_2O \qquad (1)$$

At 7 days conspicuous peaks corresponding to carbonaceous ettringite and probably tobermorite (10 Å) appear along with the former phases. However, the AFm phases disappear over time and at 180 days were converted to ettringite. The appearance of ettringite is clearly related to the change of S:Al as gypsum is added. Its late formation, however, is probably related to the nature of the alkaline solution. In this regard, Way and Shayan[10] noted that high-alkaline solutions delay ettringite formation. They attributed this retardation to the formation of sodium-substituted monosulfate.

Incorporation of fly ash yields similar XRD patterns as those for CGK. Though addition of slag has the same XRD patterns as fly ash, the rate of reactions is considerably increased, indicating that the former is more reactive than fly ash by far. Thus, in the slag specimen (Fig. 3c) peaks of ettringite and tobermorite are identifiable at 1 day. The AFm phases disappear over time, and the 180-day pattern is composed of peaks of carbonaceous ettringite, tobermorite, and a minor amount of CH. Successive addition of slag suppresses the formation of hemicarbonate, monocarbonate, and tobermorite.

The apparent formation of tobermorite deserves further consideration. The presence of tobermorite could not be unequivocally determined by the XRD pattern, as the main peaks of CSH at 3.07, 2.8, and 1.83 overlaps that of the tobermorite. Nonetheless, the distinct peak at 10 Å suggests that this phase was formed. Tobermorite is usually formed at hydrothermal conditions rather than at room temperature. However, there are evidences pointing at the formation of low Ca:Si calcium silicate at low temperatures in strongly alkaline and sulfate environments. Macphee et al.[16] have shown that aqueous NaOH lowers the Ca:Si of the C-S-H gel and consequently increases the $Ca(OH)_2$ content. Singh and Garg[17] reported on the formation of tobermorite in slag activated by $Ca(OH)_2$ and gypsum, and recently, McCarthy et al.[18] noted the appearance of tobermorite and ettringite in hydrated high-calcium coal combustion by products containing gypsum and a few percentages of alkali.

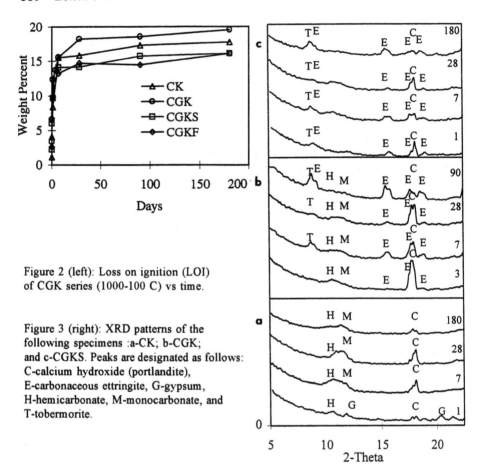

Figure 2 (left): Loss on ignition (LOI) of CGK series (1000-100 C) vs time.

Figure 3 (right): XRD patterns of the following specimens :a-CK; b-CGK; and c-CGKS. Peaks are designated as follows: C-calcium hydroxide (portlandite), E-carbonaceous ettringite, G-gypsum, H-hemicarbonate, M-monocarbonate, and T-tobermorite.

3.4 Transport Properties

In view of the limited chemical fixation capability of portland cement [19], leachability is dependent to a large extent on the physical containment of the waste. Thus, the pore microstructure is of prime importance for evaluating long-term containment. In this paper, the pore microstructure was interpreted by two independent methods: solvent exchange and impedance spectroscopy. In the former method a saturated specimen is immersed in a relatively large volume of solvent that is miscible with the pore solution. The weight loss of the specimen is caused by counter diffusion of the solvent into the specimen and the original pore fluid into the surrounding and it should be related to the transport properties of the specimens.

A typical plot of the specimen weight versus the square root of time yields an initial linear section that levels off asymptotically with time. The weight loss of the specimen in this linear section is a measure of a coarser microstructure and/or a lesser degree of tortuosity and interconnectivity of pores. Table 3 shows the weight loss in the linear section after a 625-minute immersion period in methanol. Since the cross-sectional area of all the specimens were the same, the weight loss can be directly related to the volume penetrated by methanol; the greater the weight loss the greater the accessability of pores.

Table 3: 625-minute Weight Loss

Specimen	Solution	Curing Days	C	wt. loss (g)
w:c=0.5[1]	DI H$_2$O	28	25	0.0491
w:c=0.7[2]	DI H$_2$O	28	25	0.1416
CK	Case 3	28	25	0.0524
CGK	Case 3	28	25	0.0933
CGKF	Case 3	28	25	0.1943
CGKS	Case 3	28	25	0.1054
GS 30	Case 3	28	25	0.1683
GS 40	Case 3	28	25	0.1664
GS 50[3]	Case 3	28	25	0.1207

[1] Lafarge portland cement ASTM Type I.
[2] Ashgrove portland cement ASTM Type I/II.
[3] Occurrence of some sedimentation.

Figure 4: Normalized conductivity vs time

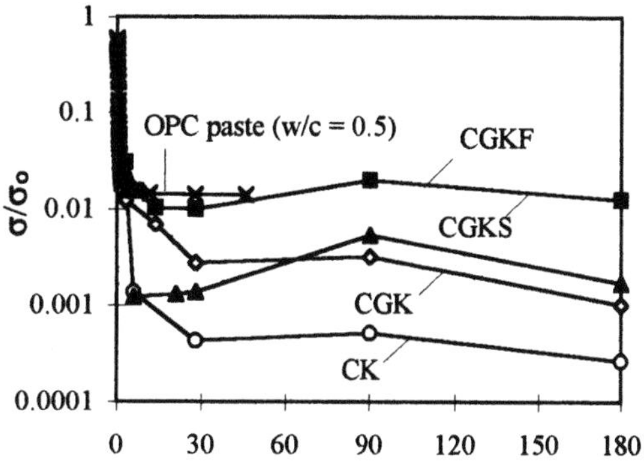

For a comparison, the weight losses of ordinary portland cement prepared at water-to-cement ratios of 0.5 and 0.7 are also tabulated. It appears that CK has the least accessible pore structure followed by CGK, CGKS, and CGKF. Despite the high w:c employed (i.e. 1.07:1), it is evident that the diffusivity of CK is similar to that of cement paste prepared at w:c 0.5. Only the diffusivity of CGKF is greater than cement paste prepared at w:c 0.7. Upon successive addition of slag the diffusivity is increased in the order CGKS < GS 30 < GS 40. Interestingly, the diffusivity of GS 50 appears to be lower than that of GS 30. This however, may be attributed to some sedimentation that took place during the preparation of the specimen that could alter the microstructure.

The normalized conductivity of the paste is directly related to the normalized diffusivity and permeability through Nernst-Einstein and Katz-Thompson equations, respectively. Further details are given in ref. [1]. Figure 4 shows the normalized conductivity of the pastes as measured by impedance spectroscopy. The main drop of the normalized conductivity occurs in the first 7 days after mixing. Yet, in agreement with the LOI data, the normalized conductivity decreases up to 180 days. Similar to solvent exchange data, CK displays the lowest diffusivity followed by CGK, CGKS, and CGKF.

Successive addition of slag considerably increases the transport properties, and the normalized conductivity is increased in the order: CGKS <GS 30 < GS 40 < GS 50. However, according to the normalized conductivity, the diffusivity of CK is about an order of magnitude lower than that of cement paste prepared at w:c 0.5, whereas that of CGKF is somewhat greater than the corresponding paste. In turn, the solvent exchange data shows that the diffusivity of CK is similar to that of paste with w:c 0.5. This discrepancy deserves further attention; the two techniques measure different properties: the impedance spectroscopy delineates transport of electric charges, whereas that of the solvent replacement is interpreted in terms of counter diffusion. Nonetheless, both methods grade the diffusivity of the specimens in the same order and indicate that the diffusivity is similar to pastes prepared at considerably lower water-to-cementitious materials ratios. That the diffusivity of CK and CGK is lower than CGKS and CGKF may be related to both, higher cement content and lower rates of reactions. This, in turn, enabled formation of a greater amount of hydrated products and refinement of the pore microstructure.

4 Concluding remarks

The system that was fabricated opens up a new avenue of stabilizing waste at high solution:cement ratios at room temperature. Addition of gypsum increases the calcium solubility in the pore fluids and facilitates early precipitation of hemicarbonate, monocarbonate, and $Ca(OH)_2$. The former two phases are converted to ettringite, and $Ca(OH)_2$ is markedly reduced over time. Additionally, it appears that tobermorite is also formed. The transport properties of the batches are comparable to cement paste with a w:c of about 0.5.

5 Acknowledgment

This work is supported by Westinghouse Hanford Company, Low-Level Waste Disposal Program.

6 References

1.   Olson, R.A, Tennis, P.D., Bonen, D., Jennings, H.M., Mason, T.O., Christensen, B.J., Brough, A.R. Sun, G-K. and J.F Young, J.F. (1996) Early Containment of High-Alkaline Solution Simulating Low-Level Radioactive Waste Stream in Blended Cement. J. Hazardous Materials (in press).

2.   Brough, A.R., Katz, A., Bakharev, T., Sun, G-K, Kirkpatrick, R.J, Struble, L.J. and Young, J.F (1995) Microstructural Aspects of Zeolite Formation in Alkali Activated Cement Containing High Levels of Fly Ash. Microstructure of Cement-Based Systems/Bonding and Interfaces in Cementitious Materials, (eds. S. Diamond, F.P. Glasser, L.W. Roberts, J.P. Skalny, and L.D. Wakeley), Proc. Mater. Res. Soc. Vol 370, Pittsburgh, pp. 199-208.

3.   Katz, A., Brough, A.R., Bakharev, T., Kirkpatrick, R.J., Struble, L.J., and Young, J.F. (1995) LLW Solidification in Cement - Effect of Dilution. Microstructure of Cement-Based Systems/Bonding and Interfaces. Cementitious Materials, (eds. S. Diamond, F.P. Glasser, L.W. Roberts, J.P. Skalny, and L.D. Wakeley), Vol 370, Pittsburgh, pp. 209-16.

4.   Bakharev, T., Brough, A.R., Olson, R.A., Tennis, P.D., Sahu, S. Diamond, S., Bonen, D., Kirkpatrick, R.J., Mason, T.O., Struble, L.J. and Young, J.F. (1996) Durability of Cement Stabilized Low Level Wastes. Mechanism of Chemical Degradation of Cement-Based Systems, Proc. Mater. Res. Soc. E & FN Spon, London (this volume, in press).

5.   Sahu, S., private communication.

6.   Young, J.F., Tong, H.S. and Berger, R.L. (1977) Compositions of Solutions in Contact with Hydrating Tricalcium Silicate Pastes. J. Am. Ceram. Soc. Vol. 60 No 5-6. pp. 193-98.

7.   Kalousek, G.L. (1944) Studies of Portions of the Quaternary System Soda-Lime-Silica-Water at 25 C. J. Res. Natl. Bur. Stand. (U.S.), Vol. 32, pp. 285-302.

8.   Diamond, S. (1975) Long-Term Status of Calcium Hydroxide Saturation of Pore Solutions in Hardened Cements. Cem. Concr. Res. Vol. 5, No. 6. pp. 607-16.

9   Moragues, A., Macias, A., Andrade, C. and Losada, J. (1988) Equilibria of the Chemical Composition of the Pore Concrete Solution. Part II: Calculation of the Equilibria Constants of the Synthetic Solutions. Cem. Concr. Res. Vol. 18, No. 3, pp. 342-50.

10.   Way, S.J. and Shayan, A. (1989) Early Hydration of A Portland Cement in Water and Sodium Hydroxide Solutions: Composition of Solutions and Nature of Solid Phases. Cem. Concr. Res. Vol. 19, No. 5. pp. 759-69.

11. W. Rechenberg, W. and S. Sprung, S. (1983) Composition of the Solution in the Hydration of Cement," Cem. Concr. Res. Vol.13, No. 1. pp. 119-26.

12. Moragues, A., Macias, A. and Andrade, C. (1987) Equilibria of the Chemical Composition of the Concrete Pore Solution. Part I: Comparative Study of Synthetic and Extracted Solutions. Cem. Concr. Res. Vol. 17, No. 2. pp. 173-82.

13. Bakharev, T. and Brough, A.R., private communication.

14. Glasser, F.P., Luke, K. and Angus, M.J. (1988) Modification of Cement Pore Fluid Compositions by Pozzolanic Additives. Cem. Concr. Res. Vol. 18, No. 2. pp. 165-78.

15 Osbaeck, B. and Jons, E.S. (1979) The Influence of Alkalis on the Strength Properties of Portland Cement. Zement-Kalk-Gibs Vol. 32, pp. 72-77.

16. Macphee, D.E., Luke, K., Glasser, F.P. and Lachowski, E.E. (1989) Solubility and Aging of Calcium Silicate Hydrates in Alkaline Solutions at 25 C. J. Am. Ceram. Soc. Vol. 72, No. 4. pp. 646-54.

17. Singh and M. Garg, M. (1995) Activation of Gypsum Anhydrite-Slag Mixtures. Cem. Concr. Res. Vol. 25, No. 2. pp. 332-38.

18. McCarthy, G.J., Longlet, J.J., Parks, J.A. and Butler, R.D. (1995) Long-Term Stability of disposed Cementitious by Products Materials. presented at 210th Am. Chem. Soc. National Meeting, August 20-24, Chicago.

19. Bonen, D. and Sarkar, S.L. (1994) The Present State-of-the Art of Immobilization of Heavy Metals in Cement-Based Materials. Advances in Cement and Concrete, Engineering Foundation Confer. (eds. M.W. Grutzeck and S.L. Sarkar), Am. Soc. Civil Engineers, New York, pp. 481-98.

PART SEVEN

DIFFUSION AND MODELLING

PART FOUR

4 RELIGION AND MORALITY

45 ACCELERATED TESTS FOR CHLORIDE DIFFUSIVITY AND THEIR APPLICATION IN PREDICTION OF CHLORIDE PENETRATION

L. TANG and L-O. NILSSON
Chalmers University of Technology, Gothenburg, Sweden

Abstract
Three different accelerated tests for chloride diffusivity are reviewed and the theoretical relationships between the diffusion coefficients from different tests are examined and discussed. The potential application of the diffusion coefficient for predicting chloride penetration into concrete structures is also discussed. An electrically accelerated test based on the non-steady state process is recommended because of its theoretical basis and simplicity. A numerical model of chloride penetration accompanied with chloride binding was used and some predicted results under different exposure conditions are presented by comparison with the measured data from the field and laboratory studies.
Keywords: Cement, chlorides, chloride penetration, concrete, diffusion, prediction, test methods, modeling.

1 Introduction

Chloride penetration into concrete is still a hot topic concerning the durability of concrete structures. For a concrete structure exposed in the submerged zone, diffusion is considered as a dominant transport process. Many rapid tests related to chloride penetration have been proposed, but only a few of them are related to chloride diffusion. As well known, complicated transport mechanisms are involved in chloride penetration. Therefore, it is of importance to clarify the theoretical basis of the test method and the definitions of the coefficient obtained from the test, before it could be applied for predicting chloride penetration into concrete structures.

Mechanisms of Chemical Degradation of Cement-based Systems. Edited by K.L. Scrivener and J.F. Young. Published in 1997 by E & FN Spon, 2-6 Boundary Row, London SE1 8HN. ISBN: 0419215700.

2 Accelerated tests for chloride diffusivity

2.1 Concentrated immersion test

A concentrated immersion test is similar to conventional immersion tests except that a solution with a very high sodium chloride concentration is used for immersion. This test usually involves 1) sealing all except one surface of the specimen to prevent multi-directional penetration; 2) immersing the specimen in a concentrated chloride solution, about 3~5 M NaCl, for a period of about one month; and 3) measuring the chloride penetration profile in the specimen. A Danish standard APM 302 [1] describes in detail the test procedures.

The basic assumption in this test is that Fick's second law should be applicable. This implies that no chloride binding or a linear chloride binding is involved in the diffusion process. The diffusion coefficient could, therefore, be calculated from the penetration profile by curve-fitting to the following equation:

$$c(x,t) = c_s - (c_s - c_i)\operatorname{erf}\left(\frac{x}{2\sqrt{D_{APM}\cdot t}}\right) \tag{1}$$

where $c(x,t)$: total chloride content at the position x, $kg_{Cl}/m^3_{material}$;
 t: immersion period, s;
 c_s: total chloride content at the exposure surface, $kg_{Cl}/m^3_{material}$;
 c_i: initial chloride content in concrete, $kg_{Cl}/m^3_{material}$;
 erf: error function, dimensionless;
 D_{APM}: diffusion coefficient according to APM 302 [1].

It can be seen from the above equation that the dimension of D_{APM} must be $\dfrac{m^2_{x\text{-coordinate}}}{s}$, or m^2_x/s in order to keep the error function dimensionless, no matter what dimension the chloride content has. In practice, however, the surface chloride content c_s changes with many factors, although the chloride concentration in an immersion solution is constant. Therefore, two parameters, D_{APM} and c_s should be taken as variables in the curve-fitting.

2.2 Non-steady state migration test

A few years ago, we proposed the first non-steady state migration test [2] to determine the diffusion coefficient, which involves applying a potential of 30~40 V across a 50 mm thick specimen for a short period, i.e. a couple of hours or days, then splitting the specimen and measuring the penetration depth of chlorides by using a colorimetric method [3]. Under the action of a constant electrical field, the diffusion function is used to be expressed as the following equation which will be discussed later in the next page:

$$\frac{\partial c}{\partial t} = -\frac{\partial J}{\partial x} = D_{nSS}\cdot\left(\frac{\partial^2 c}{\partial x^2} - \frac{zFU}{RTL}\cdot\frac{\partial c}{\partial x}\right) \tag{2}$$

where c: free chloride concentration, $kg_{Cl}/m^3_{solution}$;
 J: chloride flux through the solution, $\dfrac{kg_{Cl}}{m^2_{solution}\cdot s}$;
 D_{nSS}: diffusion coefficient from the non-steady state migration test, m^2_x/s;

z: absolute value of the ion valence, for chloride ions, $z = 1$;
F: Faraday constant, $F = 9.648 \times 10^4$ J/(V·mol);
U: absolute value of the potential difference, V;
R: gas constant, $R = 8.314$ J/(K·mol);
T: solution temperature, K;
L: thickness of the specimen, m_x.

According to the recent development [4, 5] in the Chalmers University of Technology (CTH), the following relationship between chloride diffusivity and penetration depth can be derived from the above equation:

$$D_{CTH} = D_{nss} = \frac{RTL}{zFU} \cdot \frac{x_d - \alpha\sqrt{x_d}}{t} \qquad m^2_x/s \qquad (3)$$

where D_{CTH}: diffusion coefficient according to CTH method [4, 5];
        x_d: penetration depth, m_x;
        t: test duration, s;
        α: laboratory constant,

$$\alpha = 2\sqrt{\frac{RTL}{zFU}} \cdot erf^{-1}\left(1 - \frac{2c_d}{c_0}\right) \qquad (4)$$

where c_0 is the chloride concentration in the bulk solution, c_d is the chloride concentration at which the color changes when using a colorimetric method to measure x_d, and erf^{-1} is the inverse error function.

It should be noticed that, in Eq. (2), i) the chloride binding term does not included; and ii) the flow area is related to the solution only, as shown in the dimension of the flux J. In the steady state tests, however, the flow area is often related to the material. In order to relate the diffusion coefficient in Eq. (2) to that from the steady state tests, the following equation should be used:

$$\frac{\partial c_t}{\partial t} = \varepsilon\left(\frac{\partial c}{\partial t} + \frac{\partial c_b}{\partial t}\right) = -\frac{\partial J_m}{\partial x} = D_{SSM}\left(\frac{\partial^2 c}{\partial x^2} - \frac{zFU}{RTL} \cdot \frac{\partial c}{\partial x}\right) \qquad (5)$$

where c_t: total chloride concentration, $kg_{Cl}/m^3_{concrete}$;
        ε: fraction of pore solution in concrete, or capillary porosity under the saturated condition, $m^3_{solution}/m^3_{concrete}$;
        c: free chloride concentration, $kg_{Cl}/m^3_{solution}$;
        c_b: bound chloride concentration, $kg_{Cl}/m^3_{solution}$;
        J_m: chloride flux through the material, $\dfrac{kg_{Cl}}{m^2_{concrete} \cdot s}$;

        D_{SSM}: diffusion coefficient from the steady state migration test, $\dfrac{m^3_{solution} \cdot m_x}{m^2_{concrete} \cdot s}$,

which will be discussed later in the next section.

Assuming that the binding equilibrium occurs instantaneously when chlorides penetrate into the concrete, Eq. (5) becomes

$$\varepsilon\left(\frac{\partial c}{\partial t} + \frac{\partial c}{\partial t} \cdot \frac{\partial c_b}{\partial c}\right) = D_{SSM}\left(\frac{\partial^2 c}{\partial x^2} - \frac{zFU}{RTL} \cdot \frac{\partial c}{\partial x}\right) \tag{6}$$

or

$$\frac{\partial c}{\partial t} = \frac{D_{SSM}}{\varepsilon\left(1 + \frac{\partial c_b}{\partial c}\right)} \cdot \left(\frac{\partial^2 c}{\partial x^2} - \frac{zFU}{RTL} \cdot \frac{\partial c}{\partial x}\right) \tag{6'}$$

where $\frac{\partial c_b}{\partial c}$ is called chloride binding capacity [6], which is a function of free chloride concentration. The relationship between diffusion coefficients D_{nSS} and D_{SSM} could easily be seen from a comparasion of Eq. (6') with Eq. (2), and it is in accordance with that Nilsson [6] derived a few years ago. It is difficult, however, to get an analytical solution to the above differential equation. If the diffusion term $\frac{\partial^2 c}{\partial x^2}$ could be negligible when compared with the migration term, Eq. (6') turns into

$$\frac{\partial c}{\partial t} = \frac{D_{SSM}}{\varepsilon\left(1 + \frac{\partial c_b}{\partial c}\right)} \cdot \frac{zFU}{RTL} \cdot \frac{\partial c}{\partial x} \tag{7}$$

In this case the term $\frac{\partial c_b}{\partial c}$ in Eq. (7) could then be solved [5], and will be discussed later in the next chapter.

On the other hand, if chloride binding does not immediately reach the equilibrium, the binding rate should be considered. Some researchers [7, 8] suggest to use the first order chemical reaction for describing chloride binding rate, that is,

$$\frac{\partial c_b}{\partial t} = kc \tag{8}$$

where k is the binding rate. Eq. (5) then becomes

$$\left(\frac{\partial c}{\partial t} + kc\right) = \frac{D_{SSM}}{\varepsilon}\left(\frac{\partial^2 c}{\partial x^2} - \frac{zFU}{RTL} \cdot \frac{\partial c}{\partial x}\right) \tag{9}$$

As Xu and Chandra [7] referred, an analytical solution to Eq. (9) is

$$c = \frac{1}{2}c_s e^{\frac{zFU}{2RTL}x}\left[e^{-\beta x} \operatorname{erfc}\left(\frac{x - 2\beta D't}{2\sqrt{D't}}\right) + e^{\beta x} \operatorname{erfc}\left(\frac{x + 2\beta D't}{2\sqrt{D't}}\right)\right] \tag{10}$$

where c_s: free chloride concentration at the surface on concrete, kg_{Cl}/m^3 solution;
erfc: error function complement, erfc = 1- erf;

$$\beta = \sqrt{\left(\frac{zFU}{2RTL}\right)^2 + \frac{k}{D'}} \qquad \text{and} \qquad D' = \frac{D_{SSM}}{\varepsilon}.$$

Some example profiles with different values of k are shown in Fig. 1. It is obvious that there is no remarkable change in the penetration depth with different chloride binding rates k, although the shape of chloride profiles has changed a lot. Therefore, the diffusivity determined by measuring chloride penetration depth could be considered as a diffusion coefficient without the influence of chloride binding rate.

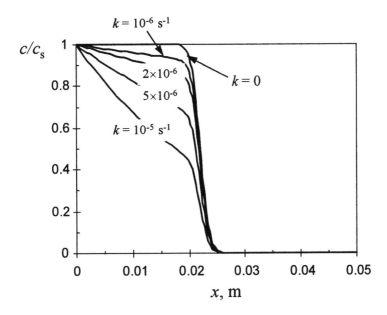

Fig. 1. Example chloride profiles with different values of chloride binding rate k.

2.3 Steady state migration test

The largest difference between the steady state and the non-steady state migration tests is that in the former test the chloride ions must be driven through the specimen and the flow rate of chloride ions must be measured until a steady state flow is reached. A number of experimental arrangements [4, 9-11] can serve for this test.

When a steady state flow of chloride ions is reached, the chloride diffusivity can then be calculated by using the equation as follows:

$$D_{SSM} = \frac{RTL}{zFU} \cdot \frac{\left(e^{\frac{zFU}{RT}} - 1 \right)}{\left(c_0 e^{\frac{zFU}{RT}} - c_1 \right)} \cdot J_m \tag{11}$$

where c_0, c_1: chloride concentrations in the cell supplying chloride ions (upstream cell) and in the cell collecting them (downstream cell), respectively.

It is obvious that the above equation can be reduced as

$$D_{SSM} = \frac{RTL}{zFU} \cdot \frac{J_m}{c_0} \tag{12}$$

if U is large enough to make $e^{\frac{zFU}{RT}} \gg 1$, and $c_0 \geq c_1$.

The chloride flux J_m can be calculated by using the following equation:

$$J_m = \frac{1}{A} \cdot \frac{\Delta q}{\Delta t} \tag{13}$$

where A: cross sectional area of concrete disc, $m^2_{concrete}$;
$\quad\quad\quad \Delta q$:   quantity of chloride ions flowed through the disc, kg_{Cl};
$\quad\quad\quad \Delta t$:   time interval, s;

Inserting Eq. (13) into (12) we obtain:

$$D_{SSM} = \frac{RTL}{zFU} \cdot \frac{1}{c_0 A} \cdot \frac{\Delta q}{\Delta t} \qquad \frac{m^3_{solution} \cdot m_x}{m^2_{concrete} \cdot s} \tag{14}$$

3 Relationships between diffusion coefficients from different tests

3.1 Relationship between diffusion coefficients D_{CTH} and D_{SSM}

A theoretical relationship between diffusion coefficients D_{CTH} and D_{SSM} could be found by comparing Eq. (6') with Eq. (2) and solving the term $\frac{\partial c_b}{\partial c}$ in Eq. (7) [5]:

$$D_0 = \frac{D_{SSM}}{\varepsilon} = D_{CTH} \cdot \left(1 + K_b \cdot \frac{W_{gel}}{\varepsilon}\right) \tag{15}$$

where D_0:   intrinsic diffusion coefficient [5], m^2_x/s;
$\quad\quad\quad K_b$:   binding constant involved in the non-steady state migration test, $K_b = 0.59 \times 10^{-3}$ $m^3_{solution}/kg_{gel}$;
$\quad\quad\quad W_{gel}$:   hydrate gel content in concrete, $kg_{gel}/m^3_{concrete}$.

Some of the measured data and the calculated results are listed in Table 1.

Table 1   Some of the measured data and the calculated results [5].

Concrete w/c	0.35	0.40	0.50	0.75
Steady state D_{SSM}, $\times 10^{-12}$ $\frac{m^3_{solution} \cdot m_x}{m^2_{concrete} \cdot s}$	1.1	1.6	2.1	3.2
Non-steady state D_{CTH}, $\times 10^{-12}$ m^2_x/s	2.8	5.4	13.6	39 7
Porosity ε	0.098	0.105	0.129	0.140
Gel content W_{gel}, kg/m^3	394	420	439	312
Binding constant K_b, $m^3_{solution}/kg_{gel}$		0.59$\times 10^{-3}$		
$D_0 = D_{SSM}/\varepsilon$, $\times 10^{-12}$ m^2_x/s	11.4	15.2	16.3	23.0
$D_0 = D_{CTH} \cdot (1 + K_b \cdot W_{gel}/\varepsilon)$, $\times 10^{-12}$ m^2_x/s	9.4	18.1	40.8	91.8

It can be seen that for the concrete with a low w/c the intrinsic diffusion coefficients obtained from both the steady state and the non-steady state tests are rather comparable, but for the concrete with a high w/c the coefficients from the steady state test are remarkably lower than those from the non-steady state test. This large difference may be attributed to many factors, e.g. 1) blocking effect of bound chlorides on pore structure, which may decrease the chloride flow, resulting in an underestimation of D_{SSM}, or overestimation of D_{CTH}; 2) effect of alkali on c_d in Eq. (4), the critical concentration for color change in the colorimetric method [12], which may result in an overestimated value of D_{CTH} for high w/c concrete due to a lower alkali content, causing a lower value of c_d, in such a concrete; 3) effect of interface potential or polarization on chloride flow, which may have more influence on steady state test due to a longer testing duration.

3.2 Relationship between diffusion coefficients D_{CTH} and D_{APM}

When there is no external electrical field, i.e. potential $U = 0$, Eq. (6') is valid for a non-steady state immersion test. Theoretically speaking, therefore, the relationship between D_{CTH} and D_{APM} should be that

$$D_{CTH} \cdot \left(1 + K_b \cdot \frac{W_{gel}}{\varepsilon}\right) = D_{APM} \cdot \left(1 + \frac{\partial c_b}{\partial c}\right) \tag{16}$$

It is difficult, however, to solve the binding capacity $\dfrac{\partial c_b}{\partial c}$ involved in the immersion test, because it changes tremendously with the free chloride concentration [6], and the determination of D_{APM} is strongly dependent on the whole profile of chloride distributions. Some experimental results have shown a fairly good correspondence between the both tests [13]. Probably under the condition of a very high chloride concentration, the average effect of chloride binding involved in the APM method, by a coincidence, becomes similar to that involved in the CTH method. On the other hand, this correspondence may also be explained by using our numerical model [14,15] as will be mentioned later.

4 Potential application

Since chloride binding usually follows a non-linear binding isotherm [16], the diffusion coefficient determined by the concentrated immersion test could not analogously be used in other immersion conditions. Some empirical parameters must be introduced to modify Fick's second law [17] before D_{APM} could be used for predicting chloride penetration into concrete structures.

Migration tests, specially non-steady state migration test, provide a unique tool to determine the diffusion coefficient, and the influence of chloride binding involved in the tests could be excluded by using Eq. (15). This coefficient can, therefore, be used for predicting chloride penetration by taking into account chloride binding capacity, mix design, hydration, porosity, etc., as well as the exposure conditions. A numerical model as well as a Windows program [14,15] has been developed for this prediction. Preliminary test results show that the predicted chloride profiles correspond fairly well with the measured data, in spite of different exposure conditions and duration. Some examples of prediction under laboratory and field exposure conditions are shown in Figs. 2 to 4.

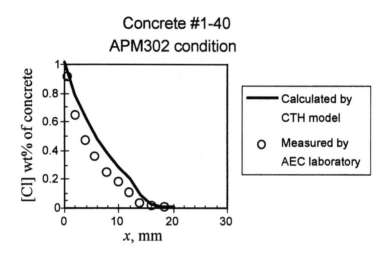

Fig. 2.   Example of the exposure in a concentrated NaCl solution in the laboratory. The concrete with w/c 0.4 and D_{CTH} 7.2 $\times 10^{-12}$ m^2_x/s was immersed in a solution of 165 g NaCl per litre for a period of 40 days.

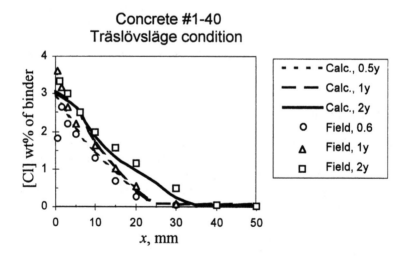

Fig. 3.   Example of the field exposure at the Träslövsläge harbour at the west coast of Sweden. The concrete with a quality similar to that in Fig. 2 was immersed in seawater with an average concentration of 14 g Cl per litre for different periods.

Ölandsbron
after 4 years exposure

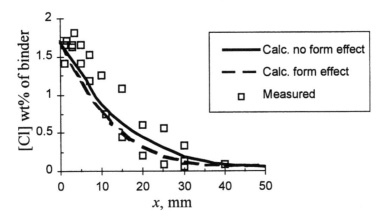

Fig. 4. Example of the exposure in the Baltic Sea. The concrete from Öland Bridge
with a quality similar to that in Fig. 2 and 3 was immersed in seawater with an
average concentration of 4.1 g Cl per litre for a period of 4 years.

5 Concluding remarks

Among the three accelerated tests mentioned above for determining chloride diffusivity,
the non-steady state migration test seems theoretically clearest and experimentally
simplest. The diffusion coefficient from this migration test have been applied for
predicting chloride penetration in concrete structures by using a numerical model taking
into account chloride binding, mix design and certain properties of concrete, as well as
the exposure conditions. The predicted results correspond fairly well to the measured
data.

6 References

1. AEC (1991) *Chloride ingress into concrete, Test method APM 302*, AEC
    Laboratory, AEC Consulting Engineers Ltd., Vedbæk, Denmark.
2. Tang, L. and Nilsson, L-O. (1992) Rapid determination of chloride diffusivity of
    concrete by applying an electric field, *ACI Materials Journal*, Vol. 49, No. 1,
    Jan./Feb., pp. 49-53.
3. Collepardi, M., Marcialis, A. and Turriziani, R. (1970) The kinetics of penetration
    of chloride ions into the concrete, *Il Cemento*, No. 4, Oct., pp.157-164.
4. Tang, L. (1995) Electrically accelerated methods for determining chloride
    diffusivity in concrete, accepted for publication in *Magazine of Concrete
    Research*.
5. Tang, L. (1995) On chloride diffusion coefficients obtained by using the
    electrically accelerated methods, to be published in the proceedings of the RILEM

International Workshop on Chloride Penetration into Concrete, Oct. 15-18, 1995, St. Rémy-lès-Chevreuse.

6.   Nilsson, L.-O. (1992) *A theoretical study on the effect of non-linear chloride binding on chloride diffusion measurements in concrete*, Division of Building Materials, Chalmers University of Technology, Publication P-92:13, 1992.

7.   Xu, A. and Chandra, S. (1994) A discussion of the paper 'Calculation of chloride diffuion coefficients in concrete from ionic migration measurements' by C. Andrade, *Cement and Concrete Research*, Vol. 24, No. 2, pp. 375-379.

8.   Gerard, B. (1995) *Ion transport mechanisms in cement-based materials—Experimental procedures*, unpublished report.

9.   Zhang, M.-H. (1989) *Microstructure and properties of high strength lightweight concrete*, Doctoral dissertation, NTH 1989:51, Division of structural engineering, The Norwegian Institute of Technology, University of Trondheim.

10.   Dhir R.K. et al. (1990) Rapid estimation of chloride diffusion coefficient in concrete, *Magazine of Concrete Research*, Vol. 42, No. 152, pp. 177-185.

11.   Andrade, C. (1993) Calculation of chloride diffuion coefficients in concrete from ionic migration measurements, *Cement and Concrete Research*, Vol. 23, No. 3, pp. 724-742.

12.   Otsuki, N. et al. (1992) Evaluation of $AgNO_3$ solution spray method for measurement of chloride penetration into hardened cemetitious matrix materials, *ACI Materials Journal*, Vol. 89, No. 6, pp. 587-592.

13.   Tang, L. (1993) *Methods for Determining Chloride Diffusivity in Concrete*, Licentiate thesis, Dept. of Building Materials, Chalmers University of Technology, Publication P-93:8.

14.   Tang, L. and Nilsson, L-O. (1995) A numerical method for prediction of chloride penetration into concrete structures, in *The Modelling of Microstructure and its Potential for Studying Transport Properties and Durability*, (ed. H. Jennings et al.), Kluwer Academic Publisher, Dordrecht, pp.537-550.

15.   Tang, L. (1995) A Windows program for the prediction of chloride penetration into submerged concrete, to be published in the proceedings of the RILEM International Workshop on Chloride Penetration into Concrete, Oct. 15-18, 1995, St. Rémy-lès-Chevreuse.

16.   Tang, L. and Nilsson, L.-O., (1995) Chloride binding isotherms—An approach by applying the modified BET equation, to be published in the proceedings of the RILEM International Workshop on Chloride Penetration into Concrete, Oct. 15-18, 1995, St. Rémy-lès-Chevreuse.

17.   Poulsen, E., (1995) Description of exposure to chloride of marine concrete structures, to be published in the proceedings of the RILEM International Workshop on Chloride Penetration into Concrete, Oct. 15-18, 1995, St. Rémy-lès-Chevreuse.

46 MATHEMATICAL MODELLING OF CHLORIDE DIFFUSION IN CONCRETE

Y. XI
Department of Civil and Architectural Engineering, Drexel University, PA, USA

Abstract
Two material parameters in equation of chloride diffusion are modeled, one is the chloride binding capacity and the other is the chloride diffusivity. The basic trends of the chloride binding capacity and the diffusivity are predicted by the present model successfully. With increasing w/c, both the diffusivity and the increment of free chloride increases. With the same w/c, longer curing time leads to more bound chloride and lower diffusivity. Since the diffusivity of aggregates are usually lower than that of cement paste matrix in concrete, with increasing volume fraction of aggregate, the effective diffusivity of concrete decreases. The present mathematical models are used in the nonlinear chloride diffusion equation which predicts profiles of the chloride penetration very well.
Keywords: concrete, transport, diffusion, chloride, microstructure.

1. Introduction

When the chloride concentration in the concrete reaches certain level, the protective environment in the concrete is destroyed, and the corrosion process of embedded steel starts. Therefore, chloride penetration into concrete is a critical issue in the process of steel bar corrosion. The problem is particularly acute in marine environment, bridges and roadways subjected to deicing salts, and parking garages into which salt is tracted from salted roadways.

When chloride ions enter concrete, a part of them will be bounded on the internal surface of cement paste and aggregate, which are called bound chloride; while the others, called free chloride, diffuse around in the concrete. Researches have shown that steel corrosion is related only to the free chloride content but not total chloride content. So, the governing equation should be expressed in terms of the free chloride content. The movement of free chloride ions in pores of the concrete is driven by three different mechanisms: diffusion, convection and ionic migration. This study concentrates on the first mechanism. The governing equation representing the diffusion process driven by concentration gradient of chloride ions can be written

$$\frac{dC_f}{dt} = \frac{dC_f}{dC_t} div\left[D_{Cl} \; grad\left(C_f\right)\right] \tag{1}$$

in which C_t = the total chloride concentration in gram of total chloride per gram of concrete, g/g; C_f = the free chloride concentration in gram of free chloride per gram of concrete, g/g; t = time in days; D_{Cl} = the chloride diffusivity; dC_f/dC_t = binding capacity.

Mechanisms of Chemical Degradation of Cement-based Systems. Edited by K.L. Scrivener and J.F. Young. Published in 1997 by E & FN Spon, 2–6 Boundary Row, London SE1 8HN. ISBN: 0419215700.

Weight ratio of chloride to concrete for Cf and Ct is used because it is the commonly used unit system for evaluation of chloride content in concrete industry.

The two material parameters, binding capacity and chloride diffusivity, must be determined first in order to solve the equation analytically or numerically. Eq. 1 is formulated based on the Fick's first law and mass conservation. It is valid in different scale levels as long as the two material parameters have correct physical meanings. In microscopic level, the equation holds at a point in space, the parameters in the equation must represent properties of a single phase when the point is located in the phase. In macroscopic level, the equation holds for representative volume of the composite, the two parameters must be effective properties of the composite. In the present study, we treat concrete as a composite material, and thus the two material parameters must be effective properties of the concrete.

Since chloride diffusion is a very complicated process, many material properties other than binding capacity and diffusivity are involved. To avoid curve fitting and lumping up all physical and chemical mechanisms into one over simplified equation, we will develop proper material model for each involved material property. These models are calibrated based on corresponding test data of their own. In this manner, each material parameter will have clear physical meaning.

2. Binding Capacity dC_f/dC_t

Binding capacity is defined as the relationship between the change of total chloride concentration and a change of free chloride concentration. The total chloride concentration is the sum of the bound and the free chloride, and therefore, the binding capacity can be expressed in terms of dC_b/dC_f since $C_t = C_f + C_b$ and $dC_f/dC_t = 1/(1 + dC_b/dC_f)$ in which C_b = bound chloride concentration in gram of bound chloride per gram of concrete. C_b is an effective parameter for concrete, which includes the bound chloride in cement paste matrix and in aggregates. Now, what we need to know is dC_b/dC_f, which is called chloride isotherm and has been obtained experimentally.

Recently, Tang and Nilsson [1] established the relationship between free chloride and bound chloride which was based on so-called Freundlich isotherm

$$Log(C_b') = A \; Log(C_f') + B \quad \text{or} \quad C_b' = (C_f')^A 10^B \tag{2}$$

in which A and B are two material constants related to chloride adsorption; C_b' and C_f' are bound and free chloride concentration, respectively, but in a unit system differing from C_b and C_f. C_b' is in milligram of bound chloride per gram of calcium silicate hydrate (C-S-H) gel, mg/g; C_f' is in mole of free chloride per liter of pore solution, mol/l. The advantage of using such a unit system is that the resulting Freundlich isotherm is independent of water to cement ratio and aggregate content of concrete because it is based on amounts of C-S-H gel and pore solution. However, for solving engineering problems, it is more convenient to use the unit system defined for C_b and C_f.

To this purpose, we need to convert Eq. 2 into the unit system adopted in Eq. 1, and then determine the binding capacity. Considering C_f' first, by introducing a conversion factor β_{sol}, we have the relationship between Cf and C_f': $C_f = 35.45 \beta_{sol} C_f'$, where β_{sol} = the ratio of pore solution to concrete in liter of the pore solution per gram of the concrete, l/g. By the same manner, introducing a conversion factor β_{C-S-H}, we have the relationship between C_b and C_b': $C_b = \beta_{C-S-H} C_b'/1000$, where β_{C-S-H} = the ratio of C-

S-H gel to concrete in gram of C-S-H gel per gram of concrete, g/g. Substituting these two equations into Eq. 2 gives

$$C_b = \frac{\beta_{C-S-H}}{1000} \left(\frac{C_f}{35.45 \beta_{sol}} \right)^A 10^B \tag{3}$$

then the binding capacity can be evaluated

$$\frac{dC_f}{dC_t} = \frac{1}{1 + \frac{A 10^B \beta_{C-S-H}}{35450 \beta_{sol}} \left(\frac{C_f}{35.45 \beta_{sol}} \right)^{A-1}} \tag{4}$$

From the definition of β_{sol}, one can write

$$\beta_{sol} = \frac{V_{sol}}{W_{conc}} = \frac{W_{sol}}{\rho_{sol} W_{conc}} = \frac{n(H,T)}{\rho_{sol}} \tag{5}$$

in which ρ_{sol} = the density of the solution in gram/liter, g/l; V_{sol} and W_{sol} = the volume and weight of the pore solution, respectively; W_{conc} is the weight of the concrete; $n(H,T) = W_{sol}/W_{conc}$, which is the ratio of the weight of pore solution to the weight of concrete. n(H,T) is a function of temperature and pressure. In the present study, we focus on saturated concrete only, and both pressure and temperature are kept constants.

From Eq. 5, one can see that β_{sol} depends on density of the solution and n(H,T). ρ_{sol} depends on chloride concentration which is the unknown we are looking for. Actually, in concrete, density of pore solution varies in a very limited range, and thus, to simplify the problem ρ_{sol} can be considered to be the density of pore water. On the other hand, n(H, T) depends on H, T, and pore structures, and in turn, the pore structure depends on H, T, as well as age of the concrete. To simplify the problem, n(H, T) can be considered to be the water adsorption isotherm instead of adsorption of chloride solution.

The concrete can be treated as a two phase material with aggregate as inclusion and cement paste as matrix. The adsorption isotherm for the two constituents may be studied separately and then combined to obtain the effective moisture adsorption isotherm of concrete. For cement paste, a model of adsorption isotherm has been developed based on BET theory, in which the three BET parameters were expressed in terms of water to cement ratio, type of cement, curing period and temperature [2,3]. The BET theory can also be used for aggregates, and the model for the three BET parameters for aggregate has also been developed, which is much more simpler because there is no aging effect for aggregate [4]. The effective adsorption isotherm for concrete may be obtained $n(H,T) = f_{cp} n_{cp}(H,T) + f_{agg} n_{agg}(H,T)$, where f_{cp} and f_{agg} are weight percentages of cement paste and aggregate, respectively; $n_{cp}(H,T)$ and $n_{agg}(H,T)$ are adsorption isotherm of cement paste and aggregate, respectively.

To determine β_{C-S-H}, a recently developed microstructural model for process of cement hydration is used in the present study [5]. The model is based on cement chemistry and analyses of micrographes of scanning electron microscope. In the model, there are five different phases in cement paste: anhydrous, inner product, outer product, capillary pores, calcium hydroxide (CH) and other AFm phases. The formulation of the model has been rewritten and rearranged to fit the need in this study and the details can be seen in [4]. When the volume fraction of C-S-H is calculated, it needs to be converted to the weight ratio of C-S-H. The typical value of specific gravity of concrete is 2.3 and the typical value of specific gravity of C-S-H gel is about 2.34, so β_{C-S-H} can be considered to be equal to f_{C-S-H} without significant loss of precision.

From above analyses on β_{C-S-H} and β_{sol}, one can see that β_{C-S-H} determines the effect of cement composition and age on volume fraction of C-S-H, while β_{sol} represents the effect of the structure of hydration products because for the same β_{C-S-H} the difference in internal structure of C-S-H makes β_{sol} different.

3. Chloride Diffusivity

The effective diffusivity of concrete can be obtained on the basis of composite theories. Christensen [6] developed a composite model for diffusivity based on the three phase model and Fick's first law

$$D_{\it eff} = D_m\left\{1 + g_i/\left[(1-g_i)/3 + D_m/(D_i - D_m)\right]\right\} \tag{6}$$

where D_i and D_m are diffusivities of inclusion (aggregate) and matrix (cement paste), respectively; $D_{\it eff}$ = effective diffusivity which depends on the configuration of the two constituent phases; g_i is the volume fraction of inclusions.

A modified Kozeny-Carmen equation [7] is used in this study for evaluation of D_m and D_i in which only the relevant porosity and the corresponding surface area are included. This means that the porosity and specific surface area in the modified Kozeny-Carmen equation are not those measured by water adsorption or mercury intrusion tests, but a part of them. Martys [7] suggested to introduce the critical porosity, V_p^c, into Kozeny-Carmen equation. V_p^c is the porosity at which the pore space first being percolated. The physical meaning of V_p^c is very much similar to the critical pore diameter, d_c, defined in Katz-Thompson theory [8,9]. d_c has been linked to the inflection point in the cumulative intrusion curve obtained from mercury intrusion test. Martys proposed [7]

$$D = 2\left(V_p - V_p^c\right)^f\left[1 - \left(V_p - V_p^c\right)\right]/S^2 \tag{7}$$

in which D = D_m or D_i in Eq. 6; V_p = porosity; S = specific surface area (surface area/bulk volume); $f \approx 4.2$. The important parameter in Eq. 7 is V_p^c. Numerical simulation on porous media composed of randomly overlapping spheres showed that the pore space becomes disconnected at V_p^c = 3% [7]. Connectivity in the microstructure of cement paste are much more complicated than that of randomly overlapping spheres, and thus, V_p^c for cement paste should be higher than 3%, but there have been no test data available in this aspect.

D_m can be evaluated from Eq. 7 in which the surface area of cement paste S can be estimated by the monolayer capacity V_m in BET theory since V_m is proportional to S; porosity V_p for cement paste can be estimated by n(H,T) at saturation (H = 1). It should be noted that n(H,T) is in weight ratio and it needs to be converted into volume fraction. The value of D_i can be estimated by Eq. 7 in a similar way or taken simply as a constant, typically 1×10^{-12} sec/ cm, which is very small compared to the diffusivity of cement paste matrix.

Chloride diffusivity is concentration dependent, which makes the governing equations, Eq. 1 nonlinear. Kozeny-Carman equation is valid basically for an undissociated diffusant, and thus the major resistant to the diffusion process is due to the wall effect, i.e. interaction of the diffusant with the pore wall. However, for diffusion of ions, especially diffusion of chloride ions in the present study, the movement of ions are restricted by a strong electrostatic field induced by the other ions presented in the solution. This restriction may be characterized by the following equation [10]

$$D_{ion} = D'\left[1 - k_{ion}\left(C_f\right)^m\right] \tag{8}$$

in which $D' = RT\Lambda_0/F^2|Z_{ion}|$; R = universal gas constant; T = temperature; Λ_0 = reference conductance; $k_{ion} = k_c/\Lambda_0$; Z_{ion} = the valency of the ion; F = the Faraday constant; and k_c and m are two positive constants.

4. Influential Parameters on Chloride Binding Capacity and Diffusivity

The present model can be used to predict the effect of various influential parameters on chloride binding capacity. Fig. 1 shows the effect of water-cement ratio on dC_f/dC_t calculated at saturation state with various curing periods. One can see that with increasing w/c, dC_f/dC_t increases. This is because with higher w/c (at fixed curing time and same type of cement), the volume fraction of capillary pore is larger, and consequently, the increment of free chloride in saturated pore solution is larger upon a change in total chloride. With the same water-cement ratio, longer curing time corresponds to higher surface area and tortuosity of the microstructure, which means more bound chloride and thus lower dC_f/dC_t.

Fig. 2 shows the effect of water-cement ratio on chloride diffusivity at various curing periods predicted by the present model. The curves are calculated at 5% of NaCl concentration in the pore solution. One can see that with increasing w/c, the diffusivity increases. This is because higher w/c (at fixed curing time and same type of cement) leads to higher volume fraction of capillary pore, and consequently higher value of the diffusivity. While with the same water-cement ratio, longer curing time leads to lower diffusivity.

Fig. 3 shows the effect of aggregate content on the diffusivity. The lowest aggregate content used in the model is taken to be 50%. The water-cement ratio in Fig. 3 is 0.55, and curing time is 7 days. The assumption used in this example is that the aggregate has much lower diffusivity than that of cement paste matrix. Therefore, with increasing volume fraction of aggregate, the ratio of the effective diffusivity of concrete to the diffusivity of cement paste decreases. For a given cement paste, the magnitude of the reduction in the effective diffusivity depends on the diffusivity of the aggregate used.

5. Numerical Solution

Once the binding capacity and chloride diffusivity are determined, Eq. 1 can be solved by a proper numerical method, such as finite element method or finite difference method. In the present study, Eq. 1 is solved for a one dimensional case by a finite difference method, the Crank-Nicolson finite difference algorithm [11]. Diffusion equation under polar coordinate system is also solved by the finite difference method and can be used to predict distribution of chloride concentration in concrete cylinders.

In the above formulation, there are not many free parameters which can be manipulated to calibrate with the chloride diffusion test data. As one can see, Λ_0, m and

k_c in Eq. 8, and V_p^c in Eq. 7 are not known for sure. In the present study, V_p^c is taken to be a constant 0.03 based on results from computer simulation as described earlier. Λ_0 suppose to be a constant, but because of many other cations exist in the pore solution, it cannot be simply estimated based on pure chloride solution. The other two parameters, m and k_c, are apparently the functions of water to cement ratio and age of cement paste. This is because the concentration dependence of molar conductance is influenced by pore size and geometry, i.e. the microstructural features, especially when the size of the pore is small.

The parameters in the expressions for chloride diffusivity that must be evaluated by calibration with chloride profiles are k_{ion}, m, and D', which are combined parameters. Computational experiences show that the profile is not sensitive to the value of m around m = 0.5. The profile of free chloride concentration in a concrete specimen predicted by the present model is shown in Fig. 4. The water to cement ratio of the concrete specimen is 0.55 and type of cement used is Type I. After three days of curing, the specimen is immersed in a C_aCl_2 solution with chloride concentration 11.8 g/l .

6. Conclusions

There are two material parameters in the governing equation for chloride diffusion. One is the chloride binding capacity and the other is the chloride diffusivity. The chloride binding capacity is modeled based on chloride adsorption isotherm. Two parameters, β_{sol} and β_{C-S-H}, are considered important for evaluation of the binding capacity. β_{C-S-H} is the ratio of C-S-H gel to concrete in gram of C-S-H gel per gram of concrete, it determines the effect of cement composition and age on volume fraction of C-S-H. It is characterized by a microstructural model based on chemical reaction of the hydration reaction. β_{sol} is the ratio of pore solution to concrete in liter of the pore solution per gram of the concrete, the amount of pore solution depends C-S-H gel structure, so, β_{sol} represents the effect of the structure of hydration products. It is estimated based on a model for moisture capacity of cement paste, and calibrated based on moisture adsorption test results.

Chloride diffusivity is modeled by composite theory for the effect of aggregates. The three phase model developed by Christensen [6] has been used in the present study. The diffusivity for cement paste is characterized by adopting a modified Kozeny-Carmen model suggested by Martys et al. [7], in which the effects of surface area and effective porosity of cement paste are included. The influence of chloride ion concentration has also been handled in the model.

The basic trends of chloride binding capacity and diffusivity are predicted by the present model successfully. With increasing w/c, both the diffusivity and the increment of free chloride increases. With the same w/c, longer curing time corresponds to higher degree of maturity of C-S-H gel, which means more bound chloride and lower diffusivity. Since the diffusivities of aggregates are usually lower than that of cement paste, with increasing volume fraction of aggregate, the ratio of the effective diffusivity of concrete to the diffusivity of cement paste decreases. The present models for chloride binding capacity and diffusivity are used in the chloride diffusion equation, which is solved by a finite difference method.

7. References

1. Tang, L., and Nilsson, L.O. (1993) "Chloride Binding Capacity and Binding Isotherms of OPC Pastes and Mortars", *Cement and Concrete Research*, Vol. 23, pp. 247-253.

2. Xi, Y., Bazant, Z.P., and Jennings, H.M. (1994) "Moisture Diffusion in Cementitious Materials: Adsorption Isotherm", *J. of Advanced Cement-Based Materials*, Vol. 1, pp. 248-257.

3. Xi, Y. (1995) "A Model for Moisture Capacities of Composite Materials - Formulation", *Computational Materials Science*, Vol. 4, pp. 65-77.

4. Xi, Y. (1995) "A Model for Moisture Capacities of Composite Materials - Application to Concrete", *Computational Materials Science*, Vol. 4, pp. 78-92.

5. Jennings, H.M., and Tennis, P.D. (1994) "A Model for the Developing Microstructure in Portland Cement Paste", *J. of Ame. Ser. Soc.*, Vol. 77, No. 2, pp. 3161- 3172.

6. Christensen, R.M. (1979) *Mechanics of Composite Materials,* Wiley-Interscience, New York.

7. Martys, N.S., Torquato, S., and Bentz, D.P. (1994) "Universal Scaling of Fluid Permeability for Sphere Packings", *Physical Review*, E, Vol. 50, No. 1, pp. 403-408.

8. Katz, A.J., and Thompson, A.H. (1986) "Quantitative Prediction of Permeability in Porous Rocks", *Phys. Rev.*, Vol. B34, pp. 8179-8181.

9. Katz, A.J., and Thompson, A.H. (1987) "Prediction of Rock Electrical Conductivity from Mercury Injection Measurements", *J. of Geophys. Res.*, Vol. 92, No. B1, pp. 599-607.

10. Chatterji, S. (1994) "Transportation of Ions through Cement Based Materials, Part I. Fundamental Equations and Basic Measurement Techniques", *Cement and Concrete Research*, Vol. 24, No. 5, pp. 907-912.

11. von Rosenberg, D.U. (1969) *Methods for the Numerical Solution of Partial Differential Equations,* American Elsevier Publishing Company, Inc., New York.

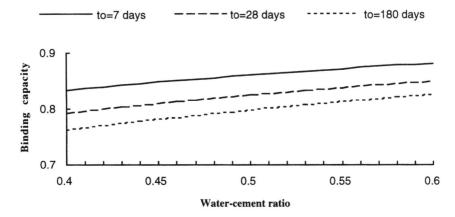

Fig. 1 Effect of water-cement ratio and curing time on binding capacity

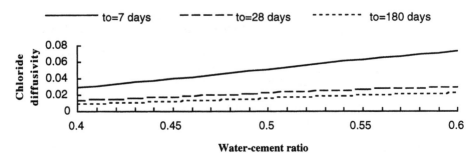

Fig. 2 Diffusivity at various water-cement ratio and curing   times

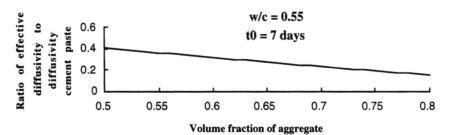

Fig. 3 Effect of aggregate content on chloride diffusivity

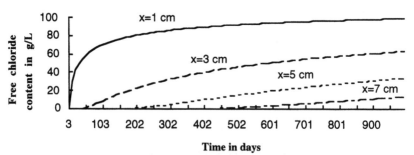

Fig. 4 Predicted profile of free chloride concentration

(x = depth from specimen surface)

47 DEGRADATION OF CONCRETE STRUCTURES DUE TO CHEMICAL ATTACK: MATHEMATICAL PREDICTION

J-P. BOURNAZEL and B. CAPRA
Laboratoire de Mécanique et Technologie, ENS Cachan, France

Abstract
Service-life prediction of a concrete structure appears as key issue in civil engineering. Realistic results are possible if material and structural considerations are taking into account in the same model. Thermodynamics of irreversible processes result in a mathematical model which involves physico-chemical reactions and site observations. Considering the macro, meso and micro levels, the structural effects, such as expansion, cracking, loss of strength, are at the macrolevel, while the chemical reactions include atomic bonds between cement hydrates and internal and external ions through the pore solution. So the model starts at the microlevel and progresses to the concrete structure.
Keywords : Durability, Modelling, Thermodynamic, cracking, permeability, AAR

1 Introduction

Mathematical service-life prediction of a concrete structure is now a key issue for researchers in material science. Nevertheless, it is a rather complex problem because a chemical reaction such as alkali-aggregate reaction is a coupled one. Chemistry, physics and mechanics strongly interfere.

Mehta [1] has proposed a holistic approach of a problem of concrete durability. This model clearly indicate that for a given phenomenon, experimentally studied, the total domain must be considered. In this way, thermodynamic of irreversible processes is a good approach [2].

In mechanical computations, this sort of model presents some defects. In fact, concrete is an heterogeneous material, the chemical reactions are local and the induced effects are on the structure itself. It is so necessary to introduce this

Mechanisms of Chemical Degradation of Cement-based Systems. Edited by K.L. Scrivener and J.F. Young.
Published in 1997 by E & FN Spon, 2–6 Boundary Row, London SE1 8HN. ISBN: 0419215700.

heterogeneous character, and, this multiscale aspect. Two problems of durability, cracking at early ages and alkali-aggregate reaction, permit to show the necessity of this incremental multiscale approach.

2 Theoritical approach

The starting postulate is that the thermodynamic state of a material, at a given time and a meso level, is completely defined by the knowledge of the state variable [3]. This postulate implies that phenomena can be described with an accuracy which depends on the choice of the nature and number of thermodynamic state variables.

Strain (ε), temperature (T) and humidity (H) are the state variables which can be observed. On the contrary, concentration of chemical species (A) and damage (D), which are internal variables are not directly measurable. Introduced in constitutive equations both internal and observable variables describe the actual state of the reaction and automatically adapt the behaviour of the material to this state.

From a general point of view, we define the free energy potential of the material. This potential depends on state variables and can be defined by (1) :

$$\rho\psi = \rho\psi(\varepsilon, T, H, A, D) \tag{1}$$

The first state law (2) gives the expression of the stress, σ, which is the associated variable of strain, ε.

$$\sigma = \frac{\partial \rho\psi}{\partial \varepsilon} \tag{2}$$

3. Cracking at early ages

3.1 Free energy potential and state laws

In this work high temperatures of curing at not considered, only hydration effects are taking into account (the higher temperature do not exceed 80 °C). In the case of mass concrete, an elementary volume can be considered as an isolated system in which shrinkage is only autogeneous and in which restrained strains induce stresses and risk of damage.

Through the concept of state variables, two major state variables are introduced into the model : a state variable called maturity (M) which described the global evolution of microstructure. A state variable called damage (D) which described the state of degradation of the material. The free energy potential is expressed as follows :

$$\psi = \frac{1}{2}K(M)(1-D)(\varepsilon - (\varepsilon^{th} + \varepsilon^{sh} + \varepsilon^{c}))^2 + \psi_M + \psi_T \tag{3}$$

ψ is the free energy, ψ_M is the part linked to maturation, ψ_T is the part of free energy linked to thermal effects, K is the tensor of elastic characteristics

Equation (4) expresses the hypothesis of strains partition :

$$\varepsilon = \varepsilon^{th} + \varepsilon^{sh} + \varepsilon^{c} + \varepsilon^{e} \tag{4}$$

where $\varepsilon^{th} + \varepsilon^{sh}$ are the volumic strains due to thermal variations and autogeneous shrinkage, ε^{c} is the maturation viscous strain and ε^{e} is the elastic strain

It is possible to deduce the expression of stress by the first state law (see equation 2):

$$\sigma = \frac{\partial \psi}{\partial \varepsilon} = K(M)(1 - D)(\varepsilon - (\varepsilon^{th} + \varepsilon^{sh} + \varepsilon^{c})) \tag{5}$$

3.2 Evolution laws

3.21 Maturity
Based on previous works ([4], [5]), the evolution law of maturity is choosen as follows :

$$M = \frac{e^{-\frac{U}{RT}} \langle t - to \rangle_{+}}{1 + e^{-\frac{U}{RT}} \langle t - to \rangle_{+}} \tag{6}$$

U represents the activation energy, T is temperature (K), t is time, to is the setting time.

Maturity function is directly linked to the Young's modulus evolution and can be expressed as :

$$E = E_{\infty} M \tag{7}$$

E_{∞} is the Young's modulus for M=1

3.22 Damage
In a unixial isotropic case, the damage variable D is related to the Young's modulus [6] by the folllowing equation :

$$E = Eo(1 - D) \tag{8}$$

with

$$D = \alpha_{c} D_{c} + \alpha_{t} D_{t} \tag{9}$$

D_i is either D_t in tension or D_c in compression

the evolution law is given as follows :

$$D_i = 1 - \frac{\varepsilon_0 (1 - A_i)}{\varepsilon} - \frac{A_i}{e^{B_i (\varepsilon - \varepsilon_o)}} \tag{10}$$

3.3 Autogeneous shrinkage and basic creep

These two phenomena can be linked to the evolution of the microstructure. The two following equations, based on experimental results, are proposed :

$$\dot{\varepsilon}^{sh} = \varepsilon^{sh}_\infty \, M \tag{11}$$

ε^{sh}_∞ is this autogeneous shrinkge for M=1

$$\varepsilon^c = ae^{-b(\frac{1-M}{M})^c} \left(\frac{M-M_\tau}{1+M-M_\tau} \right)^d \tag{12}$$

a, b, c and d are material coefficients and M_τ is the maturity at the age of loading

The heat of hydration can be linked to maturity function (see [7])

3.4 Application

This model has been applied to the case of a French arch dam. the results of computations (figure 1) show the significance of such an approach. The map cracking of the dam, due to hydration effects, is reproduced with a good precision.

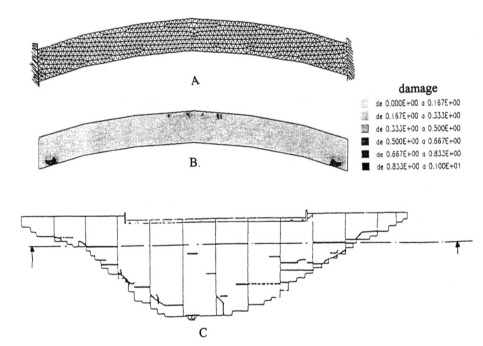

damage	
	de 0.000E+00 a 0.167E+00
	de 0.167E+00 a 0.333E+00
	de 0.333E+00 a 0.500E+00
	de 0.500E+00 a 0.667E+00
	de 0.667E+00 a 0.833E+00
	de 0.833E+00 a 0.100E+01

Figure 1 : Calculations of an arch dam : (a) mesh and boundary conditions, (b) map of damage, (c) real structure

It is possible to link this map of cracking to the distribution of permeability in the structure.

$$K = K_0 . e^{f(D)} \tag{13}$$

Based on experimental results, it is possible to express the form of the f function. For example, using the concrete of a dam, we have found :

$$k = 8.10^{-8} e^{(9,43D^{0.859})}$$

4 Alkali-aggregate reaction

In this work, only alkali silica reaction (ASR) is considered, the chemical processes are those proposed by Dent Glasser and Kataoka [8]. It is assumed that the porous medium is saturated [9] and therefore fluide transfers into concrete do not need to be considered.

The description of the state of the material is obtained by the choice of state variables associated with chemical reactions and mechanical induced effects. In the case of an isothermal process at temperature To, the following free energy potential is proposed :

$$\psi = \frac{1}{2}K(1-D)(\varepsilon - \varepsilon^{aar})^2 + \frac{1}{2}LA^2 \tag{14}$$

K is the fourth order tensor of elastic characteristics, L is a intrinsic characteristic of the material, A is the reaction rate of ASR, ε^{aar} is the ASR induced expansion

The state la permit to obtain :

$$\sigma = \frac{\partial \psi}{\partial \varepsilon} = K(1-D)(\varepsilon - \varepsilon^{aar}) \tag{15}$$

and

$$Y = -\frac{\partial \psi}{\partial A} = K(1-D)(\varepsilon - \varepsilon^{aar})\frac{\partial \varepsilon^{aar}}{\partial A} - LA \tag{16}$$

A complementary equation, permitting to consider the kinetics of reaction, is needed to solve this system. As the ASR is controlled by ions diffusion, the kinetic of the reaction is considered as one of the first order :

$$\dot{A} = k.e^{-\frac{U}{RT}}.(1-A) \tag{17}$$

where k is a parameter which can depend on the rate of damage, U is the activation energy, R is the perfect gas constant and T is the temperature (K).

If the induced swelling is considered as proportional to the reaction rate, the following relation is proposed :

$$\varepsilon^{aar} = f(A) = \varepsilon_0^{aar}(1-e^{-Kt}) \tag{18}$$

K represent the Arrhenius' law and ε_0^{aar} is the free expansion induced by AAR

This relation leads to a good agreement with experimental data. Figure 2 compare experimental results ([10]) to computational calculations.

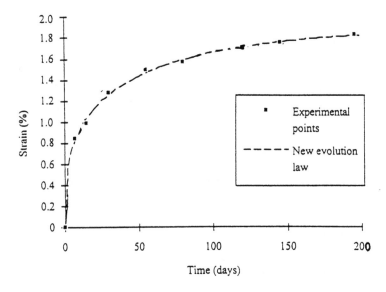

Figure 2 : Comparison between modeling and experimental data [9]

Nevertheless, this solution is not efficient in the case of a free expansion test, the heterogeneous character of concrete must be considered. Each finite element is assigned to its own evolution by a ramdom sampling of the parameter a of a simplified evolution law, proposed as follows

$$\varepsilon^{aar} = \frac{at}{1+bt} \tag{19}$$

This parameter is characterised by its mean value and its standard deviation which have been infered from expansion tests (figure 3).

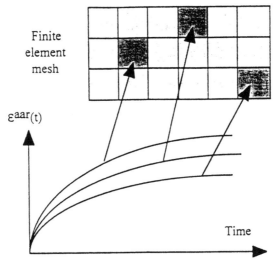

Figure 3 : Principle of simulations ([9])

The first results have shown some interesting achievements [11]. At one hand, such an approach is able to describe map cracking of concrete observed on attacked concrete structures, the damage starts at the edges of the concrete cub and besides, comes through the structure (figure 4).

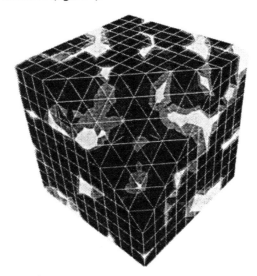

Figure 4 : Damage map of a concrete cube ([11])

If such an approach is able to describe some phenomena, there is however a major disadvantage : there should have enough simulations, with different random distributions, for a real statistical treatment of results. In reality, in the case of a site structure, this will be impossible because numerous simulations are costly. A work in progress [12], in the lab, will permit to have a more global approach which will be applied to real structures such dams.

progress [12], in the lab, will permit to have a more global approach which will be applied to real structures such dams.

5. Conclusions

Until now, chemical reactions such as ASR or cracking at early ages have brought about a lot of problems to civil engineers. It is difficult to take into account with a good agreement towards reality because of the complexity of the phenomena. In the approach proposed, each phenomenon is described by one or two state variables, and a large part of parameters introduced are physical. Nevertheless, new developments are necessary to make this approach totally operationnal.

6. References

1.   Mehta, P.K. (1994) "Concrete Technology at the Crossroads - Problems and Opportunities", Concrete Technology : Past, Present and Future, ACI, SP 144, ed P.K. Mehta, pp 1-30
2.   Lemaitre, J. and Chaboche, J.L. (1985) "Mécanique des Matériaux Solides", ed Dunod, Paris.
3.   Bournazel, J.P. and Moranville-Regourd, M. (1994) "Mathematical Modeling of Concrete Durability : The Use of Thermodynamics of Irreversible Processes" Concrete Technology : Past, Present and Future, ACI, SP 144, ed P.K. Mehta, pp 233-249
4.   Carino, N.J. (1982) "Maturity Functions for Concretes", RILEM int. Conf. on concrete at early Ages, Vol I, pp 123-128
5.   Regourd, M. and Gautier, E. (1980) "Comportement des Bétons Soumis au Durcissement Accéléré" Annales de l'ITBTP, n° 198, pp 83-96
6.   Mazars, J. "A Description of Micro and Macroscale Damage of Concrete Structures" Engineering Fracture Mechanics, Vol 25, n° 5/6, pp 729-737
7.   Bournazel, J.P. (1992) "Contribution à l'Etude du Caractère Thermomécanique de la Maturation du Béton" PhD thesis, LMT, 165 pages.
8.   Dent Glasser, L.S. and Kataoka, N. (1981) " The Chemistry of Alkali-Aggregate Reactions" Proc. of the 5th Int. Conf. on AAR, Cape Town, Paper S252/23
9.   Capra, B., Bournazel, J.P. and Bourdarot, E. (1995) "Modeling of Alkali-Aggregate Reactions Effects in Concrete Dams", Second Int. Conf. on AAR in hydroelectric plants and dams, USCOLD, pp 441-456
10.   Diamond, S., Barneyback, R.S. and Struble, L.J. (1981) "On the Physics and Chemistry of Alkali-Silica Reaction", Proceedings of 5th Int. Conf. on AAR, Cape Town, S252/22.
11.   Capra, B. and Bournazel, J.P. (1995) "Perspectives nouvelles pour la Prise en Compte des Alcali-Reactions dans le Calcul des Structures", Matérials and Structures, Vol 28, pp 71-73
12.   Capra, B. (1996) "Prise en Compte des Réactions Alcali-Granulat dans les Bétons de Barrage", thèse de l'ENS Cachan, PhD dissertation, to be published.

48 RECENT DEVELOPMENTS IN THE MEASUREMENT OF TRANSPORT PROPERTIES IN CEMENT-BASED MATERIALS

J.D. SHANE, J-H. HWANG, D. SOHN, T.O. MASON, H.M. JENNINGS
Northwestern University, Evanston, IL, USA
E.J. GARBOCZI
National Institute of Standards and Technology, Gaithersburg, MD, USA

Abstract

Fundamental to understanding and controlling the durability of cement-based systems is the relationship between microstructure and transport properties, particularly ionic diffusivity and hydraulic permeability. While our understanding of cement paste microstructure and interfacial zone contributions has improved considerably over the past few years, largely due to advances in nondestructive methods and modelling, there is a demonstrated need for improved laboratory and field techniques for determining diffusivity and permeability. This work provides an overview of existing and emerging techniques for measuring or predicting the transport properties of cement-based materials. Although a major focus will be impedance spectroscopy, alternative methods will also be considered, including solvent exchange kinetics, mercury intrusion porosimetry, nuclear magnetic resonance, microstructure-based modelling, and the more conventional permeameter and rapid chloride penetration techniques. The applicability of the Nernst-Einstein (diffusivity) and Katz-Thompson (permeability) equations to cement-based materials will be considered. Ramifications for rapid testing and field testing of concrete will also be discussed.

Keywords: Diffusivity, impedance spectroscopy, mercury intrusion porosimetry, modelling, nuclear magnetic resonance, permeability, solvent exchange, transport properties.

1 Introduction

Establishing the relationship between microstructure and transport properties is a critical step in predicting the durability of cement-based materials. Unfortunately, the conventional methods used to measure transport properties in cement-based materials, particularly ionic diffusivity and hydraulic permeability, are often time-consuming, labor-intensive, and inconsistent. Some typical measurements include ponding/profiling diffusivity, divided cell diffusivity, differential pressure permeability, and the rapid chloride penetrability test. Unfortunately, the usefulness of these methods is somewhat limited for reasons discussed subsequently.

Recently, several new techniques have emerged that measure or predict transport properties in cement-based materials. These novel techniques include impedance spectroscopy, solvent exchange kinetics, nuclear magnetic resonance, and microstructure-

Mechanisms of Chemical Degradation of Cement-based Systems. Edited by K.L. Scrivener and J.F. Young.
Published in 1997 by E & FN Spon, 2–6 Boundary Row, London SE1 8HN. ISBN: 0419215700.

based modelling. In addition to providing fundamental microstructure/transport-property relationships, the information can be used to predict important durability-related parameters using the Nernst-Einstein (diffusivity) and the Katz-Thompson (permeability) equations.

2 Conventional Techniques

2.1 Diffusion Measurements

2.1.1 Ponding/Profiling

A common way to measure the diffusion of a species into concrete is by direct exposure. For instance, the rate of ingress of chloride and sulfate ions can be determined from cores taken from structures exposed to aggressive environments. This exposure is simulated in the laboratory with ponding experiments. In these tests, specimens are immersed in a solution (e.g., 1 molar NaCl) for a specified length of time. They are then removed from the solution and the concentration of the diffusant is determined as a function of depth within the sample. The diffusion coefficient can be found by fitting the concentration profile with a modified version of Fick's second law.

When the total diffusant content is determined for profiles, the diffusion coefficient includes contributions from diffusion as well as reaction with the microstructure, which is known as binding. When binding is not accounted for, the coefficient calculated is known as the apparent diffusion coefficient (D_{app}).

The major limitation of ponding/profiling testing is the time required to make measurements, which, for many concretes, is on the order of months. For example, the AASHTO T-259 standard ponding test requires a 90 day exposure [1]. This restricts the use of ponding/profiling tests to the measurement of mature pastes, concretes, and mortars.

2.1.2 Divided Diffusion Cell

An alternative to the ponding test is known as the divided diffusion cell test. This method involves placing a disk-shaped specimen between two ionic solutions, one of which is a reservoir, the other a sink. This is shown schematically in Figure 1. Often chloride is chosen as the diffusing species because of its importance to the durability of steel-reinforced structures. Hydroxide solutions are normally used in both compartments to minimize leaching of $Ca(OH)_2$. Chloride concentration is monitored with time in both compartments, and is maintained at a constant level in the reservoir, while the increase in concentration in the sink is recorded. When steady state flow is reached, the rate of change of concentration in the downstream side is used to calculate a diffusion coefficient using Eq. (1). Because Eq. (1) requires steady state diffusion, the effect of binding is not a factor and an effective diffusion coefficient (D_{eff} in cm^2/s) is calculated.

$$D_{eff} = \frac{Jl}{\left(C_A - C_B\right)} \tag{1}$$

where J is the unidirectional flux of the species (mol/cm^{2}s), C_A and C_B are the concentration of the diffusant in cells A and B (mol/cm^3), respectively, and l is the sample thickness (cm).

The major drawback of the divided cell test is that can take weeks to achieve steady state flow. During the period of the test, the microstructure is continually changing from hydration and leaching of $Ca(OH)_2$. This makes it virtually impossible to test samples with a low degree of hydration.

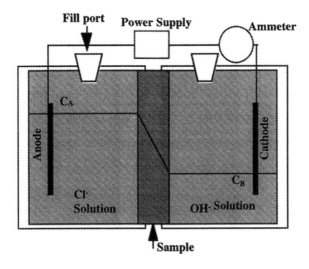

Fig. 1. Schematic of a diffusion cell used for the divided diffusion cell, the rapid chloride penetrability, and Norwegian tests. Note for the divided diffusion cell test, the electrodes, power supply, and ammeter are not present, and for the Norwegian test, the ammeter is not required.

2.1.3 Rapid Chloride Penetrability* Test

Because the time involved in long term diffusion measurements is undesirable, a faster test was developed which makes use of a strong electric field which drives ions through the concrete. This rapid chloride penetrability test, as developed by Whiting [2], has been adopted by both the AASHTO [3] and ASTM [4] as a standard test method. The test itself consists of monitoring the amount of charge passed through a cylindrical sample during a 6 hour test period. A d.c. potential difference of 60 volts is maintained across the sample, one end of which is immersed in a NaCl solution (anode), while the other end is immersed in a NaOH solution (cathode). The apparatus is shown schematically in Figure 1. The total charge passed is then related to a penetrability, which gives a qualitative measure of the durability.

The current passing through the sample can result in a temperature increase from Joule heating. This heating is especially a concern in permeable samples that pass large amounts of current, because a rise in temperature can have a profound effect on the diffusion coefficient. Moreover, Feldman *et al.* concluded that the rapid chloride penetrability test induced changes in pore structure and resistivity of the concrete specimens [5]. We have recently demonstrated that high fields can affect the microstructure [6]. Therefore, because the microstructure of the specimen is being changed, it is impossible to know if a representative specimen is being tested.

The resistance of cement-based materials is strongly dependent on the type and concentration of the ions in the pore fluid. The rapid chloride penetrability test does not discriminate between the ions that carry charge. Therefore, when comparing the results

---

* Commonly referred to as "permeability," however we prefer "penetrability" or "ion migration" and reserve permeability for fluid flow situations

from two concretes with different pore solution chemistries, the difference in charge passed cannot be assumed to be purely a result of microstructural differences.

Nonetheless, the test has become a standard adapted by both the AASHTO and ASTM and has quickly become a popular tool used by many researchers to characterize the relative durability of concrete.

2.1.4 Modified Rapid Chloride Penetrability Test (Norwegian Test)

The Norwegian test, described by Detwiler *et al.,* was designed to correct some of the shortcomings of the rapid chloride penetrability test [7]. There are two major modifications. First, the applied potential is lowered from 60 to 12 Volts. The lower voltage decreases the rate of charge passed, thus reducing Joule heating. Unfortunately, this increases the time required to perform the test. Second, instead of current passed, the Norwegian test measures the concentration of chloride ions in solution directly. This eliminates the sensitivity of the test to changes in pore solution chemistry. Unlike the rapid chloride penetrability test, the Norwegian test allows a diffusion coefficient to be calculated.

From a practical point of view, there are several limitations to using the Norwegian test. For instance, the duration of the test is both longer and less predictable than the rapid chloride penetrability test. Experiments can take anywhere from days to weeks, depending on the quality of the concrete. Also, to determine that steady state flow has been reached, the apparatus requires constant monitoring. Measurement of the chloride concentration in solution presents its own set of problems, including the need for laborious titrations or expensive analytical equipment.

2.2 Permeability Measurements

Permeability of concrete plays an important role in durability because it controls the movement and the rate of entry of water, which may contain aggressive chemicals. Banthia *et al.* called permeability, "...by far the most important property of concrete for determining its durability." [8]

Fluid flow through cement-based materials is dictated by the porous nature of the microstructure. As cement hydrates, capillary porosity is being consumed while gel porosity increases, and the permeability drops by approximately 10-12 orders of magnitude [9]. Powers *et al.* showed that the permeability in neat pastes was influenced by the overall volume of capillary porosity [10, 11]. Later, Goto *et al.* showed that the volume, distribution, connectedness, and shape of the pores control the fluid flow through the microstructure [12].

The measurement of permeability relies on Darcy's law, which states that the coefficient of permeability is proportional to the flow rate per unit area and unit pressure gradient. When studying the permeability of cement-based materials, water is by far the most commonly used fluid. Unfortunately, water will react with any unhydrated cement grains, thus changing the pore structure during the test. Since it can take several weeks to reach equilibrium flow, data for younger samples is questionable.

The general design for a device to measure permeability is relatively straightforward. A fluid under pressure is applied to one surface of a specimen disk, while the other surface is maintained at a lower pressure. The fluid flow rate on the low pressure side is monitored and, knowing the sample geometry and pressure head, a coefficient of permeability can be calculated. A simple device to measure permeability was proposed by Ludirdja *et al.*, where the flow rate induced by a column of water was monitored using a burette [13]. This had the advantage of being a low-cost, easy-to-build apparatus. Unfortunately, the design was limited by low fluid pressures and the imprecise nature of the flow rate measurements. Mills and Hearns [14], and later Hooton [15], have designed and used more advanced devices to measure the permeability. Here, the applied pressure from a weighted mechanical arm provides pressures much higher than a column of water. The flow rates are monitored using LVDT's (linear variable displacement transducers) interfaced to a personal computer. This has the advantage of a much higher sensitivity in flow rate detection, as well as continuous collection of data. The limitation of this design is the relatively high cost of construction. Regardless of design, fluid-tight sealing of the cement/concrete specimen to the apparatus can be problematic.

3 Recent Developments

3.1 Impedance Spectroscopy

Cement paste relies on the pore solution to carry electrical charge, thus the pore structure plays an important role in governing the resistance to the flow of charge. The most important factors are the capillary porosity, concentration and types of ions in the pore fluid, and the continuity of the pore network. The effect of these factors can be easily probed with various electrical measurements, particularly impedance spectroscopy.

Impedance spectroscopy (IS) is an experimental technique which is becoming increasingly prevalent in the study of materials [16]. The technique involves applying an a.c. excitation signal to the specimen of interest and recording the time-varying response. This is repeated over a large range of frequencies, and the gain and phase angle differences are monitored. This information is often represented graphically by plotting on a complex plane the negative of the imaginary part of the impedance versus the real part of the impedance. This type of representation, known as a Nyquist plot, has the advantage of representing different responses as distinct semi-circular arcs, as is shown in Figure 2. Several parameters can be determined from an analysis of IS data. The most reliable parameter obtained is the specimen resistance which is determined from the intersection of the bulk cement paste arc with the Real Impedance axis (see Fig. 2).

There are several important advantages of IS over other methods used to examine cement-based materials. First, IS is an in-situ, nondestructive technique, which, unlike many methods, does not require a dried, and therefore altered, specimen. Moreover, geometry constraints are much less severe for samples tested with IS, compared to many commonly used methods, which might allow in-field testing of concrete structures. Another useful trait of IS is the ability to test a single sample continuously, monitoring the evolution of properties over time, without the need for multiple samples.

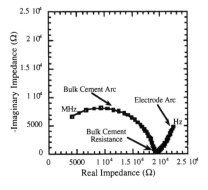

Fig. 2. Nyquist plot of water/cement = 0.4 Type I ordinary portland cement paste hydrated for approximately 24 hours. Note location of bulk cement arc, electrode arc, and the value of cement resistance, as well as the frequency values [17].

During hydration, the microstructure and pore fluid composition of cement based materials change. This change can be represented using the following relationship: [18]

$$\frac{\sigma}{\sigma_o} = \phi_{cap}\beta \qquad (2)$$

Where σ is the overall bulk conductivity $(\Omega \cdot cm)^{-1}$, σ_o is the pore solution conductivity $(\Omega \cdot cm)^{-1}$, ϕ_{cap} is the capillary porosity, and β is the connectivity (inverse tortuosity) of the capillary pore network.

The fraction, σ/σ_o, is often called the normalized conductivity, but is also known as the inverse formation factor, or the inverse MacMullin number. It is assumed that the current is carried by the pore fluid exclusively, and the contribution from the solids within the microstructure is negligible. The value of σ_o is obtained by measuring the conductivity of the pore fluid expressed from the hardened paste or estimating it based on chemical analysis of the cement powder [19].

3.2 Microstructure-Based Modelling

The pore structure dictates the transport properties of cement-based materials. Analysis of the influence of the pore structure, however, is often difficult as a result of the ongoing cement hydration process and the destructive nature of many testing techniques. Recently, microstructure-based modelling has been used to predict the transport properties in a number of cement-based systems.

Modelling the transport properties allows one to predict the characteristics of a particular microstructure. Output from the computer is compared to experimental data to validate the model and its underlying assumptions. The model can then contribute to the understanding of microstructure/transport property relationships by the use of computer simulations. Model parameters can be varied to assess their effects on microstructural development and, ultimately, transport properties.

An example of the use of microstructure-based models is the prediction of the diffusivity of cement paste. In this model, the 3D microstructure of the cement paste is simulated using a pixel based model. Each pixel will have an identity defining its location and composition. When water is "added", the surfaces of the cement particles are allowed to dissolve, and, following random walk algorithms, diffuse through the water. When the diffusing particles collide with another particle, there is a probability that they will react and form a reaction product, which is predominantly C-S-H gel or calcium hydroxide.

Once the microstructure is built up, specific conductivities are assigned to each pixel, forming a random conductor network, similar to Figure 3. The overall electrical conductivity is determined using a method outlined in reference [20]. When the conductivity has been determined, it can be related to diffusivity using the Nernst-Einstein relation, which equates electrical conductivity and ionic diffusivity. This will be discussed in more detail in Section 3.5.

The importance of this model, and computer simulations in general, is that once a model has been designed and verified, a large number of experimental variables can be examined. This is important, for instance, when predicting the diffusivity of cement pastes at early times. Experimentally it is difficult test young pastes owing to the rapid rate of hydration, coupled with the long time required to make the measurement. With microstructure-based modelling, however, any degree of hydration can be simulated, and properties such as diffusivity can be predicted. Agreement between model-based diffusivities/conductivities and those determined by IS is quite encouraging [19].

C-S-H   Capillary Pore

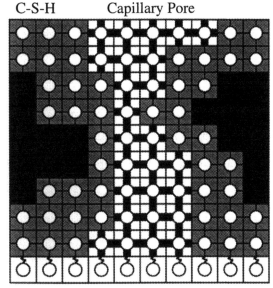

Unreacted
cement
and Ca(OH)₂

Fig. 3. Schematic diagram of the digital image for a random conductor network map used to compute electrical conductivity of the cement paste model. Conductances of different bonds are predetermined [20].

Electrode

3.3 Nuclear Magnetic Resonance

Analysis of nuclear magnetic resonance (NMR) relaxation has been used with considerable success to study porous materials such as sol-gel glasses, sandstones, and borosilicate glasses. Only recently, however, has this technique been applied to cement-based materials [21]. The advantage of using NMR is that it allows the in-situ analysis of the evolving pore structure during hydration.

Although complete description of the theoretical background for this technique is beyond the scope of this text, a few basic aspects will be discussed. First, NMR analysis of cement-based materials relies on the spin-echo relaxation from protons in the free water contained in the capillary pore network. It is established that NMR relaxation rates are enhanced near a solid-liquid interface. By applying a magnetic field, and measuring the spin decay rates when the field is removed, one can determine the distribution of bulk and near-surface water. This information can then be used to calculate a pore size distribution of the cement-based material [22].

This technique also has the potential to measure effective diffusion coefficients (e.g., protons or chloride ions). In order to increase the magnitude of the magnetic field, which increases the sensitivity of the measurement, a pulsed NMR signal is used. With this method, direct information about the evolution of the mean square displacement of water molecules with time is obtained. In particular, it can be determined if classical diffusion occurs during the time frame of the NMR experiments. If this is the case, the mean-square displacements are linearly proportional with time, and the relationship $<r(t)^2> = 6D_{eff} \, t$ holds. Here $<r(t)^2>$ is the mean-square displacement, D_{eff} is the effective diffusion coefficient, and t is time. The mean-square displacements can be determined by fitting the decay of the spin echo from the specimen. The technique has been successfully employed to monitor diffusion in other porous media, but not, to date, in cement-based materials.

3.4 Katz-Thompson Equation

In 1986 A.J. Katz and A.H. Thompson proposed the relationship shown in Equation (3) to predict the hydraulic permeability of porous rocks using indirect measurements [23].

$$K = cd_c^2\left(\frac{\sigma}{\sigma_o}\right) \tag{3}$$

where K is the permeability (m^2), d_c is the critical pore diameter (m), σ is the electrical conductivity of the sample $(\Omega\cdot cm)^{-1}$, σ_o is the conductivity of the pore fluid $(\Omega\cdot cm)^{-1}$, and c is a constant.

To predict the permeability of a material using the Katz-Thompson equation, the critical pore diameter (d_c) and the normalized conductivity (σ/σ_o) must be measured. Katz and Thompson related the critical pore diameter to the inflection point in mercury intrusion curves. The physical interpretation of the critical diameter is that it is the smallest continuous pore size that percolates through the sample. The normalized conductivity is determined using electrical methods, particularly impedance spectroscopy, and is a measure of the heterogeneity of the pore network. For the electrical measurements, it is necessary that the pore structure be saturated with a conductive pore fluid (e.g., brine).

The main advantage of the Katz-Thompson relation is that the time required to make these measurements is much shorter and much less labor intensive than is required to make permeability measurements. Also, unlike permeameters, there is no limit of permeability that can be predicted, once set has occurred, and d_c can be measured. This allows specimens of any age, or composition, to be evaluated.

The Katz-Thompson relation has been shown to predict the permeability of porous rocks quite accurately. Christensen *et al.* tested this procedure on young cement pastes and found agreement with permeameter data to be quite encouraging [19].

3.5 Nernst-Einstein Equation

Ionic diffusion is a critical issue for the design of structures, such as those with steel reinforcement and those used for waste containment. There can be numerous difficulties in accurately measuring diffusion coefficients. However, using IS and the Nernst-Einstein equation, shown in Equation (4), an estimate of the diffusion coefficient can be easily determined.

$$\frac{\sigma}{\sigma_o} = \frac{D_{eff}}{D_o} \tag{4}$$

where σ and σ_o are as defined previously, D_{eff} is the effective diffusivity of a given ion in the porous medium, and D_o is the intrinsic diffusivity of that ion in 100% pore fluid. Intrinsic diffusivities are readily available in the literature, and show no more than a 15% change from pure water to concentrated (0.5 molar) solutions. Using Eq. (4), determining the diffusion coefficient of any ion in the bulk is simply a matter of measuring the normalized conductivity, and multiplying by D_o, the intrinsic diffusion coefficient. Agreement between σ/σ_o and D_{eff}/D_o is quite good [19].

There are several advantages to using the Nernst-Einstein equation to predict diffusivities. First, it is possible to make estimates of the diffusivities for young pastes (approx. 1 day), something that is not possible with the Norwegian test, ponding/profiling, or divided diffusion cell techniques. This allows the continuous monitoring of the diffusivity versus time for a particular sample. Also, because of the shorter time involved in making individual measurements, a larger number of samples can be tested in a given period.

3.6 Solvent exchange Kinetics

Cement-based materials have a water-filled, porous microstructure. When immersed in an organic solvent (e.g., isopropanol, methanol), the pore fluid is replaced by the solvent. The rate of this replacement can be investigated with methods such as weight loss or electrical conductivity changes, and provides insight concerning the pore network (e.g., amount and connectiveness). The use of impedance spectroscopy allows continuous, in-situ measurements of the conductivity during the exchange process. Specimens that have been hydrated to different times show different rates of exchange, as shown in Figure 4.

As originally suggested by Feldman *et al.* [24], solvent exchange kinetics could potentially be used to establish parameters such as diffusivity and permeability. We are currently investigating the solvent exchange rate as an alternative parameter to d_c^2 in the Katz-Thompson equation (Eqn. (3)). The advantage of this method is the ease of the test, in terms of both time and labor. Also, although the pore fluid is being exchanged with an organic solvent, the microstructure is perturbed very little, especially when compared with mercury intrusion porosimetry and BET, which both require drying of the samples prior to measurement.

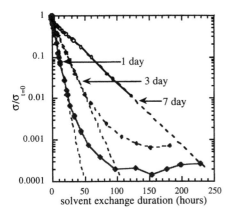

Fig. 4. Solvent exchange data for a Type I OPC, water:cement = 0.4 hydrated for 1, 3, or 7 days prior to solvent exchange in isopropanol. Conductivity, normalized by pre-immersion conductivity is plotted as a function of exchange duration [25].

4 Conclusion

In summary, the measurement of transport properties is important in predicting the durability of cement-based materials. Unfortunately, many established techniques are time-consuming, labor-intensive, and inconsistent. These techniques include ponding/profiling, divided diffusion cell, rapid chloride penetrability test, and the Norwegian test, as well as the differential pressure permeameter.

A number of techniques have recently emerged that have the potential to measure or predict transport properties in an accurate and efficient manner. Often, the tests are nondestructive, fast, and relatively easy to perform. Some examples of these techniques include impedance spectroscopy, microstructure-based modelling, nuclear magnetic resonance, and solvent exchange kinetics. Additionally, the information gathered in these novel methods can be used to predict important durability-related parameters, such as diffusivity (by the Nernst-Einstein equation) and permeability (by the Katz-Thompson equation).

Acknowledgments:

This work was supported by the National Science Foundation (DMR-91-20002) through the Science and Technology Center for Advanced Cement-Based Materials.

5 References

1. AASHTO-T259-80 (1993) Resistance of Concrete to Chloride Ion Penetration; American Association of State Highway Transportation Officers
2. Whiting, D. (1981) Rapid Measurement of the Chloride Permeability of Concrete. *Public Roads*; Vol. 45, No. 3, pp. 101-112.
3. AASHTO-T277-83 (1993) Electrical Indication of Concrete's Ability to Resist Chloride; American Association of State Highway Transportation Officers
4. ASTM-C1202-91 (1991) Standard Test Method for Electrical Indications of Concrete's Ability to Resist Chloride Penetration; American Society for Testing and Materials.
5. Feldman, R.F., Chan, G.W., Brousseau, R.J., Tumidajski, P.J. (1994) Investigation of the Rapid Chloride Permeability Test. *ACI Materials Journal*; Vol. 91, No. 2, pp. 246-255.
6. Sohn, D. (1995) Unpublished data
7. Detwiler, R.J., Kjellson, K.O., Gjorv, O.E. (1991) Resistance to Chloride Intrusion of Concrete Cured at Different Temperatures. *ACI Materials Journal*; Vol. 88, No. 1, pp. 19-24.
8. Banthia, N., Mindess, S. (1989) Water Permeability of Cement Paste. *Cement and Concrete Research*; Vol. 19, No. 5, pp. 727-736.
9. Mindess, S., Young, J.F. (1981)*Concrete*. Prentice Hall, Inc., Englewood Cliffs, NJ
10. Powers, T.C., Copeland, L.E., Hayes, J.C., Mann, H.M. (1954) Permeability of Portland Cement Paste. *Journal of the American Concrete Institute*; Vol. 26, No. 3, pp. 285-298.
11. Powers, T.C. (1958) Structure and Physical Properties of Hardened Cement Paste. *Journal of the American Ceramic Society*; Vol. 41, No. 1, pp. 1-6.
12. Goto, S., Roy, D.M. (1981) The Effect of W/C Ratio and Curing Temperature on the Permeability of Hardened Cement Paste. *Cement and Concrete Research*; Vol. 11, No. 4, pp. 575-579.
13. Ludirdja, D., Berger, R.L., Young, J.F. (1989) Simple Method for Measuring Water Permeability of Concrete. *ACI Materials Journal*; Vol. 86, No. 5, pp. 433-439.
14. Hearn, N. (1990) A Recording Permeameter for Measuring Time-Sensitive Permeability of Concrete. *Ceramic Transactions -- Advances in Cementitious Materials*; Vol. 16, No. pp. 463-475
15. El-Dieb, A.S., Hooton, R.D. (1994) A High Pressure Triaxial Cell with Improved Measurement Sensitivity for Saturated Water Permeability of High Performance Concrete. *Cement and Concrete Research*; Vol. 24, No. 5, pp. 854-862.
16. MacDonald, J.R. (1987)*Impedance Spectroscopy: Emphasizing Solid Materials and Systems*. Wiley Interscience, New York
17. Shane, J.D. (1995) Unpublished data
18. Garboczi, E.J. (1990) Permeability, Diffusivity, and Microstructural Parameters: A Critical Review. *Cement and Concrete Research*; Vol. 20, No. 4, pp. 591-601.
19. Christensen, B.J., Coverdale, R.T., Olson, R.A., et al. (1994) Impedance Spectroscopy of Hydrating Cement-Based Materials: Measurement, Interpretation,

and Application. *Journal of the American Ceramic Society*; Vol. 77, No. 11, pp. 2789-804.

20. Garboczi, E.J., Bentz, D.P. (1992) Computer Simulations of the Diffusivity of Cement-Based Materials. *Journal of Materials Science*; Vol. 27, pp. 2083-2092.

21. D'Orazio, F., Bhattacharja, S., Halperin, W.P., Eguchi, K., Mizusaki, T. (1990) Molecular Diffusion and Nuclear-Magnetic-Resonance Relaxation of Water in Unsaturated Porous Silica Glass. *Physical Review B*; Vol. 42, No. 16, pp. 9810-9818.

22. Bhattacharja, S., Moukwa, M., D'Orazio, F., Jehng, J.-Y., Halperin, W.P. (1993) Microstructure Determination of Cement Paste by NMR and Conventional Techniques. *Journal of Advanced Cement Based Materials*; Vol. 1, pp. 67-76.

23. Katz, A.J., Thompson, A.H. (1986) Quantitative Prediction of Permeability in Porous Rock. *Physical Review B*; Vol. 34, No. 11, pp. 8179-8181.

24. Feldman, R.F. (1987) Diffusion Measurements in Cement Paste by Water Replacement Using Propan-2-ol. *Cement and Concrete Research*; Vol. 17, No. 4, pp. 602-612.

25. Hwang, J.-H. (1995) Unpublished data

PART EIGHT

CEMENTITIOUS WASTE FORMS FOR NON-NUCLEAR APPLICATIONS

49 DEGRADATION MECHANISMS OF CEMENT-STABILIZED WASTES BY INTERNAL SULFATE ASSOCIATED WITH THE FORMATION OF THE U PHASE

G. LI
Laboratoire de Mécanique et Technologie
P. LE BESCOP
Commissariat à l'Énergie Atomique, DCC/DESC/SESD, C.E. Saclay, France

Abstract

In cement-stabilized wastes that contain high amounts of sulfate, the U phase, a sodium-substituted AFm phase, is pointed out. This phase is shown to be responsible for the degradation of the simulated samples. Two types of expansion processes which lead to the degradation have been characterized: 1. secondary formation of the U phase, in samples $C_3S+C_3A+Na_2SO_4$. 2. the U phase $\rightarrow$ ettringite transformation through leaching tests on samples C_3S+U. The U phase, which is not detected in traditional cements, appeared as deleterious in our studied cement-stabilized wastes.
Keywords: Cement-stabilized wastes, sodium sulfate, U phase, ettringite, expansion

1 Introduction

Among wastes produced by nuclear industry, liquids are the most important fraction by volume. Evaporation is widely applied to reduce their apparent volume, and it leads to solutions characterized by high concentrations of different salts. In some cases, the concentration of sodium sulfate may reach 200 to 250 g/L. Large quantities of these wastes are solidified in cement matrices. It is necessary to investigate the behavior of this kind of cemented wastes because of the potential waste-binder interactions.

Considering cement-stabilized wastes containing high amounts of Na_2SO_4 (10~15%), the first section of this paper presents a preliminary study which reveals the formation of the U phase and its deleterious effects on the studied system. From experimental evidences, two possible degradation mechanisms associated with the U phase are proposed and elucidated in the other two sections: the secondary formation of the U phase, in samples $C_3S+C_3A+Na_2SO_4$, and the transformation of the U phase into ettringite through leaching tests on samples C_3S+U phase.

Mechanisms of Chemical Degradation of Cement-based Systems. Edited by K.L. Scrivener and J.F. Young. Published in 1997 by E & FN Spon, 2–6 Boundary Row, London SE1 8HN. ISBN: 0419215700.

2 U phase formation and its deleterious effects

In the characterization of a cement paste, the mineralogy of the solid phases has a particular importance, and mineralogical analysis by X-rays diffraction constitutes a useful starting point Corresponding to one of the methods used in cementing the waste solutions with high concentration in Na_2SO_4, some simulated samples were prepared according to the French norm NF EN-196-3, using the French OPC cement (CPA55R) mixed with 20% wt Na_2SO_4 solution at a solution/solid ratio = 0.50 and sealed adiabatic curing for the first seven days. The XRD analyses were carried out at different ages. Fig. 1 shows a typical XRD pattern at the age of two years, obtained using Cu-Kα radiation (3s count time and 0.02° step size).

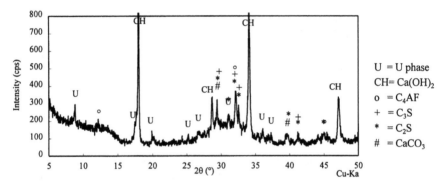

Fig. 1. XRD pattern of the sample under sealed curing for two years

The identification of the peaks indicated not only the presence of some classic hydrates such as $Ca(OH)_2$, and some anhydrous cement phases, but revealed especially the formation of an AFm sodium-substituted phase called "U phase", first observed by Dosch and zur Strassen [1] in studying the chemical system $CaO-Al_2O_3-SO_3-Na_2O-H_2O$. The plausible composition of the U phase was established as $4CaO.0.9Al_2O_3.1.1SO_3.0.5Na_2O.16H_2O$, belonging to the group of hexagonal or pseudo-hexagonal layered structures like AFm, but differing from the latter in the fact that it contains sodium between the layers and possessed a higher interlayer distance. Its diffraction lines corresponded to 10.00, 5.00, 4.46, 3.53, 3.33, 2.88, 2.77, 2.76Å etc.

The similar cement-stabilized systems with other compositions were also studied in [2] and the U phase was systematically observed in these special cement media. In addition, the observations of the samples on the scanning electronic microscope showed clearly that the U phase was formed in hexagonal platelets.

Although the U phase was mentioned in some publications [3] [4] [5] [6] [7] and detected in some experimental studies [8] [9], no information is available in the literature on the potential effects of the U phase. It is thus interesting to examine the immersion behavior of the samples, so that samples identical with the previous XRD analyses (Fig. 1) have been set in pure water after sealed adiabatic curing for 7 days.

Generally, the samples underwent obvious swelling which led to many cracks or even complete destruction. Fig. 2 shows a sample immersed for one year. For an earlier age, some cracks appeared which crossed the sample, and then many microcracks were observed on the surface of the fragments which even resulted in some debris.

Fig. 2. Severe degradation of the sample immersed in water for one year

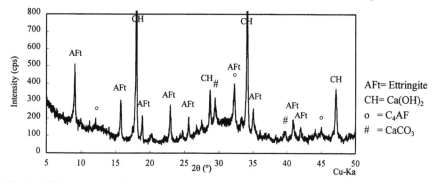

Fig. 3. XRD patterns of the debris resulted from the cracking of the immersed sample

In order to compare the mineralogical composition to this at the initial state, XRD analyses were systematically carried out on the immersed samples at different ages. No change of mineralogical composition was observed before the samples were cracked, except that the intensity of the peaks due to the U phase had a tendency to increase. On the contrary, large differences appeared in the debris resulted from the cracking at the later ages: the U phase was not detected, but ettringite was formed abundantly, and no other known deleterious phase was detected (Fig. 3).

These experimental evidences led us to suppose that the degradation of the studied cement-based systems should be associated with the U phase formation. The following two mechanisms can be envisaged:

- secondary formation of additional U phase can take place during the immersion if the original cement paste mixing water is insufficient for clinker hydration and U phase formation. This reaction may be expansive as the case of the secondary formation of ettringite [10] or gypsum [11].

- the U phase-ettringite transformation may be caused by a decrease of alkaline concentration, according to [1]. During the immersion, which leads to the decrease of alkaline concentration due to leaching, the U phase may be destabilized, and it is possible that the conversion process to ettringite induces an expansion.

3 Expansion by secondary formation of the U phase

3.1 Experimental procedure

To explore the first assumption, the choice of a system in which the U phase can be formed and remains always stable, is necessary. Taking into account the nature of the hydraulic material and the high content of Na_2SO_4 in the previously mentioned cement-stabilized wastes, a system composed of $C_3S+C_3A+Na_2SO_4$ was chosen.

It was indicated in a preliminary study that in such a system, C_3A could react directly with Na_2SO_4 and form the U phase with the composition proposed by Dosch and zur Strassen $4CaO.Al_2O_3.1.1SO_3.0.5Na_2O.16H_2O$ [1]. A possible reaction is shown below (1), with the necessary $Ca(OH)_2$ supplied by C_3S hydration (2):

$$2.7\ C_3A + 3.3\ Na_2SO_4 + 45.9\ H_2O + 3.9\ Ca(OH)_2 \rightarrow 3.0\ U + 3.6\ NaOH \qquad (1)$$
$$1.0\ C_3S + 3.8\ H_2O \rightarrow 1.0\ C_{1.5}SH_{2.3} + 1.5\ Ca(OH)_2 \qquad (2)$$

The C-S-H from C_3S hydration was taken in the form of $C_{1.5}SH_{2.3}$ [12], a C/S value smaller in the alkaline media than those in normal conditions.

To envisage the secondary formation of the U phase in the mixture, the proportion of the constituents was determined in such a way that: 1. the available amount of $Ca(OH)_2$ from C_3S hydration (2) is higher than that necessary to the formation of the U phase (1), not only at the stage of primary hydration, but also during the secondary reactions. 2. a large part of mixing water should be consumed by the primary hydration of C_3S and C_3A. The remaining anhydrous compounds may undergo a secondary hydration under an external water supply. Based on these considerations, the mixture should be 38.5% C_3S + 38.5% C_3A + 23.0% Na_2SO_4 and the water/solid ratio = 0.345.

Taking into account the high basicity in the cement-stabilized wastes [2] and the stability of the U phase [1], a 3M NaOH solution was chosen as the immersion solution.

Cylindrical specimens ϕ30mm x h30mm were prepared at 20°C. They were demoulded after three days and then immersed in the 3M NaOH solution in a specially designed device, which measures continuously the linear dimensional variation. The samples were analyzed by XRD using Cu-Kα radiation under 40kV/25mA and 3s/.01°.

3.2 Results and discussion

The tests were conducted for about four months till the dimensional variation became stable. The expansion data are plotted in Fig. 4 as a function of time.

The mineralogical analyses of the specimens by XRD at the beginning of the tests are shown in Fig. 5(a). These at the end of the tests, e.g. after the immersion for about four months in 3M NaOH solution, are shown in Fig. 5(b).

It is noted that the specimens underwent an enormous swelling (5%), which lead to a large amount of cracking during the immersion and even complete destruction at the end of the tests. In fact, this expansion phenomenon can be explained only by the further formation of the U phase, because the chemical conditions in the studied system are compatible only with the U phase formation and its accompanying C_3S hydration.

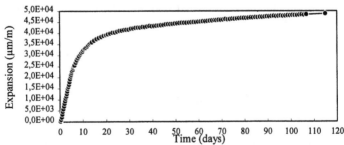

Fig. 4. Linear expansion of the sample immersed in the 3M NaOH solution

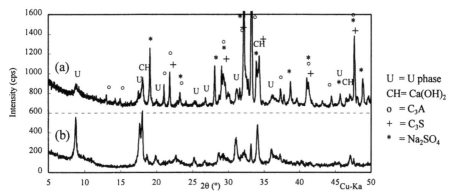

Fig. 5. Typical XRD patterns at the beginning of the tests and after ~4 months
(a) at the beginning of the tests   (b) after ~4 months in 3M NaOH solution

The envisaged secondary formation of the U phase was confirmed by XRD analyses (Fig. 5a and Fig. 5b). It is first interesting to remark that the U phase is the only stable hydrate formed from C_3A hydration with Na_2SO_4, both at the beginning and at the end of the tests; no trace of others deleterious phases is detected. Moreover, the amount of the U phase increases considerably in the immersed samples in comparison with the samples before the immersion, evidently due to the secondary formation of the U phase.

Because the U phase is the only hydrate that forms at the expense of C_3A and Na_2SO_4, U phase formation should be accompanied by the disappearance of C_3A and Na_2SO_4. Taking the peaks of C_3A at d = 2.7Å and Na_2SO_4 at d = 4.66Å, the decrease of their amounts can be clearly observed, and the secondary formation of the U phase is therefore confirmed.

Concerning $Ca(OH)_2$, the characteristic peaks after the immersion became stronger although the secondary formation of the U phase consumed a part of the $Ca(OH)_2$, because the hydration of C_3S in the mixture could supply more $Ca(OH)_2$ than this necessary for U phase formation.

3.3 Conclusions

These results confirm that the secondary U phase can be formed by an external water supply and the correspondent reaction process can induce a large expansion.

4 Expansion by U phase → ettringite transformation

4.1 Experimental procedure

The experiments by Dosch and zur Strassen [1] have been carried out in alkaline solutions of different concentrations. They have described the U phase → ettringite transformation, with some intermediate compounds, when the NaOH concentration is progressively decreased: U phase → C_4AH_{13} → Monosulfoaluminate → ettringite, and this was confirmed by our preliminary study on the pure U phase in water. However, no more information is available for hydrated cement media. Can this transformation occur in cement-based system during immersion and result in an expansion phenomenon? If yes, what is the expansion process?

The proposed methodology consists in leaching the alkalis from a system containing beforehand the U phase. Taking into account the compatibility with the OPC cement-stabilized wastes, and in order to eliminate the potential secondary formation of the U phase which has just been identified as deleterious, a mixture C_3S+U phase was chosen.

Supposing that in the long term, C_3A+C_4AF can be hydrated into the U phase, and $C_3S +C_2S$ into C-S-H, for the OPC cement-stabilized wastes, the U/C_3S ratio = 0.69 for the C_3S+U mixture was selected. The amount of C_3S in the mixture is equivalent to C_3S+C_2S in OPC cement, and that of the U phase is equivalent to C_3A+C_4AF. To guarantee the stability of the U phase in the sample before the leaching tests, the mixture was hydrated by a 1.5M NaOH solution with solution / C_3S ratio = 0.70 wt.

Two leaching devices were used: one keeps the leaching solution deionized by means of a continuous circulation through ion exchange resins, another keeps the pH at 7.0 by nitric acid addition with a self-acting burette, and renewing the leachant frequently to avoid excessive concentrations of different ions. These two devices are considered to be compatible one with the other [13]. A special cell was also combined with the first device to measure the linear expansion.

Cylindrical specimens with the dimensions ~ϕ30mm x h30mm were sealed cured for one month, then put into the leaching solution. At the end of the tests, the layer by layer XRD analyses were performed. The sample was carefully scratched from the surface to the core and the powders collected in different thin 0.2~0.4mm zones were analyzed under 40kV/25mA and 3s/0.02° using Cu Kα radiation. The proportion of phases was approximately estimated from the height of the peaks: d=10Å for the U phase, d=9.72Å for AFt, d=4.92Å for Ca(OH)$_2$ and d=8.91Å for AFm. SEM observations of the samples were also carried out.

4.2 Results and discussion

After about fifteen days, the sample underwent a swelling which then increased rapidly. At one month, many cracks induced were well observed at the surface of the sample. The appearance of the sample and the measured linear expansion are shown in Fig. 6.

To avoid an excessive cracking, the samples for XRD analyses were taken out after one month. The periphery of ~2.0mm was nearly separated from the main sample, so the analyses could be performed only on the surface and then the whole debris. For the main sample, more than twenty thin zones on the ~3.0mm external layer were successively scratched and analyzed, the mineralogical distribution is illustrated in Fig. 7 and the SEM observations related to the mineralogical zoning are presented in Fig. 8.

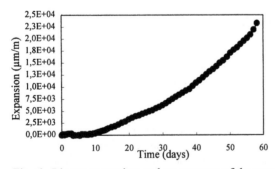

Fig. 6. Linear expansion and appearance of the sample after leaching test

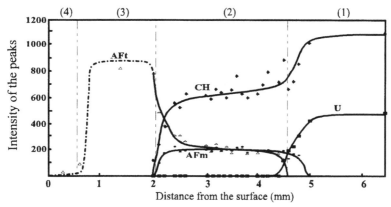

Fig. 7. Mineralogical distribution in the sample under leaching tests for one month

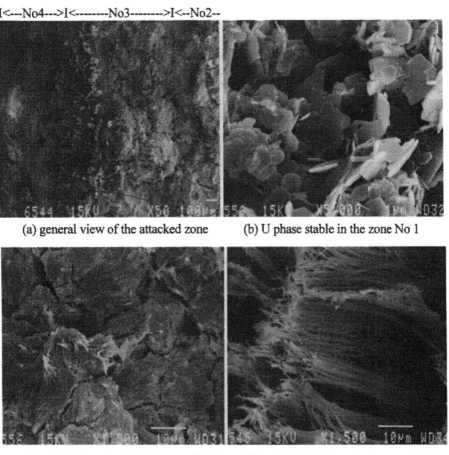

(a) general view of the attacked zone   (b) U phase stable in the zone No 1

(c) colloidal ettringite in the zone No 2   (d) ettringite in fibers in the zone No 3
Fig. 8. SEM observations related to the zoning characterized by XRD analyses

Several zones can be distinguished from the mineralogical distribution (Fig. 7) and the apparent observations of the sample (Fig. 8a):

• Zone (1): The central nucleus (~ϕ20mm x h20mm) of the sample keeps the initial mineralogical composition: $Ca(OH)_2$ is very abundant, the U phase is still stable (Fig. 8b), and the presence of anhydrous C_3S is not negligible. The microstructure is compact and homogeneous; no degradation character is present.

• Zone (2): Outside of the non attacked core, there exists a quite large zone with a thickness of ~2.5mm, in which the U phase disappeared, but $Ca(OH)_2$, the monosulfoaluminate and ettringite are present with constant quantities. This zone is first characterized by the simultaneous existence of ettringite and $Ca(OH)_2$, the later remaining still quite abundant. Moreover, the SEM observations show that the ettringite massively formed in this zone is amorphous or very poorly crystallized (Fig. 8c); it really resembles a sponge that can inflate in water, and the sample is very microcracked. We think that the ettringite found here originates directly from the U phase conversion caused by the decrease of the alkali concentration; it is formed as a colloid in the presence of $Ca(OH)_2$ and presents the expansive character.

• Zone (3): A very little outer part of the main sample and a large part of the separated debris constitute a zone enormously cracked (thickness of ~1.5mm). No trace of the U phase is detected, nor monosulfoaluminate, nor $Ca(OH)_2$. It contains only ettringite. The morphology of the ettringite was identified by the SEM observations: It is perfectly crystallized in fibers with a length of more than 50µm and those in cracks are often perpendicular to the crack surfaces (Fig. 8d). We are convinced that the ettringite found here would be formed from the interstitial solution by recrystallisation, preceding a presolubilisation of the ettringite formed in the zone (2), and that this type of ettringite is not expansive. The enormous cracking of the sample must have resulted from the microcracks generated previously by the swelling stress of the expansive ettringite and developed later by acceleration of the degradation kinetics. The complete disappearance of $Ca(OH)_2$ suggests a weak alkali concentration in the interstitial solution.

• Zone (4): Both the XRD analyses and the SEM observations show that the thin periphery of the sample (thickness of ~0.5mm) is carbonated despite the protection, but no other mineral is detected except the trace of ettringite due probably to scratching too deeply. Moreover, the EDS analyses indicate that the Ca content is very low, but those of Si and Al are relatively high. According to one study [14], only a gel of alumina and silica persists in this zone.

It is noted that some intermediate zones between two successive zones described previously are observed, by both SEM observations and the mineralogical distribution.

These experimental results demonstrate that during the immersion of a system containing the U phase, the decrease of the alkali concentration resulting from diffusion leads to the destabilization of the U phase. Related to the stability of different compounds, a zoning phenomenon is produced. It is revealed that the U phase transforms first into ettringite, probably with monosulfoaluminate as intermediate compound, being colloidal in the presence of $Ca(OH)_2$ and presenting an expansive character. The later would be dissolved in the interstitial solution, as the alkali concentration is low, and then recrystallized in perfectly formed fibers, which are not expansive. Indeed, these physical, chemical and mechanical processes are closely linked one to the other, and the apparent degradation results from their simultaneous interactions.

4.3 Conclusions
A progressive transformation of the U phase into ettringite can occur with some intermediate compounds appearing as the alkali concentration progressively decreases. In the presence of $Ca(OH)_2$, the first formed ettringite issued from this transformation is colloidal and presents the expansive character, in which the origin of the apparent degradation of the sample lies.

5. General conclusions
In cement-stabilized wastes containing high amounts of alkali sulfate, the U phase can occur. Generally, this phase is only stable at high alkali concentrations, conditions not compatible with traditional cement pastes.

For the simple systems chosen, $C_3S+C_3A+Na_2SO_4$ and C_3S+U phase, the U phase can lead to the degradation on contact with water by both the secondary formation of the U phase and the U phase $\rightarrow$ ettringite transformation.

Taking into account the compatibility of the studied systems with the previous cement-stabilized wastes, and for reasons of safety in storing this type of radioactive wastes, the U phase should be considered as deleterious, although it is not detected in traditional cement media.

6. References
1. Dosch W. and zur Strassen H. (1967), Ein alkalihaltiges calciumaluminatsulfat-hydrat (Natrium - Monosulfat), *Zement-Kalk-Gips*, Vol. 20, No. 9, pp. 392-401.
2. Li G. (1994), Ph. D. Thesis, *Etude du phénomène d'expansion sulfatique dans les bétons: Comportement des enrobés de déchets radioactifs sulfatés*, ENPC, Paris.
3. Schwiete H. E. and Ludwig U. (1968), Crystal Structures and Properties of Cement Hydration Products, *5th ISCC* Tokyo, Session II-2, Vol. 2, pp. 37-?.
4. Lea A. F. (1970), *The Chemistry of Cement and Concrete*, 3th edition, Arnold.
5. Taylor H. F. W. (1973), Cristal Structures of Some Double Hydroxide Minerals, *Mineralogical Magazine*, Vol. 39, No. 304, pp. 377-?.
6. Taylor H .F. W. (1990), *Cement Chemistry*, Academic Press, London.
7. Brown P. W. and Bothe Jr J. V. (1993), The stability of ettringite, *Advances in Cement Research*, Vol. 5, No. 18, pp. 47-?.
8. Seligmann P. and Greening N. R. (1968), Phase Equilibia of Cement-Water, *5th ISCC,* Tokyo, Session II-3, Vol. 2, pp. 179-200.
9. Way S. J. and Shanyan A. (1989), Early hydration of portland cement in water and sodium hydroxide solutions, *Cement and Concrete Research.*, Vol. 19, pp. 759-?.
10. Cottin B. (1979), Hydratation et expansion des ciments, *Ann. Chim. Fr.*, Vol. 4, pp. 139-144.
11. Regourd M. (1982), La résistance du béton aux altérations physiques et chimiques, in *Le Béton Hydraulique*, ENPC Press, Paris.
12. Atkins M. and al. (1991), Characterization of radioactive waste forms, *research rapport of the Commission of the European Communities,* EUR 13542 EN.
13. Adenot F. (1992), Ph. D. Thesis, *Durabilité du béton: Caractérisation et modélisation des processus physiques et chimiques de dégradation du ciment*, University Orléans, Orléans, France.
14 Revertégat E., Richet C. and Gegout P. (1992), Effet de pH on the durability of cement pastes, *Cement and Concrete Research*, Vol. 22, pp. 259-272.

50 PERMEABILITY AND LEACH RESISTANCE OF GROUT-BASED MATERIALS EXPOSED TO SULPHATES

M.L. ALLAN and L.E. KUKACKA
Brookhaven National Laboratory, Upton, USA

Abstract
Cementitious grouts have uses in such geotechnical and environmental applications as soil stabilisation, containment barriers, and immobilisation of hazardous wastes. Ingress of sulphate ions can threaten the durability of grouts and soil cements. The influence of sulphate attack on permeability and leach resistance was investigated. Low permeability was maintained in a sulphate environment for up to 12 months when grouts were prepared from Type V (sulphate resistant) cement or with 80% blast furnace slag/20% Type I (ordinary Portland) cement. Degradation of 100% Type I cement-based materials caused a marked increase in permeability. Replacement of Type I cement with 30% slag or 5% silica fume improved performance over the first 6 months of accelerated testing, but was less effective than Type V cement or 80% slag. Exposure to sulphate ions significantly increased the leachability of hexavalent chromium from grout treated soil unless slag was incorporated in the mix.
Keywords: Chromium, containment barriers, contaminated soil, grout, leaching, permeability, sulphate attack, waste management.

1 Introduction

Cementitious grouts are being investigated for subsurface barriers to isolate hazardous waste plumes and for treatment of contaminated soil. Barriers and treated soil are at risk of attack by sulphate-bearing groundwater. The degree of sulphate attack is usually measured in terms of expansion or loss of strength. However, such tests do not indicate how hydraulic properties are affected. Data on permeability changes throughout sulphate attack would contribute to performance prediction. In addition, the potential for release of contaminants from cementitious waste forms exposed to

Mechanisms of Chemical Degradation of Cement-based Systems. Edited by K.L. Scrivener and J.F. Young. Published in 1997 by E & FN Spon, 2–6 Boundary Row, London SE1 8HN. ISBN: 0419215700.

sulphates is an important factor when assessing the suitability of stabilisation agents.

The purpose of the research presented was to address the issues of permeability transitions and leachability of contaminants in a sulphate environment. The research was conducted in two parts. The first part involved monitoring the permeability of grouts and grout-treated soils (soil cements) exposed to a sodium sulphate solution. The effects of cement type, blast furnace slag, and silica fume were investigated. The second part of the research examined the effect of sodium sulphate on leachability of hexavalent chromium from soil stabilised with grout. The role of slag in controlling leach resistance was studied. Previous research has evaluated the use of slag-modified grouts for stabilisation of chromium contaminated soil [1].

2 Experimental procedure

2.1 Permeability experiments

The effect of sulphate attack on permeability of grout and grout-treated soil (soil cement) was determined. The grout mixes were as follows: (1) Type I cement (ordinary Portland); (2) Type V cement (sulphate resistant); (3) 30% blast furnace slag/70% Type I cement; (4) 80% blast furnace slag/20% Type I cement; and (5) 5% silica fume/95% Type I cement. The ground granulated blast furnace slag conformed to ASTM C 989-89 Grade 100 and had an alumina content of 7.5%. The liquid superplasticizer (SP) used contained 42% naphthalene sulphonate formaldehyde condensate. Table 1 presents the mix proportions of the grouts. The water/cementitious material ratio was held constant at 0.45 to enable comparison of the different cement types and supplementary cementing materials.

The soil cements were produced by mixing grout with fine-grained sandy soil from Albuquerque, New Mexico. The soil contained calcium carbonate deposits, in addition to gypsum. Three different grouts were used. These were: (1) Type I cement; (2) 40% blast furnace slag/60% Type I cement; and (3) 80% blast furnace slag/ 20% Type I cement. The water/cementitious material ratio of all the soil cements was 0.72. The mix proportions for the soil cements are presented in Table 2.

Table 1. Mix proportions of grouts

Mix No.	Cement (kg/m³)	Slag (kg/m³)	Silica Fume (kg/m³)	Sand (kg/m³)	Water (kg/m³)	Bentonite (kg/m³)	SP (L/m³)
1	819	0	0	983	369	17.7	16.4
2	819 (V)	0	0	983	369	17.7	16.4
3	563	241	0	965	362	18.0	16.1
4	159	634	0	952	357	17.8	15.9
5	771	0	41	975	365	7.2	16.2

Table 2. Mix proportions of soil cements

Mix No.	Cement (kg/m³)	Slag (kg/m³)	Water (kg/m³)	Bentonite (kg/m³)	SP (L/m³)	Soil (kg/m³)
1S	537	0	388	7.7	10.7	1075
2S	322	215	388	7.7	10.7	1075
3S	107	430	388	7.7	10.7	1075

The grouts and soil cements were cast as cylinders 75 mm diameter and 150 mm long. Three specimens per batch were cast. The cylinders were demoulded after 24 hours and cured in a water bath for 28 days. After curing, the cylinders were trimmed to a length of 105 mm and vacuum saturated with de-aired tap water.

The coefficient of permeability (hydraulic conductivity) of the grouts and soil cements under saturated conditions was measured in a flexible wall triaxial cell permeameter. This type of permeameter gives uniaxial flow. Use of a latex membrane to seal the specimens permitted testing of degraded materials since the membrane conforms to the surface. The permeant was de-aired tap water. The applied pressure gradient was 207 kPa over the length of the specimen. The confining pressure applied to seal the latex membrane to the wall of the specimen was 310 kPa. The experimental set-up followed ASTM D 5084-90. Measurements that were conducted over pressure differentials from 69 to 207 kPa indicated that Darcy's law was valid. However, this may not remain so if cracking is severe.

Following measurement of the initial 28 day permeability, the ends of the specimens were dipped in molten wax and the specimens were placed in a bath of 50 g/L (0.35 M) Na_2SO_4. The purpose of wax coating the ends of the specimens was to promote radial diffusion of sulphate ions and prevent the formation of "shoulders". This resulted a "shrinking core" type mode of attack while keeping the cylinder ends at a right angle. The Na_2SO_4 solution was replaced on a monthly basis to ensure maintenance of the SO_4^{2-} concentration. The pH was not controlled. The specimens were removed from the sulphate bath after a period of 6 months and the waxed ends were cut off. The permeability test was repeated and the specimens were re-waxed and returned to the bath. This procedure was repeated after 12 months of exposure to sulphates. The average cross sectional area after exposure was used in the calculation of permeability, rather than the original value.

2.2 Leach experiments

The leachability of hexavalent chromium from soil cement exposed to Na_2SO_4 was measured following a procedure modified from ANS 16.1. The mixes tested were 1S and 3S and this permitted determination of the effect of slag on leachability. Hexavalent chromium contaminated soil was produced by adding dissolved CrO_3 to give an initial soil concentration of 1000 mg/kg (ppm). Grout was then mixed with the contaminated soil and cylinders 65 mm high and 31 mm diameter were cast. The Cr concentration of the soil cements following dilution with grout was 532 ppm. The specimens were maintained in a plastic bag for 28 days prior to testing. Three cylinders per batch were tested. The cylinders were placed in individual containers and exposed to a 0.35 M Na_2SO_4 solution. The ratio of leachate volume to surface area was 10.0 cm³/cm². The leachate was sampled and replaced at a 14 day interval. Atomic absorption spectroscopy was used to measure total chromium concentration in the leachate. Aliquots of leachate with standard additions of chromium were analysed to check for interference by other anions or cations. The concentration of SO_4^{2-} in the leachate was measured using a UV spectrophotometer and the method was based on precipitation of barium sulphate.

3 Results

3.1 Permeability experiments
The effect of exposure to 0.35 M Na_2SO_4 on permeability of grout is depicted in Figure 1. Figure 2 shows the response of soil cements. Both plots also give the permeability after 28 days of wet curing. The error bars represent one standard deviation. A *t*-test was conducted at the 5% significance level to compare the sample means of the permeabilities. It was assumed that the permeabilities were normally distributed.

Surface cracking and expansion were evident on the soil cement prepared with Type I cement (Mix 1S) after 3 months of exposure. The Type I cement grout (Mix 1) exhibited surface deterioration after 4 months. Attack progressively worsened. The grouts and soil cements containing supplementary cementing materials were not visibly deteriorated at 6 months, nor was the grout prepared with Type V cement.

After 12 months the soil cement with 40% slag/60% cement (Mix 2S), the 30% slag/70% cement grout (Mix 3) and the silica fume-modified grout (Mix 5) exhibited slight cracking. Of these three materials, deterioration was greatest for Mix 2S and least for Mix 3. The cracks on the trimmed ends were concentric with the outer circumference. No visible deterioration was observed on the mixes with 80% slag or Type V cement after 12 months. However, it is possible that microcracking occurred in these mixes.

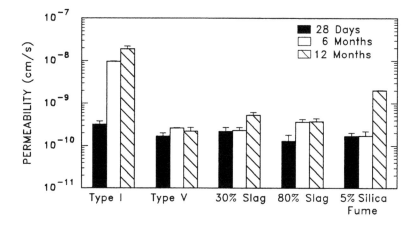

Fig. 1. Permeability of grouts after 28 days wet cure and exposure to sulphates.

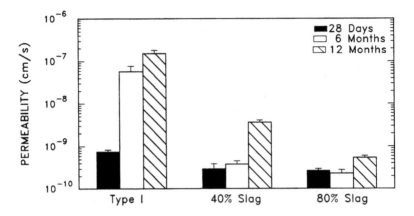

Fig. 2. Permeability of soil cements after 28 days wet cure and exposure to sulphates.

3.2 Leach experiments

The mean cumulative fractions leached for chromium in the Na_2SO_4 solutions are plotted against time in Figure 3. The Type I cement based material (1S) showed a high release of Cr in the initial 2 week period. Although the leachate was not analyzed until 2 weeks, the solution turned yellow within hours of immersion and this indicates almost immediate release of Cr. Concentration of SO_4^{2-} in the leachate decreased throughout the sampling periods. Severe cracking, softening and expansion occurred and the specimens broke in half after 8 weeks of exposure. The specimens disintegrated by 10 weeks and the leachate was left in place until 24 weeks. The final cumulative fraction leached from the Type I cement based material was 0.91 and this indicates that virtually all of the added Cr had leached out.

The slag modified-soil cement (3S) did not exhibit any visual indications of attack and SO_4^{2-} concentration in the leachate remained constant. The final cumulative fraction leached for Cr was 0.033 for this mix.

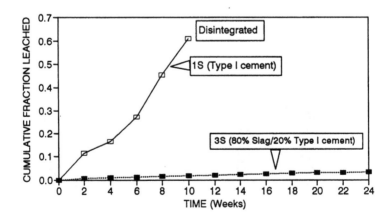

Fig. 3. Effect of sulphate attack on leachability of Cr from soil cement.

4 Discussion

4.1 Effect of sulphates on permeability

Of the tested grouts, the 100% Type I cement mix was the most susceptible to sulphate attack under the aggressive experimental conditions and demonstrated a severe loss of integrity. A potential temporary decrease in permeability associated with filling of pore space or cracks with gypsum and/or ettringite was not detected for this mix. Such a permeability decrease may be more favored in confined conditions where expansion is restrained. Testing at more frequent intervals may reveal more information on the permeability transitions in the early stages of attack.

Application of a *t*-test at the 5% significance level indicated that the mean permeability of the Type V cement grout increased after 6 months. However, the 12 month permeability was not significantly different from the 28 day value. It is possible that slight microcracking or leaching occurred in the first 6 months. Microcracks may have become partially blocked with reaction products by 12 months. Type V cement is not permanently immune to sulphate attack, as shown in microstructure studies [2], and integrity may eventually be lost after prolonged exposure.

The grout containing 30% slag retained the same permeability after the first 6 months, but developed visible cracking with a consequent increase in permeability at 12 months. The permeability of the 80% slag grout increased after 6 months of exposure and remained stable after 12 months. The 80% slag mix behaved in a similar manner to the Type V cement grout. Slag is known to improve sulphate resistance [3,4]. The relatively low alumina content (7.5%) of the slag used also contributed to the observed sulphate resistance since high alumina slags may be more prone to attack [3,4].

The silica fume-modified grout maintained the same permeability over the first 6 months. Sulphate attack was evident at 12 months and the mean permeability increased from 1.7×10^{-10} to 2.0×10^{-9} cm/s . Higher replacement levels of silica fume may enhance sulphate resistance [5,6]. The replacement level of 5% was chosen on the basis of rheological studies since higher levels significantly increase apparent viscosity. However, 5% replacement appears inadequate for sulphate resistance. In conditions where exposure to excessive concentrations of SO_4^{2-} is expected, the silica fume replacement level must be balanced to achieve both workable rheological properties and long-term durability.

The higher initial permeability of the Type I cement-based grout partially contributed to the degree of sulphate attack. Other properties controlling attack and subsequent cracking are cement chemistry and fracture resistance. Crack propagation occurs once expansive reaction products generate sufficient stress at a crack tip to cause the strain energy release rate to equal or exceed the critical value (toughness). Hence, the resistance to expansion-induced cracking is controlled by the toughness of the material. Also important are the size of pre-existing flaws and the magnitude of any confining stress.

The Type I cement-based soil cement exhibited the greatest susceptibility to sulphate attack of all the materials tested. A permeability increase of two orders of magnitude was measured after 6 months. By 12 months the mean permeability was 1.5×10^{-7} cm/s. The high water/cementitious material ratio and greater initial permeability contributed to the increase in permeability concurrent with sulphate attack. Although not measured, the toughness of the soil cement as compared to the grouts is expected to be influential. Microstructure studies prior to exposure to SO_4^{2-} showed that the Type I cement-based soil cement had greater amounts of ettringite needles than the corresponding grout and that these were concentrated in pores. The initial presence of ettringite is derived from gypsum in the soil, in addition to the cement. Since the soil cement already contains a significant amount of ettringite in the pores, the amount of space available for expansion without inducing stress is less than that for grout. Consequently, the time to cracking is reduced. Another potential reaction product is thaumasite $(CaSiO_3.CaSO_4.CaCO_3.15H_2O)$ since the soil contained calcium carbonate.

Partial replacement of cement with slag improved sulphate resistance of the soil cements and decreased the rate of attack. The mix with 40% slag did not exhibit a statistically significant increase in permeability after 6 months. The mean permeability increased from a 28 day value of $3.0x10^{-10}$ to $3.6x10^{-9}$ cm/s at 12 months. The soil cement with 80% slag retained relatively constant low permeability throughout the first 6 months, followed by an increase to $5.4x10^{-10}$ cm/s by 12 months.

It is important to compare the experimental conditions with those likely to be encountered in service. Lower sulphate concentrations in groundwater may result in a different reaction mechanism. Confining pressure may also alter the degree of cracking. When exposed to sulphate ions, the specimens were in an unconfined state and free to expand and crack. The extent of expansion, cracking and permeability increase may not be as pronounced for a subsurface barrier effectively confined by surrounding soil. Therefore, the specimens underwent accelerated deterioration due to both the high sulphate concentrations and the lack of confining pressure during exposure. The permeability tests were conducted under a confining pressure to simulate subsurface conditions and the pressure acted to close cracks partially. The measured permeability change is indicative of the relative performance of the different materials under accelerated conditions.

4.2 Effect of sulphates on chromium leachability

The leaching tests demonstrated that accelerated sulphate attack results in significant release of hexavalent chromium from Type I cement grout-stabilized soil and that this can be countered by partial replacement of cement with slag. Slag-modified grouts provide chemical immobilization by reduction of Cr(VI) to the trivalent state [1,7], in addition to physical encapsulation and sulphate resistance. Type I cement does not reduce Cr(VI) and relies solely on physical encapsulation. Sulphate attack compromises the physical integrity and functionality of a Type I cement-based waste form.

The observed decrease in SO_4^{2-} concentration and increase in Cr concentration in the leachate for the Type I cement-based soil cement is consistent with ion exchange between SO_4^{2-} and CrO_4^{2-}. Reactions to form ettringite, gypsum, and, possibly, thaumasite induced cracking and increased the surface area exposed to the leachate.

Reduction of Cr(VI) to Cr(III) by slag negates the ion exchange between SO_4^{2-} and CrO_4^{2-} that occurs in the Type I cement-Cr(VI) system. The decreased $Ca(OH)_2$ content of the slag-modified soil cement limits the reaction with SO_4^{2-} to produce gypsum. Therefore, slag is beneficial in two ways: (1) decreased sulphate attack and (2) decreased exchange with CrO_4^{2-}.

5 Conclusions

Accelerated tests enabled comparison of the relative performance of grout-based materials exposed to sulphate ions. The durability and functionality of cementitious grouts used for subsurface containment barriers and for stabilization of hazardous wastes can be severely compromised in the presence of high concentration Na_2SO_4 solutions. Type I (ordinary Portland) cement based materials will not maintain low permeability in sulphate environments, nor will Type I cement materials with 5% replacement with silica fume or 30-40% replacement with blast furnace slag. Type V (sulphate resistant) cement or 80% replacement of Type I cement with slag provides superior performance and decreased rate of permeability change. Leach resistance of hexavalent chromium from grout-stabilized soil exposed to a Na_2SO_4 solution is significantly improved by partial replacement of Type I cement with slag. Long-term testing under realistic field conditions of groundwater chemistry and confining stress is necessary.

6 Acknowledgement

This work was performed under the auspices of the U.S. Department of Energy, Washington, DC, under Contract No. DE-AC02-76CH00016.

7 References

1.   Allan, M.L. and Kukacka, L.E. (1995) Blast Furnace Slag-Modified Grouts for In-Situ Stabilization of Chromium Contaminated Soil, *Waste Management*, Vol. 15, pp. 193-202.
2.   Gollop, R.S. and Taylor, H.F.W. (1995) Microstructural and Microanalytical Studies of Sulfate Attack. III. Sulfate Resisting Portland Cement: Reactions with Sodium and Magnesium Sulfate Solutions, *Cement and Concrete Research*, Vol. 25, pp. 1581-1590.
3.   Lea, F.M. (1971) *The Chemistry of Cement and Concrete*, Chemical Publishing Co., NY.
4.   Biczok, I. (1972) *Concrete Corrosion and Concrete Protection*, Publishing House of the Hungarian Academy of Sciences, Budapest.
5.   Hooton, R.D. (1993) Influence of Silica Fume Replacement of Cement on Physical Properties and Resistance to Sulfate Attack, Freezing and Thawing, and Alkali-Silica Reactivity, *ACI Materials Journal*, Vol. 90, pp. 143-151.
6.   Torii, K. and Kawamura, M. (1994) Effects of Fly Ash and Silica Fume on the Resistance of Mortar to Sulfuric Acid and Sulfate Attack, *Cement and Concrete Research*, Vol. 24, pp. 361-370.
7.   Kindness, A., Macias, A. and Glasser, F.P. (1994) Immobilization of Chromium in Cement Matrices, *Waste Management*, Vol. 14, pp. 3-11.

51 RESPONSE OF Pb IN CEMENT WASTE FORMS DURING TCLP TESTING

J.H. BOY, T.D. RACE and K.A. REINBOLD
U.S. Army Corps of Engineers, Construction Engineering Research Laboratories, Champaign, IL, USA
J.M. BUKOWSKI, A.R. BROUGH and X. ZHU
University of Illinois, Center for Cement Composite Materials, Urbana, IL, USA

Abstract
The use of cement based systems for solidification and stabilization of Pb contaminated paint blast media waste was evaluated. The EPA's Toxicity Characteristic Leaching Procedure (TCLP) was used to evaluate the Portland cement waste form. The high alkaline conditions of Portland cement consumed the acetic acid of the TCLP extraction solution resulting in a final pH greater than 12 and the measured Pb concentrations were found to be below 1 ppm. Despite similar alkalinity, much lower levels of Pb were detected in the TCLP extraction solutions than in expressed pore solutions. Additional processes beside pH control of the metal solubility must be occurring. The titration of Pb solutions into the TCLP extraction solution of Portland cement were found to precipitate $Pb_3(CO_3)_2(OH)_2$. PbS was precipitated from TCLP extraction solutions obtained from waste forms containing 6 vol. % blast furnace slag (BFS). However, when water was used as the extraction solution, PbS was only precipitated from waste forms containing 100% BFS. It is proposed that the mechanism for the decrease in the Pb content in the TCLP extraction solution is precipitation of insoluble phases during TCLP testing. However, these precipitation reactions were not found to occur in the cement waste forms not subjected to interaction with the TCLP extraction solution.
Keywords: Acid extraction, expressed pore solutions, lead, Portland cement, TCLP.

1 Introduction

Previous work to investigate separation, recycling, and treatment options for paint blast media wastes included a systematic investigation of the stabilization of hazardous waste in Portland Cement [1]. The paint blast media wastes were

Mechanisms of Chemical Degradation of Cement-based Systems. Edited by K.L. Scrivener and J.F. Young. Published in 1997 by E & FN Spon, 2–6 Boundary Row, London SE1 8HN. ISBN: 0419215700.

generated at Army maintenance facilities. We have reported results [2-5] obtained using both the EPA Method 1311 Toxicity Characteristic Leaching Procedure (TCLP) [6] and the expression and analysis of the cement pore solutions [7].

The TCLP [6] is the regulatory standard in the United States for the classification of a waste stream as hazardous or non-hazardous. It was designed to simulate the response of a waste disposed in a municipal landfill in the presence of organic waste. The decomposition of the organic matter in these landfills is known to generate organic acids which are simulated in the TCLP by the use of acetic acid. However, the highly alkaline nature of Portland cement solidified waste forms can quickly neutralize the acid present in the TCLP extraction solution and shift the pH of the test solution to significantly higher values [8]. This shift in the pH can result in a significant change in the solubility of the hazardous metal species [9]. In the past work we concluded that the TCLP may not accurately reflect what occurs in the cement waste form [2-5].

We utilized the expression and analysis of pore solutions from Portland cement to directly measure the chemical response of the waste species in the cement waste form. This technique has been used previously to provide insight into the cement hydration process [7]. Pore solutions recovered after set are typically found to be concentrated solutions of alkali hydroxides with modest amounts of other species. With pH>13, the pore solutions are extremely basic. This is the chemical environment that any foreign species, introduced from the hazardous waste, would encounter during stabilization.

2 Background

Simulated pore solutions consisting of 0.1 M and 1.0 M solutions of KOH + NaOH (3:1 mole ratio) were utilized to isolate the effect of pH on the waste [2]. The concentrations were selected to correspond to the lower and upper limits of the alkali content of commercially available cements. It was found, as expected, that Pb was highly soluble and that the solubility of Pb increased with increasing pH.

Waste form samples containing cement and Pb containing waste were previously prepared and allowed to hydrate for 1 and 28 days [3,4]. Both a commercial low alkali (LAC) and a high alkali cement (HAC) were evaluated. The pore solutions were extracted and analyzed. The OH⁻ concentrations were measured by titration to pH = 7 and were found to be highly basic with calculated pH >12. At this pH, Pb was found to be soluble in the expressed pore solutions. The concentration of Pb was found to be higher in the low alkali cements than in the high alkali cements despite the higher pH of the fluid for the high alkali cement. We concluded that additional processes besides simple OH⁻ control of the metal solubility were occurring [3].

Cement waste forms were previously tested using the TCLP [3-5]. Solution 2, pH = 2.88, was used for the extractions. Both the initial and final pH of the TCLP were determined. The high pH of the cement completely neutralized the acid present in the TCLP test. The final pH of the TCLP extraction fluid were in the identical range as the pH of the expressed pore solutions. However, the measured Pb concentrations in the TCLP extraction solutions were below the regulatory limit of 5 ppm. All samples passed TCLP for Pb. Despite high Pb solubility in simulated and expressed pore solutions, Pb concentrations were low in the TCLP extraction solution [3]. A series of experiments was designed to determine a mechanism to explain these results.

3   Experimental Procedures

TCLP is an 18 hour acid extraction of a granulated waste or waste form. One of two nonbuffered acetic acid extraction solutions is used: solution 1 with a pH of 4.93 ± 0.05, or Solution 2, with a pH = 2.88 ± 0.05. The TCLP requires that a small sample of the waste be stirred into solution 1. If the pH raises to > 5, than solution 2 must be used. There is no other requirement on the pH of the TCLP extraction solution. As Bishop [8] has demonstrated, the final pH may be substantially different. This is especially important for cement stabilized waste forms in which the high alkalinity of the cement quickly neutralizes all the acid present in the TCLP extraction fluid.

To determine what chemical processes may be occurring during the TCLP extraction of cement waste forms, simulated extractions were performed. 100 g mixtures of cement and sand, or cement, sand, and blast furnace slag (BFS) were introduced into the TCLP acid extraction solution. The high purity silica sand was used as a non-reactive substitute for the waste. Mixtures consisting of 75 vol. % sand and 25 vol. % cement or cement plus slag were used to duplicate the high waste loading of the original stabilization work.

As a control, extractions were also conducted using CO_2-free deionized water. The water extractions were used to isolate the response of the cement waste form from any interference with the acid of the TCLP extraction process.

The simulated extractions were conducted for either 10 min. or 18 hours with the bottles rotated at 30 ± 2 rpm. During the simulated extraction, the solution would be able to react with the cement and dissolve into solution any soluble or absorbed species present in the cement. The solids were filtered and removed, and a solution of 10 gm of Pb dissolved in 75 ml of 0.1 molar nitric acid was then immediately titrated into the resulting extraction solution. Precipitation was observed to occur during the titration. The resulting solution was filtered and the precipitate analyzed by x-ray diffraction. Both a low and high alkali cement as well as cement and BFS mixtures were evaluated. The chemical analysis of the cements and the BFS are shown in Table 1.

4   Results

Simulated extractions were performed using deionized water as the extraction solution, Table 2. For all cement extractions solutions the final OH⁻ concentration was determined by titration to pH = 7 and the calculated pH > 12.5. This showed that the cement reacted with the water and that the base in the cement was released into solution. The final pH of the LAC extraction solutions were slightly higher than the HAC extraction solutions. Although the alkalis are important in expressed pore solutions, they are insignificant in the simulated extraction solutions due to dilution. The subsequent titration of a solution of Pb dissolved in nitric acid resulted in the

Table 1. X-ray florescence analysis of low and high alkali cements

Oxide Equivalents	Low Alkali Cement	High Alkali Cement	Blast Furnace Slag
SiO_2	24.04	20.45	37.55
Al_2O_3	2.58	5.41	7.45
Fe_2O_3	0.28	2.00	0.18
CaO	68.90	64.21	39.07
MgO	1.07	2.72	11.32
K_2O	0.03	1.07	0.36
Na_2O	0.14	0.24	0.30
TiO_2	0.13	0.27	0.37
P_2O_5	0.10	0.13	0.01
MnO	0.02	0.044	0.55
SO_3	2.31	2.93	2.80
Total	99.60	99.47	99.96
LOI 105° -1000°C	1.06%	0.76%	

precipitation of basic Pb nitrate, indicating that other more favorable precipitation reactions did not occur. We believe that this is an artifact of the nitric acid solution used in the titration and does not represent reactions with species dissolved from the cement. No differences in the precipitates detected were observed when part of the cement was replaced by blast furnace slag (6% total by vol.). However, the extraction of 100% blast furnace slag and subsequent titration of the Pb solution resulted in the precipitation of Pb sulfide. The PbS was formed from sulfur compounds present in the granulated blast furnace slag. The use of slag additions to cement to immobilize hazardous wastes has been reported by Macphee and Glasser [9] and by Conner [10].

Simulated extractions were performed on high alkali cement and high alkali cement and blast furnace slag mixtures using the TCLP acid extraction solution No. 2, Table 3. The final OH⁻ concentrations were determined and the calculated pH > 11, indicating reaction of the cement with the TCLP extraction solution. For high alkali cement extractions, the subsequent titration with a solution of Pb dissolved in nitric acid resulted in the precipitation of a basic Pb carbonate. For the cement and BFS mixtures, in addition to basic Pb carbonate, PbS was also precipitated. The relative

Table 2. H_2O extraction of low and high alkali cements followed by Pb titration

Sand Vol. %	Cement Vol. %		Slag Vol. %	Time	pH End TCLP	Principal Precipitates
---	LAC	100	---	18 hr.	12.76	$Pb_3(NO_3)(OH)_5$
---	HAC	100	---	18 hr.	12.64	$Pb_3(NO_3)(OH)_5$
75%	LAC	19	6	18 hr.	12.54	$Pb_3(NO_3)(OH)_5$
75%	HAC	19	6	18 hr.	12.40	$Pb_3(NO_3)(OH)_5$
---	---		100	18 hr.	11.48	PbS

Table 3. TCLP extraction of high alkali cements followed by Pb titration

Sand Vol. %	Cement Vol. %	Slag Vol. %	Time	pH Start TCLP	pH End TCLP	Principal Precipitates	Color
75	25	---	10 min.	2.85	11.05	$Pb_3(CO_3)_2(OH)_2$	Light Yellow
75	25	---	18 h.	2.90	12.05	$Pb_3(CO_3)_2(OH)_2$	Light Yellow
75	19	6	10 min.	2.85	11.16	PbS (more) $Pb_3(CO_3)_2(OH)_2$ (less)	Black
75	19	6	18 h.	2.88	11.86	$Pb_3(CO_3)_2(OH)_2$ (more) PbS (less)	Brown

intensity of the PbS was lower for the 18 h. extraction than for the 10 min. extraction.

Simulated extractions were performed on LAC as well as LAC and BFS mixtures using the TCLP acid extraction solution No. 2, Table 4. The final OH⁻ concentrations were determined and the calculated pH > 10.9, again indicating reaction of the cement with the TCLP extraction solution. For low alkali cement extractions, the precipitates identified from the subsequent titration with a solution of Pb dissolved in nitric acid was found to depend on the length of the extraction time. Extractions for 10 minutes resulted in basic Pb carbonate precipitates while the precipitation of basic Pb nitrate occurred after the 18 h. extraction. Similarly, for the cement and blast furnace slag mixtures, in addition to a basic Pb carbonate, basic Pb nitrate and PbS were also identified after the 10 min. extractions, but only the later two after the 18 h. extraction. The precipitation of PbS was due to the release of sulfides from the granulated blast furnace slag. Sulfur is typically present in the cement as calcium sulfate, $CaSO_4$.

Table 4. TCLP Extraction of Low Alkali Cements Followed by Pb titration

Sand Vol.	Cement Vol. %	Slag Vol. %	Time	pH Start TCLP	pH End TCLP	Principal Precipitates	Color
75	25	---	10 min	2.85	10.95	$Pb_3(CO_3)_2(OH)_2$	Light Brown
75	25	---	18 hr	2.90	12.20	$Pb_3(NO_3)(OH)_5$	White, V. Fine
75	19	6	10 min	2.85	11.11	$Pb_3(CO_3)_2(OH)_2$ $Pb_3(NO_3)(OH)_5$ PbS	Brown
75	19	6	18 hr	2.87	12.09	$Pb_3(NO_3)(OH)_5$ PbS (less)	Light Yellow

5   Discussion

Water extractions were used to simulate the hydration reactions of cement, cement plus BFS, or BFS. This would isolate the response of the waste form from any interference with the acid of the TCLP extraction process. The cement was found to react with the water extraction solution as shown by the rapid increase in the pH. The only precipitate identified was basic Pb nitrate which was attributed to being an artifact of the nitric acid solution used to dissolve the Pb. No other precipitates were identified. This was also the case when cement was replaced by BFS (6 % total vol.). When 100% of the BFS was subjected to 18 h. water extraction followed by Pb titration, PbS precipitate was detected.

For the extractions using the TCLP acid extraction solution No. 2, precipitation of additional phases were identified. For high and low alkali cements basic lead carbonate was generally identified with there being a time effect for the low alkali cement. When part of the cement was replaced with blast furnace slag, PbS precipitates were also identified, but not with the water extraction. Glasser [11] has shown that sulfur can be released into the pore solution of cement and slag waste forms, but higher concentrations of slag (>50%) are necessary for this reaction to occur. This is consistent with our water extraction results which showed no PbS formation with similar slag concentrations (6% total vol.). Glasser [11] also showed that the kinetics of the process are relatively slow, requiring several months to obtain significant reductions, with the reaction continuing for several years. We conclude that the presence of acid is necessary to facilitate the release of sulfur species into the extraction solution in as little as 10 minutes. In the high pH pore solution of the cement waste form, no acid is present. Therefore, the chemical processes that occur during the acid extraction are different than those that occur in the cement or cement plus BFS waste form.

Similar processes may be occurring in the cement extraction solutions.   The formation of basic Pb carbonate from the titration of Pb into cement extraction solutions indicated the presence of dissolved CO_2 in the extraction solutions. The extractions were performed in accordance with standard TCLP procedures without any special control of the atmosphere in the extraction bottle. The CO_2 could have entered into solution from the air in the extraction bottle. Alternately, the loss on ignition (LOI) data, Table 1, indicated the presence of small amounts (<1%) of unreacted or absorbed species in the cement. Soluble CO_2 could have been released from the cement by the acid of the TCLP extraction solution. Independent of the specific source, the extractions were performed in accordance with the TCLP extraction procedure, and the response of the cement waste form would be similar to that which occurs during TCLP testing. The determination of a mechanism for the apparent time effect observed in the low alkali cements would require further investigation.

This is shown schematically in Figure 1. The expression of the pore solution gives a direct measure of the chemistry of the waste species in the cement matrix without interference from the extraction solution. The metal concentrations in the high pH of the pore solution are directly measured. However, when the cement waste form is placed in the TCLP extraction solution, the cement and acid react, the acid of the extraction solution is consumed and the pH rapidly increases. The reaction of the

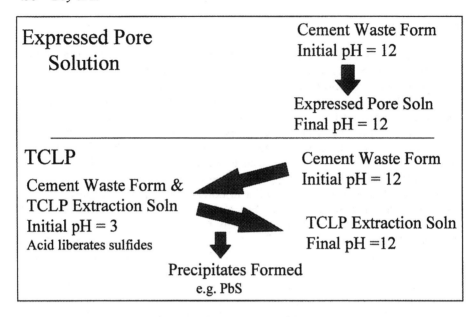

Fig. 1. Schematic comparison of expressed pore solution and TCLP testing.

acid and the cement waste form may also liberate other species which then precipitate out with the increase in pH, such as the release of S^{2-} and subsequent precipitation of PbS. Thus the analysis of the resulting extraction solution may not represent the chemical processes occurring in the cement waste form.

6 Conclusions

Pb was found to be highly soluble in simulated and expressed pore solutions. Pb concentrations were found be lower in TCLP extraction solutions than would be expected due only to dilution. This indicated that additional chemical processes were occurring during the TCLP extraction process. During the TCLP test the cement waste form and extraction solution were found to react, changing the chemistry of the extraction solution. These included changes in the pH and the solubility of the hazardous metals species. The extraction of cement and slag mixtures using the TCLP acid extraction solution and the subsequent titration of a Pb solution was found to precipitate PbS and other phases. PbS precipitates were formed by titration into acid extraction solutions and not in the water extraction solutions, except for the case of 100% BFS. Therefore, additional reactions occurred during TCLP testing that do not occur in the cement waste form. Similar processes may be occurring to account for the formation of basic lead carbonate precipitates from titration into cement extraction solutions. We conclude that the expression of pore solutions from cement waste forms gives a direct measure of the chemistry of the waste species in the cement matrix, without interference from the extraction solution. The expression and analysis of pore solutions, or the water extraction of waste forms, may provide a more direct determination of the immobilization of hazardous chemical species in cement waste forms.

7 References

1. Jeffrey H. Boy, Timothy D. Race, and Keturah A. Reinbold, "Hazardous Paint Blast Media Waste: Separation, Recycle, and Treatment Options," U.S. Army Corps of Engineers, Construction Engineering Research Laboratories (CERL), Technical Report, In Press.
2. J. Boy, T.D. Race, K.A. Reinbold, J. Bukowski, and X. Zhu, "Portland Cement Stabilization of Metal Contaminated Paint Blast Media Wastes," Proceedings of 18th Army Environmental R & D Symposium, Williamsburg, VA (June 1993).
3. J.M. Bukowski, J.H. Boy, X. Zhu, T.D. Race, and K.A. Reinbold, "Immobilization Chemistry in Portland Cement Stabilized Paint Blast Media Wastes," In the Proceedings of the Am. Ceramic Society Symposium: Environmental and Waste Management Issues in the Ceramic Industry II, Ceramic Trans., Vol. 45, Ed. by D. Bickford et. al., Am. Cer. Soc., Westerville, OH, p. 155-164, 1994.
4. J.H. Boy, T.D. Race, K.A. Reinbold, J. Bukowski, and X. Zhu, "Stabilization of Metal Contaminated Paint Removal Waste in a Cementitious Matrix Containing Blast Furnace Slag," In the Proceedings of the 87th Annual Air & Waste Management Conference, Cincinnati OH, June 1994.
5. J.H. Boy, T.D. Race, K.A. Reinbold, J. Bukowski, and X. Zhu, "Cr Stabilization Chemistry of Paint Removal Waste in Portland Cement and Blast Furnace Slag," Hazardous Waste & Hazardous Materials, 12, No. 1, 1995.
6. United States Environmental Protection Agency, *Toxicity Characteristic Leaching Procedure*, Federal Register, 51, 21672-21692 (No. 114, Fri. June 13, 1986)
7. R.S. Barneyback and S. Diamond, "Expression and Analysis of Pore Fluids from Hardened Cement Paste and Mortars," Cem. and Con. Res., **11**, 279-285 (1981).
8. P.L. Bishop, "Leaching of Inorganic Hazardous Constituents from Stabilized/ Solidified Hazardous Wastes," Hazardous Wastes and Hazardous Materials Vol. 5 No 2, 129-143 (1988).
9. Macphee, D.E., and Glasser, F.P., "Immobilization Science of Cement Systems," MRS Bulletin, 18, No. 3, 66-71 (1993).
10. Conner, J.R., *Chemical Fixation and Solidification of Hazardous Wastes*, Van Nostrand Rienhold, New York, (1990).
11. Glasser, F.P., "Reactions Between Cements and Wastes; Chromium, Molybdenum and Uranium," presentation at symposium "Cement and Concrete Science," Institute of Materials Conference, Oxford, UK, Sept. 25-27, 1994

Key word index

Also Available from E & FN Spon

Cement-based Composites: Materials, Mechanical Properties and Performance
A.M. Brandt

Chemical Admixtures for Concrete
Third edition
R. Rixom and N. Mailvaganam

Concrete Mix Design, Quality Control and Specification
K.W. Day

Durability of Concrete in Cold Climates
M. Pigeon and R. Pleau

Freeze-thaw Durability of Concrete
Edited by M. Pigeon, J. Marchand and M.J. Setzer

High Performance Fiber Reinforced Cement Composites 2
Edited by A.E. Naaman and H.W. Reinhardt

Integrated Design and Environmental Issues in Concrete Technology
Edited by K. Sakkai

Production Methods and Workability of Concrete
Edited by P.J.M. Bartos, D.L. Marrs and D.J. Cleland

Performance Criteria for Concrete Durability
Edited by H. Hilsdorf and J. Kropp

Prediction of Concrete Durability
Edited by J. Glanville and A. Neville

Steel Corrosion in Concrete
A. Bentur, S. Diamond and N. Berke